JOHN VON NEUMANN
AND THE FOUNDATIONS OF QUANTUM PHYSICS

VIENNA CIRCLE INSTITUTE YEARBOOK [2000]

8

VIENNA CIRCLE INSTITUTE YEARBOOK [2000]
8

Institut 'Wiener Kreis'
Society for the Advancement of the Scientific World Conception

The titles published in this series are listed at the end of this volume.

JOHN VON NEUMANN
AND THE FOUNDATIONS OF
QUANTUM PHYSICS

Edited by

MIKLÓS RÉDEI
Eötvös University, Budapest

MICHAEL STÖLTZNER
Institute Vienna Circle, Vienna
University of Salzburg

KLUWER ACADEMIC PUBLISHERS
DORDRECHT / BOSTON / LONDON

Library of Congress Cataloging-in-Publication Data

ISBN 0-7923-6812-6
Series ISSN 0929-6328

Published by Kluwer Academic Publishers,
P.O. Box 17, 3300 AA Dordrecht, The Netherlands.

Sold and distributed in North, Central and South America
by Kluwer Academic Publishers,
101 Philip Drive, Norwell, MA 02061, U.S.A.

In all other countries, sold and distributed
by Kluwer Academic Publishers,
P.O. Box 322, 3300 AH Dordrecht, The Netherlands.

Printed on acid-free paper

Gedruckt mit Förderung des Österreichischen Bundesministeriums
für Bildung, Wissenschaft und Kultur
Printed with financial support of the Austrian Ministry for Education, Science and
Culture

In cooperation with the *University of Vienna, Center for Interdisciplinary Research (CIR)*

Printed in the Netherlands

PREFACE

The present volume is the result of a three-year project (1997-1999), which aimed at bringing together and coordinating philosophically motivated historical research on John von Neumann's activity in the area of foundations of quantum mechanics. The project was jointly organized by the Institute Vienna Circle in Vienna, Austria, and by the Department of History and Philosophy of Science (HPS) in the Faculty of Sciences of Loránd Eötvös University, Budapest, Hungary. Financial support for the Project has been provided by the Stiftung Aktion Österreich-Ungarn, by the Austrian Science and Research Liaison Office, Austrian Institute of Eastern and South Eastern European Studies, Budapest, and by the Hungarian National Science Foundation (OTKA, contract number T 025841). We owe thanks to Béla Rásky, Elke Schmidt and Agnes Schnaider in this respect.

As part of the project a two-day workshop was held at the HPS Department in Budapest in February 1999. The lectures delivered by the participants of the workshop were based on essays analyzing different areas of von Neumann's work on foundations of quantum mechanics. The essays are published in the first part of this volume. The second part of the present volume contains hitherto unpublished documents related to von Neumann and works by him, including the typescript of his "Unsolved problems in mathematics", which was von Neumann's invited address to the International Congress of Mathematicians, Amsterdam, The Netherlands, September 2-9, 1954 and his unpublished manuscript "Quantum mechanics of infinite systems". All these documents are held in the Von Neumann Archive of the Manuscript Division of the Library of Congress, Washington, D.C., U.S.A.; they were found during a study conducted by one of the editors (M. Rédei) in the Archive. The archival search, and a substantial part of preparing the documents for publication was done during M. Rédei's appointment in the academic year 1997-1998 as a Resident Fellow in the Dibner Institute for the History of Science and Technology in the Massachusetts Institute of Technology, Cambridge, U.S.A. We wish to thank Jed Buchwald, director, and Evelyn Simha, executive director of the Dibner Institute for their support of the project. The von Neumann documents are published here with the kind permission of Marina von Neumann-Whitman, the daughter of John von Neumann; she holds the copyright of all the documents. The editors wish to thank Marina von Neumann-Whitman and all institutions and founding agencies for their cooperation and generous support of the project.

The editors also express their thanks to the Series Editor Friedrich Stadler for his embracing the idea of a von Neumann volume in the Vienna Circle Institute Yearbook Series. Special thanks go to Hartwig Jobst and Erich Papp for their technical work that has produced a book out of a bunch of electronic files and to Camilla Nielsen for upgrading the linguistic quality of some contributions.

As in each Vienna Circle Institute Yearbook the special topic is rounded off by papers emerging from the Institute's annual lecture series and by a review section. Adolf Grünbaum's paper was delivered as the seventh Vienna Circle Lecture which was generously supported by the Bank Austria. Wesley Salmon gives an overview of the history of scientific understanding and explanation during the twentieth century, in which he himself has played an important part. Henrique Jales Ribeiro introduces the classic Russellian interpretation of Carnap's *Aufbau* in a new light that is likely to open up new perspectives for present debates. The work of the Institute's staff (Angelika Rzihacek, Daria Mascha and Robert Kaller) in connection with these activities has been much appreciated.

Just before finishing this volume the Institute received the sad message that Stephan Körner, Honorary Consulting Editor of this Yearbook series, has passed away. An obituary will appear in the following Yearbook.

Budapest and Vienna   *Miklós Rédei*
September 2000   *Michael Stöltzner*

TABLE OF CONTENTS

A. JOHN VON NEUMANN AND THE FOUNDATIONS OF QUANTUM PHYSICS

B. GENERAL PART

REPORT – DOCUMENTATION

REVIEW ESSAY

REVIEWS

ACTIVITIES OF THE IVC

INTRODUCTION

John von Neumann was, undoubtedly, one of the true scientific geniuses of the 20th century. The main fields to which he contributed include different disciplines of pure and applied mathematics, mathematical and theoretical physics, logic, theoretical computer science and computer design. Von Neumann was also actively involved in politics, science management, served on a number of commissions and advisory committees and had a major impact on U.S. government decisions during, and especially after, the Second World War.

Due to the extremely wide range of his activities and the very technical nature of his scientific achievements the available literature on von Neumann falls into two categories:

1. Popular works, essays and reminiscences that aim at reviewing his career and personality without attempting to carry out a substantial analysis of his scientific work.

2. Works aiming at reviewing and analyzing in depth his achievements on an advanced technical level – these latter ones have been mainly authored by experts concentrating on narrow fields.

The major works in the first category are: the recent popular biography written by the economist N. Macrae [8], which is the first and to date only comprehensive biography of von Neumann; Heims' work on Wiener and von Neumann [6]; Halmos' and Ulam's reminiscences [5], [10], [11]; the collection [7] and Nicolas Vonneumann's personal memories of his brother and of the Neumann family circle [13].

The most significant collections of expert papers and essays on von Neumann are: the one that appeared shortly after his death [10]; the comprehensive work of Aspray [1] on von Neumann and the computer; the volume with papers focusing on his contribution to computer design [2]; the one on von Neumann's impact on economics [3] and the recent collection [4] reviewing the legacy of von Neumann in the field of pure mathematics.

The present volume falls into this second category, but also includes important unpublished sources from the von Neumann Archive of the Library of Congress, documents appearing here for the first time. The essays presented in the first part of this volume provide in-depth analyses of von Neumann's achievements in the area of foundations of quantum mechanics and show the significance of his results in current research. This volume was motivated in part by the recognition that no

1

M. Rédei and M. Stöltzner (eds.),
John von Neumann and the Foundations of Quantum Physics, 1–4 .
© 2001 *Kluwer Academic Publishers. Printed in the Netherlands.*

collection of expert essays discussing von Neumann's work in this field has ever been published. The editors thought that such a volume was long overdue since von Neumann's contribution to the field is widely acknowledged as being crucial for quantum mechanics's becoming a mature and firmly established theory. Also, von Neumann's results in this direction form a major part of his scientific accomplishments, of which he was especially proud. That the issues related to the foundations of quantum theory remained close to von Neumann's heart is also reflected by the fact that in his address at the 1954 Amsterdam International Congress of Mathematicians in which he spoke on "Unsolved problems in mathematics", he focused exclusively on operator theory and its relation to physics, logic and probability theory – a field in which he saw major changes as still being necessary. Von Neumann's lecture, published in this volume, represents the only detailed and systematic formulation of his late views on quantum theory after the development of Hilbert space formalism (1926-1932) and after the subsequent creation of von Neumann algebras. Sadly enough, he did not live to pursue these ideas further.

Today, von Neumann's name figures most prominently in two areas of quantum physical research. First, von Neumann algebras have become increasingly important in contemporary quantum mechanics of large systems and in quantum field theory. From the perspective of current research in mathematics and in mathematical physics, these relatively recent developments are more interesting than ordinary Hilbert space quantum mechanics. (Thus, the title of the volume.) Unfortunately von Neumann did not live to see algebraic quantum mechanics thrive from the late fifties to the present. As his only lecture on quantum field theory (published here for the first time) shows, von Neumann, however, clearly entertained ideas related to further applications of algebraic methods to large quantum systems.

In recent years the algebraic formalism has attracted notable attention in another field that reaches back to von Neumann's 1932 *Mathematical Foundations of Quantum Mechanics*: the interpretations of quantum mechanics. His theory of measurement and his critique of hidden variables became the touchstone of most subsequent debates. The papers analyzing von Neumann's work on these topics give the present volume a more philosophical character than the above-mentioned collections in other fields. Scholarly analyses of this specific aspect of von Neumann's contribution to the foundations of quantum theory seem to be especially appropriate because von Neumann's motivations, intentions and results are all too often misunderstood and distorted in the literature. Apart from farming out papers on all the particular aspects of von Neumann's approach to the foundations of quantum mechanics, the editors have also included two papers on von Neumann's views about mathematical physics and their historical origin.

THIRRING's paper sets the tone for the volume by briefly identifying some of the main areas in which von Neumann broke new ground. The paper not only represents the personal survey of one of the founding fathers of present-day mathematical physics, but it also highlights some intricacies related to von Neumann's proof of uniqueness of the Schrödinger representation of the Canonical Commutation Relation (CCR). Thirring's paper also gives a condensed motivation for (and discovery of) different types of von Neumann algebras, recalls the idea of quantum logic and quantum statistical mechanics and points out the significance of quantum entropy for quantum statistical mechanics. These topics are taken up one by one in the subsequent papers.

MAJER'S contribution puts von Neumann's work on the foundations of quantum mechanics in the historical context of the axiomatic approach to physics – an idea that was advocated especially by Hilbert as early as in his 1900 Paris lecture, and which dominated the Göttingen scene, where von Neumann started his work on quantum mechanics. STÖLTZNER draws the picture of von Neumann as a scientist who was fully conscious of methodological issues and whose views concerning rigor and the relation of mathematics and theoretical physics are still interesting for present debates. He argues, in particular, that the virtues of axiomatization for the context of discovery are generally underestimated by philosophers of science.

BUB'S paper outlines von Neumann's theory of measurement as it was formulated in von Neumann's 1932 book and links this treatment to some current investigations aiming at characterizing sub-lattices of a Hilbert lattice that can be viewed as representing a maximal set of propositions that can have a definite truth value simultaneously. BREUER'S paper draws a parallel between Gödel's incompleteness theorems and the unsolvability of the measurement problem in quantum theory, and conjectures that von Neumann had been strongly influenced by Gödel's result and method.

PETZ'S contribution recalls von Neumann's reasoning leading to the notion of entropy in quantum mechanics, and it traces the development of that notion up until recently, showing the power, the non-triviality and fruitfulness of different notions of entropy. KÖHLER'S paper analyses the difference between Carnap's and von Neumann's approach to, and interpretation of, entropy in a broader philosophical context, arguing that the clash in 1952 between Carnap and von Neumann over entropy and other concepts related to information is a manifestation of two opposite views concerning the modal status of a theory: factual and normative.

SUMMERS'S contribution provides a brief history of the Stone-von Neumann theorem of uniqueness of the Schrödinger representation of CCR and its impact on the later developments in some areas of analysis; the review displays the richness of the topic and shows the importance of paying attention to technical details. Based partly on the archival material published in the present volume the first time, RÉDEI reconstructs the Birkhoff-von Neumann concept of quantum logic and shows how and why the Birkhoff-von Neumann concept was intimately related to the theory of continuous, finite von Neumann algebras and to the relative frequency interpretation of probability. GIUNTINI and LAUDISA carefully and critically reconstruct von Neumann's notorious 1932 proof of the impossibility of hidden variables, emphasizing neglected aspects of von Neumann's proof, such as the intimate relation of his reasoning to the relative frequency interpretation of quantum probability.

The focus of MITTELSTAEDT'S paper is probability: utilizing infinite tensor product techniques developed by von Neumann in the 1930s Mittelstaedt argues that the probabilistic part of quantum mechanics, understood as limit frequencies of yes-no propositions on infinite quantum ensembles, can be derived from a few qualitative assumptions. SZABÓ discusses the issue of interpretation of probability in quantum mechanics, arguing in favor of a frequency view of probability, which can only be upheld in connection with "quantum probabilities" if quantum probabilities are assumed to be represented as classical conditional probabilities, which is always possible, as this paper shows.

REFERENCES

[1]   W. Aspray: *John von Neumann and the origins of modern computing*, Cambridge, Mass., MIT Press, 1990.
[2]   Jean R. Brink and C. Roland Haden (eds.): "The computer and the brain: An international symposium in commemoration of John von Neumann (1903-1957)", *Annals of the History of Computing* **11** (1989) 159-201.
[3]   M. Dore, S. Chakravarty and R. Goodvin (eds.): *John von Neumann and modern economics*, Oxford, Clarendon Press, 1989.
[4]   J. Glimm (ed.): *The Legacy of John von Neumann* (Providence, R.I., American Mathematical Society, 1990) Proceedings of Symposia in Pure Mathematics; 50.
[5]   P. R. Halmos: "The legend of John von Neumann", *American Mathematical Monthly* **80** (1973) 382-394.
[6]   Steve J. Heims: *John Von Neumann and Norbert Wiener: From mathematics to the technologies of life and death*, Cambridge, Mass., MIT Press, 1980.
[7]   T. Legendi and T. Szentivanyi (eds.): *Leben und Werk von John von Neumann: Ein zusammenfassender Überblick*, Mannheim, Bibliographisches Institut, 1983.
[8]   N. Macrae: *John von Neumann*, New York, Pantheon, 1992.
[9]   D. Petz, M. Rédei: "John von Neumann and the theory of operator algebras", in F. Brody, T. Vámos (eds.): *The Neumann Compendium*, World Scientific Series of 20th Century Mathematics, Vol. I., World Scientific, Singapore, 1995, 163-181.
[10] S.M. Ulam: "John von Neumann 1903-57", in S.M. Ulam (ed.): "John von Neumann 1903-1957", *Bulletin of the American Mathematical Society* **64** (1958), 1-49.
[11] S.M. Ulam: *Adventures of a Mathematician*, Charles Scribner's Sons, New York, 1976.
[12] J. von Neumann: *Collected Works Vol. I-VI*, A.H. Taub (ed.), Pergamon Press, 1961-1962.
[13] N.A. Vonneuman: *John von Neumann as seen by his brother*, Meadowbrook, PA, The author, 1987.

WALTER THIRRING

J. V. NEUMANN'S INFLUENCE IN MATHEMATICAL PHYSICS

John von Neumann was one of the great scientists of this century. It is a particular pleasure for me to comment on his contribution to mathematical physics since I had the privilege to know him personally. My wife and myself had the pleasure of being invited to one of his splendid dinner parties in his house in Princeton. So I first got to know him as a charming host and only gradually became familiar with some of his many ideas which he spewed like a volcano. Since they contain a legion of facets I shall focus on four points where I think he laid his fingers onto the heart of physics. For a long time his importance for physics was underrated, Pauli once told me that he had said to von Neumann: "If a mathematical proof is what matters in physics you would be a great physicist". I disagree with this statement, I think he had the right vision of what will become important in physics.

1. OPERATORS IN $\mathcal{B}(\mathcal{H})$

It is remarkable how quickly quantum mechanics was assimilated by young people in the 20th century. After Heisenberg's first proposal (1925) and Schrödinger's formulation (1926), Dirac soon recognized the abstract scheme behind it and in 1927 von Neumann at the age of 24 was already an accomplished master of the mathematical machinery. Schrödinger had realized that Heisenberg's commutation relations for position x and momentum p,

$$xp - px = i\hbar \qquad (1_H)$$

can be satisfied by x acting as multiplication and p as derivative $\frac{\hbar}{i}\frac{\partial}{\partial x}$ on suitable functions. So the question arose whether this is the only possibility. It was clear to von Neumann that the relevant function space was the Hilbert space $L^2(\mathbf{R})$ and it was better to concentrate on the unitary version

$$e^{i\alpha p}e^{i\beta x}e^{-i\alpha p}e^{-i\beta x} = e^{i\hbar\alpha\beta}, \qquad \alpha,\beta \in \mathbf{R} \qquad (1_W)$$

proposed by Weyl. Indeed von Neumann could show that all representations of (1_W) are (quasi-) equivalent to Schrödinger's. The non-triviality of von Neumann's result is illustrated by the following facts:

1. (1_H) has many inequivalent representations.

2. On $L^2(T)$ that is $(\beta \in \mathbf{Z})$ there is a 1-parameter family of inequivalent representations.

M. Rédei and M. Stöltzner (eds.),
John von Neumann and the Foundations of Quantum Physics, 5–10.
© 2001 *Kluwer Academic Publishers. Printed in the Netherlands.*

3. If instead of (x, p) (or finitely many of them) one considers infinitely many (x, p) then there are uncountably many inequivalent representations.

4. One may even question whether von Neumann's proof was conclusive as he did not spell out the hypothesis of strong continuity to his theorem. Without that there are many other representations of the unitaries in (1_W) but then x (or p) do not exist as operators. With the understanding that (1_W) means that (x, p) have to exist as selfadjoint operators von Neumann's uniqueness theorem has stood the test of time.

The non-uniqueness of (1_H) stems from the fact that these operators are unbounded and not defined on all of $L^2(\mathbf{R})$. Thus to give to them a definite meaning one has to add some (boundary) conditions specifying their domain. That this may lead to amusing consequences he has shown by an example of a repulsive potential with a bound state. It does not completely defy our intuition since the potential is so repulsive that a particle reaches infinity after a finite time. There von Neumann imposes reflecting boundary conditions so that the particle goes immediately back to the origin. So in the time mean it spends most of the time near the origin and its wave function is in $L^2(\mathbf{R})$. Von Neumann realized the subtle difference between hermiticity and selfadjointness (maximal symmetry) which arises for unbounded operators. Only if H is selfadjoint e^{iHt} gives a group of unitaries and therefore the time evolution is defined for all times. For $H = p^2 + V(x)$, V real, one can always find a domain $D(H) \subset L^2(\mathbf{R})$ such that H is selfadjoint. This is not trivial because for singular potentials it is already in classical mechanics not clear how to continue the time evolution to arbitrary times if the collision orbits are dense. Even if the time evolution is definable its uniqueness remains a puzzle. One might say that it is unique if on a convenient domain like $D(H) = C_0^\infty$, H is essentially selfadjoint. This happens if $V(x) = c|x|^{-2+\varepsilon}$, $\varepsilon > 0$. In this case also classically the orbits which hit the origin are of measure zero in phase space. This is no longer true for $\varepsilon < 0$ (or $\varepsilon = 0$, $c < -1/4$) and there are extensions of $D(H)$ such that e^{iHt} is no longer unitary. This may be the physically more reasonable description because it means that the orbits which spiral in a finite time into the origin are not reflected but absorbed. However for Coulomb systems ($\varepsilon = 1$) the time evolution of H (and its generalization for many particles) is not only definable but defined. This was recognized only in the past decades and allegedly von Neumann in his later days was still sceptical whether quantum theory of Coulomb systems can be rigorized. It was however to become one of the gems of mathematical physics.

2. INFINITE TENSOR PRODUCTS AND VON NEUMANN ALGEBRAS

For finite tensor products the dimension of the spaces is multiplicative and for infinite tensor products it is uncountable even if the individual spaces have only dimension = 2. This casts some doubt on whether there is a mathematically valid description of infinite quantum systems. Schrödinger once told me that the corresponding non-separable Hilbert space did not make sense to him. To determine N components of his ψ-function one needs N experiments and in a non-separable space one would need an uncountable number of measurements and this is nonsense. However such an opinion means that Schrödinger did not get the main message of von Neumann's celebrated paper on infinite tensor products. There he shows that the corresponding operator algebras are highly reducibly represented in this vast non-separable space and there are many (inequivalent) subrepresentations which act in a separable subspace. Thus one is led to classify subalgebras $\mathcal{M}$ of $\mathcal{B}(\mathcal{H})$. Whereas this is easy if $\mathcal{H}$ is finite-dimensional some interesting new features appear when $\mathcal{H}$ is infinite-dimensional. Von Neumann (with F.J. Murray) recognized that it is useful to study $\mathcal{M}$'s which are weakly closed which can be also characterized algebraically by $\mathcal{M} = \mathcal{M}''$ (the double commutant). Now such $\mathcal{M}$'s bear the name of von Neumann. For physics it is convenient also to include weak limits (bounded sequences always have weak limit points) since the wild objects which physicists are cooking up usually do not converge in norm. However physicists had to accept that the product of weak limits is not necessarily the limit of the products. Next von Neumann recognized that the elementary building blocks of subalgebras are the so-called factors which have a trivial center $\mathcal{Z} = \mathcal{M}' \cap \mathcal{M}'' = c \cdot \mathbf{1}$. Since $\mathcal{Z}$ can be considered as the classical part of $\mathcal{M}$, factors can be considered as purely quantal objects since to each classical quantity they assign a numerical value. Their classification is aided by the fact that for them a trace is essentially unique. Thus von Neumann arrived at the first crude classification of factors of type I_N (or I_∞) where image under the (suitably normalized) trace of projections is $\{1, 2, \ldots, N\}$ (or Z^+), of type II_1 (or II_∞) where it is $[0, 1] \subset \mathbf{R}^+$ or type III where it is $\{0, \infty\}$. He succeeded in constructing examples of II_1 and thought that they are of special importance to physics since there $\mathcal{M}$ assumes the structure of a Hilbert space. Though this frequently brings some simplifications today I and II are only considered in physics as the limiting cases of systems with temperature $T = 0$ or ∞ whereas $0 < T < \infty$ corresponds to III. In fact it was quantum statistics which provided examples and relations which eventually also led to a finer classification of type III factors. Though it was achieved only after von Neumann's time he certainly had laid the foundation to our understanding of the phenomena one encounters in infinite quantum systems.

3. QUANTUM LOGIC

Von Neumann algebras are generated by their (orthogonal) projections p_i. He (and G. Birkhoff) realized that in addition to their algebraic structure they have also the structure of a lattice. $p_1 \vee p_2$ projects onto the union of $p_1\mathcal{H}$ and $p_2\mathcal{H}$, $p_1 \wedge p_2$ onto their intersection and $p_\perp = 1 - p$ onto the orthogonal complement. $p_1 \wedge p_2$ can be expressed algebraically as the weak limit $n \to 0$ of $(p_1 \cdot p_2)^n$ and since $p_1 \vee p_2 = (p_{1\perp} \wedge p_{2\perp})_\perp$ all operations can be formulated algebraically. However not all the rules of their set-theoretical (and thus logical) analogues are valid. The distributive law $p_1 \vee (p_2 \wedge p_3) = (p_1 \wedge p_2) \vee (p_1 \wedge p_3)$ fails in general. Thus we have here a new logical structure which was later termed "quantum logic". Followers of the Copenhagen interpretation refuted that one needs in quantum mechanics a new type of logic with the argument that eventually experiments have to be interpreted in classical terms and thus with classical logic.

The failure of the distributive law is not a pathology of an infinite dimensional Hilbert space but appears already in two dimensions. Thus we can illustrate its significance by one spin which is represented by the 2×2 Pauli matrices $\vec{\sigma}$. The non-trivial projections are $p_{\vec{n}} = \frac{1}{2}(1 + \vec{\sigma} \cdot \vec{n})$, $\vec{n}^2 = 1$. The rules are

a.) $p_{\vec{n}\perp} = p_{-\vec{n}}$,

b.) $p_{\vec{n}} \vee p_{\vec{n}'} = 1$,

c.) $p_{\vec{n}} \wedge p_{\vec{n}'} = 0$ for $\vec{n} \neq \vec{n}'$.

Their meaning is

a.) We are only sure that $\vec{\sigma}$ does not point in the direction $\vec{n}$ if it points to $-\vec{n}$.

b.) The sharpest proposition which is implied by both, $\vec{\sigma}$ points to $\vec{n}$ and $\vec{\sigma}$ points to $\vec{n}'$, is the tautology 1 ("the spin points somewhere") which is always true.

c.) The proposition $\vec{\sigma}$ points to $\vec{n}$ and $\vec{n}'$ is always false.

In b.) one could imagine that a proposition like "$\vec{\sigma}$ is in the plane spanned by n_1 and n_2" is sharper than 1, however, such a proposition does not exist in the formalism. Otherwise the rules seem logical but they imply already the breakdown of the distributive law: For pairwise different $\vec{n}_i$, $i = 1, 2, 3$ we get

$$p_{\vec{n}_1} \wedge (p_{\vec{n}_2} \vee p_{\vec{n}_3}) = p_{\vec{n}_1} \wedge 1 = p_n,$$

but

$$(p_{\vec{n}_1} \wedge p_{\vec{n}_2}) \vee (p_{\vec{n}_1} \wedge p_{\vec{n}_3}) = 0 \vee 0 = 0.$$

Thus the difference to the classical situation "a particle is in a region G_i" can be traced to the following. Classically the sharpest proposition which is implied by either p_{G_1} (the particle is in G_1) or p_{G_2} (the particle is in G_2) is $p_{G_1} \vee p_{G_2} = p_{G_1 \cup G_2}$. This proposition in turn implies that the particle is either in G_1 or in G_2. For spins however $p_{\vec{n}_1} \vee p_{\vec{n}_2} = 1$ but this does not imply that $\vec{\sigma}$ points either to $\vec{n}_1$ or to $\vec{n}_2$, thus $\vee$ does not mean either or (or both). The virtue of quantum logic is that it specifies the conclusions which are legitimate within the quantum mechanical formalism and which classical reasoning may lead to contradictions or inequalities which are violated in nature (Bell's inequality).

4. QUANTUM STATISTICAL MECHANICS

Already in 1927 v. Neumann derived the key formulas of quantum statistics. He defined the entropy $S(\rho)$ of a state described by a density matrix ρ as

$$S(\rho) = -\operatorname{Tr} \rho \ln \rho \qquad (0.1)$$

and proved the so-called variational principle. It says that among the states with the same energy $\operatorname{Tr} \rho H$ the canonical state $\rho_c = e^{-\beta H}/\operatorname{Tr} e^{-\beta H}$ has the highest entropy

$$\operatorname{Tr} \rho H = \operatorname{Tr} \rho_c H \Rightarrow -\operatorname{Tr} \rho \ln \rho \le -\operatorname{Tr}\rho_c \ln \rho_c. \qquad (0.2)$$

Here β appears as Lagrange multiplier. This was the origin of many new developments. It can also be expressed by saying that ρ_c has the lowest free energy $F(\rho) = \operatorname{Tr} \rho H - TS(\rho)$, $T = \beta^{-1}$. It turns out that here the assumption $\operatorname{Tr} \rho H = \operatorname{Tr} \rho_c H$ is redundant

$$F(\rho_c) = -T \ln \operatorname{Tr} e^{-\beta H} = \operatorname{Tr} \rho_c H + \operatorname{Tr} \rho_c \ln \rho_c \le \operatorname{Tr} \rho H + \operatorname{Tr} \rho \ln \rho \qquad (0.3)$$

holds for any ρ. This inequality is a special case of the positivity of the relative entropy

$$S(\rho_1|\rho) = \operatorname{Tr} \rho(\ln \rho - \ln \rho_1) \ge 0. \qquad (0.4)$$

For $\rho_1 = \rho_c$ (4.4) reduces to the previous inequality (4.3) and $S(\rho_c|\rho)$ is the difference $F(\rho) - F(\rho_c)$. This quantity has some useful classical properties not shared by $S(\rho)$ and thus helps to generalize classical ergodic theory. It may stay finite for infinite systems where $S(\rho)$ is infinite and it is monotonic with respect to reduction to subsystems. This means that when $\mathcal{N} \subset \mathcal{M}$ is a subalgebra, ρ a state over $\mathcal{M}$, and $\rho|_\mathcal{N}$ the corresponding reduced density matrix $S(\sigma|_\mathcal{N}|\rho|_\mathcal{N}) \le S(\sigma|\rho)$ whereas if ρ is pure $S(\rho) = 0$ but $S(\rho|_\mathcal{N})$ may be > 0. This opens up new aspects of the problem of entropy increase in time. Both $S(\rho)$ and $S(\sigma|\rho)$ are unitarily invariant and thus do not change under a time evolution $\rho_t = U_t^{-1} \rho U_t$, $U_t =$ unitary. However the time evolution of $\rho|_\mathcal{N}$ is not just unitary but of the form

$$(\rho|_\mathcal{N})_t = (U_t^{-1} \rho_\mathcal{N} \otimes \rho_{\mathcal{N}^c} U_t)|_\mathcal{N} \qquad (0.5)$$

if $(\rho|_\mathcal{N})_0 = \rho|_\mathcal{N} \otimes \rho|_{\mathcal{N}^c}$. Whereas $S(\rho|_\mathcal{N})$ can increase or decrease under this time evolution $S(\sigma|_\mathcal{N}|\rho|_\mathcal{N})$ does not change under amplification $\rho|_\mathcal{N} \to \rho|_\mathcal{N} \otimes \rho|_{\mathcal{N}^c}$ or U_t but decreases under $|_\mathcal{N}$. Thus we get

$$S((\sigma|_\mathcal{N})_t|(\rho|_\mathcal{N})_t) \le S(\sigma|_\mathcal{N}|\rho|_\mathcal{N}). \qquad (0.6)$$

This fact has a simple physical explanation. We cannot expect the entropy of a subsystem to always increase with time. Upon thermalization the hotter parts of a system out of equilibrium will decrease their entropy. However we learned at school that for a system in thermal contact it is not its entropy but its free energy which behaves monotonically. This is exactly what (4.6) says. Yet the time evolution (4.5) does in general not form a (semi) group. For infinite systems it may and only then we can conclude that the free energy decreases monotonically.

Summarizing one can say that von Neumann's ideas pointed far into the future and helped mathematical physics attain its present state of perfection.

Institut für Theoretische Physik
Universität Wien
Boltzmanngasse 5
A-1090 Wien
Austria

ULRICH MAJER

THE AXIOMATIC METHOD AND THE FOUNDATIONS OF SCIENCE: HISTORICAL ROOTS OF MATHEMATICAL PHYSICS IN GÖTTINGEN (1900-1930)

The aim of the paper is this: Instead of presenting a provisional and necessarily insufficient characterization of what mathematical physics is, I will ask the reader to take it just as that, what he or she thinks or believes it is, yet to be prepared to revise his opinion in the light of what I am going to tell. Because this is precisely, what I intend to do. I will challenge some of the received or standard views about mathematical physics and replace them by a more sophisticated picture, which takes into account the methodological and philosophical roots of mathematical physics in Göttingen.

The first question, with which one is confronted as historian, is the elementary question, where to begin the story. We could, of course, begin with the foundation of the university of Göttingen in the eighteenth century (1737) and follow up the line of famous physicists and mathematicians, beginning with Lichtenberg and Gauss, followed by Riemann, Dirichlet, Weber and many, many more until we come to the golden twenties and eventually to the early thirties, when the Nazis took over political power in Germany and within few month erased the world's leading center of mathematical physics. Fortunately, for the story, I am going to tell, I don't need to go back as far as Riemann and Gauß, although they were presumably the first mathematicians in modern times who had the revolutionary idea of a "physical" geometry. For my purpose it's sufficient to start in 1900. This is the year, in which the second international congress of mathematicians took place in Paris. At this congress, Hilbert held his famous lecture, in which he challenged the mathematical world by presenting a list of 23 unsolved problems and questions from all areas of pure and applied mathematics. Among the latter, the sixth problem bears the demanding headline:

6. Mathematical Treatment of the Axioms of Physics

Let me quote the first section of this problem; it reads as such:

Through the investigations of the foundations of geometry we become confronted with the task *to treat those physical disciplines axiomatically, according to the model of geometry, in which already today mathematics plays a prominent role: these are in the first line the calculus of probability and mechanics.*[1]

11

M. Rédei and M. Stöltzner (eds.),
John von Neumann and the Foundations of Quantum Physics, 11–33.
© 2001 *Kluwer Academic Publishers. Printed in the Netherlands.*

I take this announcement and its eventual piecemeal realization in the following decades, as the birth of mathematical physics. I am fully aware that this assertion is problematic for at least two reasons. First, it seems obvious that this stipulation is rather arbitrary: Why shouldn't we set the beginnings of mathematical physics some decades earlier, for example with the works of Kirchhoff, Maxwell, Hertz and Boltzmann, or even later with the prolific works of Einstein and Minkowski, or von Neumann's book *Mathematische Grundlagen der Quantenmechanik*, which is still a paradigmatic example of mathematical physics. I admit I have no real objections against these proposals and would be happy to accept them, if there were not one particular circumstance, which is in favor of my position, and which I will explain in a moment.[2] But first let me come to the second and far deeper reason, why most physicists would presumably reject my decision. They would, I suspect, not accept the identification of mathematical physics with axiomatics, that is, with a program to axiomatize all (or most) physical theories, at least the most fundamental ones. Axiomatics, they would probably argue, is something much more narrow and special than mathematical physics, as it is usually practiced. (This is fully borne out, so they would continue, by the obvious fact that axiomatization in physics has not been terribly successful, neither in its performance nor in its results, whereas mathematical physics celebrated many triumphs in the twentieth century.)

These objections have to be taken seriously, in particular the second one. But they rest, as I shall show, in part on a misunderstanding of Hilbert's sixth problem, and in part (but to somewhat lesser extent) on a hopelessly naive and oversimplified idea of what mathematical physics *really* is about. These are much-debated questions, of course, in which a lot of confusion prevails. Hence, the first aim of my paper is to correct both misunderstandings. The first misconception I will reject explicitly by explaining, first, what Hilbert's original intentions were, when he announced the Sixth Problem in 1900, and second, by showing how the problem was attacked and solved, at least in part by Hilbert himself, in the next twenty years (or so), when he and his many excellent disciples embarked into the foundations of physics. The second misunderstanding (leading to skepticism with regard to axiomatics), rests upon an oversimplified picture of the *relation* between mathematics and physics. Here I shall only point to the fact that axiomatics is far from taking mathematical concepts and theories as mere "uninterpreted" formal languages, which have to be interpreted by correspondence rules in order to become meaningful, as it is often rather naively assumed[3], but axiomatics understands them in the very same way as normal mathematicians do: as concepts and theories of certain complex domains of abstract objects, called structures, which often have their only or, at least, their most important *instantiation* in the domain of physics. Once the prejudice has been abandoned that mathematics is nothing but a sheer *formalism*, the way is opened up to recognize axiomatics as the core of mathematical physics.

HILBERT'S ORIGINAL INTENTIONS

To begin with Hilbert's original intentions, let's go back to the Paris lecture and ask, what does Hilbert actually mean, when he suggests to treat those physical disciplines "axiomatically" *according to the example of geometry*, in which mathematics already plays an important role? As you know, Hilbert had just published the year before a book on the "Foundations of Geometry", which was to become one of the biggest successes of modern science since Newton's *Principia*. It is this very book, to which Hilbert alludes in the quotation. Consequently, we have to ask, what is the content or, perhaps better, what is the "message" of this book in order to grasp Hilbert's original intention.

In order to grasp Hilbert's intentions correctly one has to understand above all that he makes a fundamental distinction between two kinds of theoretical considerations. The first kind concerns the choice of a set of axioms for a certain field, such as geometry or arithmetic, the second a certain kind of reflection about an axiom or an axiom system of a given field. The reflection about axioms is called the "axiomatic method" and has to be sharply distinguished from the axioms themselves, respectively the choice of an appropriate set of axioms for a certain field of inquiry.

Now, the reader may perhaps wonder why is this distinction so important and why do I put such a big emphasis on it? Is it not just the well-known distinction between axioms as meaningful sentences and axioms as objects of considerations of a "formal" kind, as we have learned to accept it, last not least, from Hilbert's work on the foundations of mathematics? Maybe. However, the important point here isn't just to accept it, but to understand precisely, what this *implies* and what not. A clear and unambiguous understanding, however, what the axiomatic method implies and perhaps more important, what it does not imply, is to my mind still missing. Instead I see a lot of confusion and contradicting claims. Therefore let me be as clear as possible in a short essay like this.[4]

My first simple and substantial point is this: The axioms of a field like geometry (or any other field from mathematics or physics) are not chosen arbitrarily but with respect to the field in question. The general aim is to present a field of inquiry as perspicuously and well-ordered as possible, and, I should add, not for God almighty, but for "finite" human beings, who think "discursively" and grasp "intuitively". Consequently, the descriptive symbols of an axiomatic system of a given field are not without reference or even sense, as it is often claimed quite thoughtlessly, but they have their regular meaning, at least up to a certain point of precision, to which I shall come to in a moment. To put this beyond doubt let us take the case of geometry:

For Hilbert geometry is a "natural science", the most fundamental of all natural sciences. This was, by the way, his firm conviction from the very beginning, when he lectured for the first time on geometry in 1891, and not only after the

invention of the general theory of relativity in 1915, as one might presume.[5] This
means that geometrical expressions like "point", "line" and "between" have their
usual "sense", as we recognize it in our "spatial intuition" of extended objects.
When Hilbert says, as it is often quoted, that "it must be always possible to sub-
stitute the expressions 'point, straight-line and plane' by 'Tische, Stühle und
Bierkrüge'", this belongs to a different context to be discussed below. Before,
however, let me quote from Hilbert's first lecture about geometry in 1891, in order to make it plain and obvious that geometrical terms are not meaningless, according to Hilbert's view, but have a certain intuitive "sense" (in the terminology
of Frege's distinction between "sense" and "meaning") [6], which we normally
connect with these expressions under ordinary conditions.

Geometry is the doctrine of the properties of space. It is essentially different from the pure
mathematical areas of knowledge as, for example, the theory of numbers, algebra and
function-theory. Results of these areas can be captured by pure thinking in the way that
the asserted facts are reduced to more and more simpler ones by clear logical inferences
until one eventually needs only the concept of integer numbers.

The situation in geometry is completely different. I can never ever recognize the properties of space by mere thinking, as little as I can recognize the basic laws of mechanics,
the law of gravitation or any other physical law, in this way. Space is, of course, not the
product of my thinking, but it is given to me through my senses. I need therefore my
senses in order to recognize its properties. I need intuition and experiment like in the exploration of physical laws, in which case furthermore the matter is given through the
senses.[7]

You will find quite similar remarks in all of Hilbert's lectures on geometry and
related issues. The remarks become more and more sophisticated over time until
1927, the year, in which Hilbert gave his last lecture on geometry. But he never
changed his basic attitude that geometry is a natural science. This means in the
present context that the geometrical expressions have their usual sense, as we
recognize and presuppose it not only in our daily life, but also use it in science.[8]
This does, of course, not determine the choice of the basic concepts and axioms;
here we have still a remarkable degree of "freedom", to which I will come later.

Now, let me turn to the second aspect of the distinction between the axioms
of a field and the axiomatic method as a reflection *about* the axioms of a field
and ask: What is the axiomatic method? How are we to understand it as the essential means of mathematics? This last question leads us immediately to the
core of our theme, the historical roots of mathematical physics, because for Hilbert there is no doubt that geometry becomes a *mathematical* discipline only if
and insofar as it is subjected to an axiomatic treatment according to the "axiomatic method". Otherwise it will remain a natural science like others – with no
particular relation to mathematics. Hence, if we understand in this case what it
means to apply the axiomatic method, we have a model how a natural science
becomes part of mathematical physics.

The first point one has to come to grips with is the circumstance that the application of the axiomatic method presupposes that the theory is "well devel-

oped", at least in the practical sense that most of the basic facts of the respective field are already known. This is doubtlessly the case in geometry, as Hilbert points out. Whether this is also the case in other disciplines like mechanics or electrodynamics is a serious question which has to be clarified, *before* one submits them to the axiomatic method. Otherwise, the results of an axiomatic treatment could be very misleading because the axiomatic presentation of a field, in which the facts are insufficiently known, puts the theory in a more *finished* state than it really is, and this in turn could mean that one cannot decide which axioms are necessary and which are superfluous. This will become clear, as soon as we have explained what the core of the axiomatic method is.[9]

Therefore the next question to be answered is this: What does it mean to subject geometry to an axiomatic treatment? What's the goal of an axiomatic treatment? We all seem to know this, yet, according to Hilbert, we don't really know, at least not explicitly. It took Hilbert himself almost ten years to find a convincing answer. Therefore, let me proceed very carefully and begin with the negative part of the answer: The task of an axiomatic treatment is not merely to set up a system of axioms for geometry, as Euclid did it for the first time and geometers have done since. This is at best part of the whole enterprise, and not even the most important one. The task of an axiomatic treatment is much more, and much more profound.

It is an investigation of certain *meta-logical* properties and relations among various axioms and axiom systems. Before elaborating on this point, let me first point out why the axiomatic method enters the scene precisely at this junction. The answer, although not simple, is nevertheless straightforward. In order to investigate meta-logical properties like *consistency* and *completeness* of axiom systems or meta-logical relations such as dependence or independence of axioms, one needs certain means because these properties (in general) cannot be grasped immediately neither by our finite intellect nor by our sensual intuition. The question of the logical independence of the axiom of parallels from the other axioms of Euclid is a typical case in point. Quite unusual and sophisticated means are required to recognize its logical independence from the other axioms of Euclid because in our spatial intuition "parallelism" of straight lines seems inevitably connected with the other properties of Euclidean space like congruence and the Archimedian axiom of continuity.[10]

Today we see a certain tendency to view this and similar questions not only as settled, but also as equally easy and straightforward to answer. The latter evaluation, however, rests on an illusion. The question of the logical dependence or independence of the axiom of parallels only seems simple for *us*, because we are accustomed to its solution by model-theoretic means[11], but for the geometers of the time these questions were very tricky and hard to solve. The main problem was connected with the logical function or *role* of the so-called non-Euclidean models in the *proof* of the logical independence of the axiom of parallels from the rest of Euclidean axioms: What did the construction of such "non-standard" models really show? The consistency of the respective non-Euclidean geometry, for which they are models? The answer to this question is, of course, no, at least not in an "absolute" sense of consistency, because the possibility of such models or, more precisely, their internal consistency is itself in question. Consequently, one should first prove the consistency of these models, before one uses them in turn to prove the consistency of a non-Euclidean geometry. However, the proof of the consistency of a "non-Euclidean-model" is a difficult task, if one wants to avoid an *infinite regress* or a *vicious circle*.[12] Fortunately, I do not have to go into this further because there exists an alternative approach to the logical analysis of geometry, which avoids the problem of completely constructing consistent geometrical models for non-Euclidean geometry.

This approach is the interpretation of geometrical theories, of their terms and relations by arithmetical concepts. Although the basic idea of this approach is rather old (it dates back to Descartes' analytical geometry), it has been mainly worked out and elaborated by Hilbert between 1894 and 1899. Indeed, Hilbert was able to raise this method of "arithmetical interpretation" – as I will call it – to such a level of sophistication that he was able to prove the logical independence not only of the axiom of parallels but also of the axiom of Archimedes and many other axioms from the respective remainders of the Euclidean axioms. In this way Hilbert was able to show that the axioms and sentences of Euclidean geometry formed a complex logical network of mutual dependencies and independencies, instead of a simple *linear* chain of deductive reasoning, as it is commonly taken. The fundamental schema, by which Hilbert achieved these results, is the following:

In order to prove that sentence B is logically independent of the sentences A_1 to A_n $(=A)$, one forms two sets of sentences: *A and B* and *A and not-B*. If both sets are consistent (in the sense that, taken as axioms together with a proper logic of deduction, no contradiction can be deduced) then B is logically independent of A, in the sense that neither B nor its negation $\neg B$ can be deduced from A_1 to A_n.[13] In order to prove that both sets of sentences are consistent, Hilbert constructs a pair of arithmetical interpretations, one interpretation for every one of the two sets, which differ only in one respect. In the one interpretation B is valid, in the other $\neg B$, the interpretation of the remaining sentences $A_1 - A_n$ being the same. The fundamental trick, by which Hilbert achieves this, is the construction of appropriate number fields, such that in the first number field B is valid and in the second *not-B*, whereas all other axioms $A_1 - A_n$ are valid in both number fields. Although I cannot deal with this in great detail, let me just point out the two main aspects of what Hilbert had achieved:

1) The question of the logical independence of a sentence B from a system of axioms A_1 to A_n has been reduced to the question of the consistency of both axiom systems *A and B* and *A and $\neg B$*. This is a purely *conceptual* reduction, which fixes the meaning of the "meta-logical" concepts of the independence-dependence distinction.

2) The question of the consistency of a *geometrical* axiom system has been reduced to the question of the existence of a consistent *arithmetical* interpre-

tation for the axiom system in question. This is a *semantical* reduction in the following relational sense. If we suppose that the theory of arithmetic, which was used in the construction of the corresponding number field, is consistent, the proof is completed with the stipulation of an appropriate number field over which the geometrical axioms can be interpreted as formally correct and true. (I shall elaborate on this below.) Otherwise, one has to prove that the arithmetical interpretation, i.e. the theory, that was used in the construction of the number field, is consistent.

Although Hilbert had no proof that arithmetic is consistent, the reduction of the consistency of geometry to that of arithmetic was nonetheless a big advantage for two reasons: First, there was not the faintest hint that arithmetic would turn out to be inconsistent. Second, Hilbert hoped already at the turn of the century that it would be possible to prove the consistency of arithmetic in a completely new way, in which one had not to rely upon the consistency of any theory whatsoever. Let me call such a proof an "absolute" proof of consistency in distinction to the relative consistency proofs of modern model theory. (By the way, the first problem in Hilbert's Paris list is in fact an absolute consistency proof of arithmetic). However, this hope for an absolute proof of consistency of arithmetic turned out to be futile thirty years later, when Gödel proved the deductive incompleteness of the Peano-arithmetic. Yet quite independent from this, relative consistency proofs remain a valuable tool in the logical analysis of scientific theories.

Now, let me return to my main theme, the roots of mathematical physics, and answer two questions: (1) What precisely is the axiomatic method? (2) In which sense is geometry a model for the axiomatic treatment of other physical disciplines, such as mechanics or the kinetic theory of gases?

The answer to the first question is relatively easy; it consists of two parts: First, the goal of the axiomatic method is the *logical analysis* of a scientific discipline such as geometry or mechanics. Second, the means of the axiomatic method are three: First of all, the discipline has to be axiomatized. Because this is possible in several ways, one has to choose the most "appropriate"[14] kind of axiomatization. But this is only the first necessary step. The second step is more important; it is a *deliberate* variation of the axiom system, as outlined above, in order to study the logical dependencies as well as independencies among the different axioms and concepts of a given field of inquiry. The third and final step is the proof of the consistency of the various axiom systems by specifying an arithmetical interpretation. This is, by the way, the context into which the quotation with the "Tische, Stühle and Bierkrüge" belongs. This reminds me of a problem, which I have neglected thus far.

I believe that most readers know that Frege had vehemently criticized Hilbert's *Foundations of Geometry*. Part of the critique was that the notion of interpretation was notoriously unclear. Although, I think Frege's critique was excessively exaggerated, I have to admit that he was right in one specific respect. Hilbert had failed to explain in the *Foundations*, what he meant by an interpretation of an axiom system. In particular, he did not specify the distinction between the logical and the non-logical, i.e. the descriptive vocabulary of a theory, and in this way he failed to determine the limits of admissible interpretations of a theory.

It is, however, important to note that a theory cannot be interpreted *arbitrarily*, but only within certain limits determined by the meaning of the logical expressions as well as the axioms of the field. The axioms of the "original" theory have to remain true – besides their syntactical correctness – in the new interpretation; otherwise, the interpretation is no interpretation at all, but just a different theory! What this implies is best illustrated by an example. Take the sentence: $\forall x \; \exists y: xRy$. This sentence can be interpreted in a variety of ways – in particular over quite different domains of (individual) objects. Let me just mention two: (1) If we take the natural numbers 1, 2, 3, ... as the domain of objects, over which the variables x, y run, then we *cannot* interpret R as the predecessor relation, because 1 has no predecessor! (2) If we take human individuals as domain of objects, then we can interpret R as "y is father of x", because the sentence "each human individual has a father" is true to all what we know about human beings.[15] The formal sentence could, however, not be interpreted in precisely the same way over the domain of bees, because *male* bees have no father, but only a mother. Consequently, if we choose bees as the domain of individuals, we have to interpret R as "y being the mother of x". This example has made clear, I hope, in which way the meaning of the logical constants restricts the range of the *admissible* interpretations of the descriptive expressions. Now, let me turn to the second question: In which sense is geometry a model for the axiomatic treatment of other physical disciplines?

The main goal is, of course, precisely the same as in geometry: a logical analysis of the concepts and sentences of a certain discipline. But the "application" of the axiomatic method to a proper physical discipline, like mechanics or electrodynamics, is more or less different because of the fact that these disciplines have not the same degree of maturity as geometry. For example, according to Hertz, there was a certain conceptual confusion regarding the concept of *force* in mechanics, which had to be removed before mechanics could be reasonably subjected to an axiomatic inquiry. Furthermore, there are other differences between geometry and mechanics, which had to be respected. The most obvious is the considerably greater complexity of mechanics, in the trivial sense that geometry is a proper part of mechanics. What I have just said applies, mutatis mutandis, to other disciplines like electrodynamics and thermodynamics. Hilbert was completely aware of this and modified the axiomatic method correspondingly, as I will show in two steps.

First, I shall return to the Sixth Problem and point out that Hilbert's formulation of it already "reflects" the differences between geometry and the other sciences. Second, and more important, I will at least outline that Hilbert and his school tried to solve the Sixth Problem on the basis of *modified* standards. In other words, I will show that Hilbert (together with Minkowski and Born) took serious efforts to axiomatize physics and, thereby, to contribute to mathematical physics, as they understood it.

In order to avoid a common misunderstanding of the axiomatic method, let me add a clarifying remark. It's a widespread prejudice that axiomatics contributes little (or nothing) to the progress of physics as a research discipline; that is as a discipline which tries to recognize new fundamental laws of nature. It is primarily a mental exercise, so this prejudice continues, which is of interest mainly for the mathematician and not for the working physicist. This may be the case (at least I will not deny it). But if this view is intended as an objection against an axiomatic approach to physics, it rests on a plain category mistake. The aim of the axiomatic method – at least as Hilbert saw it – is not the recognition of *new* laws, but the conceptual and logical "clarification" of *existing* ones! Of course, sometimes both aims come together, for example, when Einstein and Hilbert discussed the foundations of general relativity in 1915 in Göttingen.[16] But normally they are kept apart and do not compete with each other. To put it directly, the aim of the axiomatic method within physics is much more similar to the aim of the logical empiricists (and analytical philosophy in general) to reconstruct *existing* theories according to logical standards, than to physics as a *research discipline*. Indeed, the logical empiricists of the Vienna Circle (in particular H. Hahn, M. Schlick and R. Carnap) borrowed the idea of a "logical reconstruction" of physics to a large extent from Hilbert's work on the *Foundations of Geometry* [17]. Now, let me come back to the statement of the Sixth Problem in Paris.

First, let me stress that Hilbert talks of geometry only as a model for the axiomatic treatment of other physical disciplines, such as mechanics or electrodynamics. This already reflects, as I view it, the need to *modify* the axiomatic method according to the higher complexity of the physical disciplines. That such a liberalization is needed and that Hilbert was indeed aware of it, is confirmed by the fact that he refers to the works of E. Mach, H. Hertz, L. Boltzmann, and P. Volkmann as *examples* of foundational investigations of mechanics, which come close to his own intentions.

In respect to the foundations of mechanics already significant inquiries from the side of physics exist. I point to the works of Mach, Hertz, Boltzmann, and Volkmann; it is, therefore, very desirable that the foundations of mechanics be discussed also by the mathematicians.[18]

Now, with respect to this remark two questions arise: First, in which sense come these works close to Hilbert's own axiomatic intentions? Second, why is it, nonetheless, desirable that the foundations of mechanics be discussed *also* by mathematicians, if the physicists have already done a good job? In other words, what is still missing in the works of these physicists that is so indispensable for Hilbert? Let me take the first question first and then move on to the second, more complicated question.

In spite of Hilbert's uniformly "positive" reference to the works of Mach, Hertz, Boltzmann, and Volkmann a first trivial yet important point could be easily overlooked. These treatises are very different in style as well as in the arrangement of the content and the principles of mechanics. Mach's treatise, for example, follows the historical development of mechanics, whereas Hertz's book on the *Principles of Mechanics* is a very abstract and systematic investigation of the foundations of mechanics, which pays little attention to the applications of mechanics.[19] Both books are indeed so different that the question arises: what do they have in common that could arouse Hilbert's interest? It is not easy to find a plausible answer. It can't be the mathematics because Mach's book contains almost no mathematics. Nor is it an axiomatic treatment of mechanics, because again Mach is loathe to such a treatment and, consequently, avoids any reference to axioms. In fact he presents Gauß' "principle of least force" and other minimal principles of mechanics as expressing no more than "the sentence of the conservation of work".

The best answer to the question "What have both books in common?" is, to my mind, their "critical attitude" towards the conceptual foundations of mechanics. Both authors were not content with the orthodox way of presenting mechanics and tried instead to find a better, more coherent conceptual basis for mechanics. In Mach's case this lead to an impressive critique of some of the traditional concepts of mechanics, such as mass and absolute rotation. Hertz, on the other hand, developed a completely new conceptual basis for mechanics from the geometrical idea of a "straightest path", which avoided the notoriously unclear notion of force altogether. Although both books are conceived in very different ways (the first historically with many concrete examples and descriptions of experiments, the second in an almost axiomatic form with no concrete example at all) it was not too difficult for Hilbert to recognize their common root, to wit, a critical attitude with respect to the conceptual foundations of mechanics. This attitude is, however, already very similar to Hilbert's own intentions in the axiomatic treatment: a thoroughgoing *logical analysis* of mechanics. Indeed, it is so close that Hilbert does not only recommend the books of Mach, Hertz and Boltzmann[20], but reproduces the principal content of the last two, more or less, in his lectures on mechanics from 1898 to 1911. I will return to this point below. Before, I shall discuss why the mathematicians should at all enter the foundations of mechanics, if some physicists have already done the job? Hilbert's explicit answer is this: Because the physicists didn't solve all problems regarding the foundations of mechanics.

Boltzmann's book on the Principles of Mechanics, for example, stimulates us to develop a strong mathematical justification for the limit processes, indicated above, which lead from the atomistic conception to the laws of motion of continua. Conversely, one could try to deduce the laws of motion of rigid bodies by limit processes from a system of axioms, which rest on the idea of a medium filling the entire space and whose states, defined by some parameters, vary continuously [after a dash Hilbert adds] because the question of the equivalence of different axiom systems is always of the highest principle interest. [21]

This is, of course, a fundamental problem of the first rank – one which has not yet been solved satisfactorily – all assertions to the contrary not withstanding. Nonetheless, Hilbert's hope to solve this problem during his lifetime, was one of the primary motives for his search for a truly "universal" theory, which could serve as the foundation for all physical theories of a more special kind. There are still two subtler and intimately related reasons why not only the physicist but also the mathematician should investigate the foundations of physics. The first concerns the inner structure of physics, the second its consistency. As Hilbert points out, physical theories become *extended* again and again by new laws and the double question arises as to what the logical order or hierarchy of these extensions is and, furthermore, whether the different extensions are consistent with each other. One wants to know which new assumption depends on which, and which are logically independent of the previously accepted laws. Both inquiries need special means and skills, means and skills which the mathematician is already acquainted with from his logical investigations of geometry. As in geometry, the inquiry of axiom systems, which are possibly not true, or only presumably true, is a meaningful task also in physics. It may help to reveal the logical structure of a particular physical theory or even of physics as a whole.

MATHEMATICAL PHYSICS IN GÖTTINGEN SINCE 1900 UNTIL 1930

Unfortunately, there isn't enough time to tell the whole history of mathematical physics in Göttingen from 1900 (our point of departure) until 1932, the year in which von Neumann published his *Mathematical Foundations of Quantum Mechanics*. I can only point to the main steps of the joint efforts of Hilbert and his many excellent disciples to *axiomatize* physics, and then choose one or two concrete examples from this development, which at least show, what it means to inquire the logical structure and consistency of a physical theory or a whole discipline like mechanics or electrodynamics. As far as I can see, one has to distinguish at least seven topics, in which Hilbert dealt with particular problems or domains in physics. Of course, these topics are not sharply separated, neither in time nor with respect to their content. In part they overlap, in part they are separated by a kind of transition phase belonging to no particular period or field.

1) Hilbert's lectures on mechanics as the foundation of all physics (1898–1911).[22]

2) Hilbert's and Minkowski's concern with the kinematics of the special theory of relativity and Minkowski's four-dimensional theory of space-time (1905-1911).

3) Hilbert's inquiries of the dynamics of the electron within the framework of special relativity and Born's theory of "Rigid Motion" (1908-1911).

4) Hilbert's (published) axiomatic treatment of Kirchhoff's "Laws of Radiation".

5) Hilbert's lectures on "Kinetic Theory of Gases" and "Statistical Mechanics" – and his critique of Boltzmann's alleged deduction of an irreversible equation, the so-called H-theorem, from statistical mechanics (1911-1922).

6) Hilbert's approach to general relativity on the basis of Mie's field theory and Einstein's theory of relativity and his concern with the "principle of causality" (1915-1918).

7) Hilbert's lectures on "Mathematical Methods of Quantum Theory" in 1922 and 1926 until the appearance of von Neumann's masterpiece *Mathematische Grundlagen der Quantenmechanik* in 1932.

As already said before, there are many further topics which are not listed, for example the axiomatic treatments of classical and relativistic thermodynamics, lectures on the molecular theory of matter including early quantum mechanics, treatment of stability problems, Noether's theorem etc., not to forget the many popular lectures about the new physics. Of the listed topics, points 2), 3) and 4) are relatively well-known, at least among historians of science; point 6) has been the focus of a *priority debate* quite recently, of course, not between Hilbert and Einstein, but between some historians of physics.[23] Hence, there remain only the topics 1), 5) and 7), about which relatively little is known. Because the last topic has not yet been studied sufficiently, I will confine my remarks to the first and fifth point. Both points exemplify very well and in different ways, what I intend to show, namely, how a logical inquiry of a physical discipline by axiomatic means can combine with an interest into the foundations of that same discipline and vice versa. One should notice that both aspects are not identical; they can occur quite independently from each other, and they often do. But under favorable conditions they can also come together and in these lucky cases it is extremely interesting to recognize in which ways they work together, how they influence and stimulate each other.

In this context let me mention once more Hilbert's research in the foundation of mathematics. It is well known that Hilbert was a kind of *reductionist* regarding the foundations of arithmetic and analysis, but – and this is too often neglected – a very peculiar one. Hilbert does not propose a kind of ontological reductionism in the sense that the *actual* infinite, occurring in analysis in the very large and the very small, should be reduced to some finite domain of elementary arithmetic – this is, of course, impossible. Instead, he proposes a kind of epistemological reductionism: The infinite, as it occurs as an *idea* in analysis, has to be tamed and controlled from a "finite perspective". It's very important to understand that this does not mean, that the most elementary arithmetic, simple counting, has to be finite in the strict sense of negating the possibility of the "and so on and so forth". This would be a lethal misunderstanding and make whatever kind of reduction totally impossible. What Hilbert really proposes is something much more modest. On the level of a meta-theory, which has to be "finite" in a certain reasonable sense, we should make sure that the iterated process of theory extensions (resp. domain expansions) by "ideal" elements does not lead to contradictions. This can be achieved, according to Hilbert's view, by controlling the deductive structure of the formalized object theories in question and their stepwise extensions by new axioms and ideal elements. Of course, the "means" by which we control the consistency of the object theories, have to be finite in some intelligible sense. The deep mystery is, however, what "finite" on the meta-level of investigation means in spite of an undeniable "infinity" of the domains of most object theories. I will not discuss this old controversy but only remark that, whatever it's final solution will be, it does not imply an ontological reduction of consecutive object theories. It only erects a certain epistemological limit to domain expansions resp. new theories. They have to be *consistent* with respect to the already "established" theories respectively with the experimental results and observations. This doesn't exclude the introduction of "new" entities (objects, properties and relations), whose existence is not implied by the former theories, and it permits a new interpretation or better a re-interpretation of old experimental results and observations in the light of new theories.

A SHORT HISTORY OF HILBERT'S LECTURES ON CLASSICAL MECHANICS

Hilbert held between 1898–1911 seven lectures on classical mechanics. Except for the last one, their structure is quite similar. Hilbert organizes the whole field of mechanics into three classes of problems and tasks: (i) systems with *one* particle, (ii) systems with many, but a *finite* number of particles and, last but not least, (iii) systems with an *infinite* number of particles. The latter embraces roughly what is traditionally called continuum mechanics. This looks, I confess, not very interesting and seems to confirm the usual prejudice that Hilbert's approach to physics is rather formal – typical for a mathematician, so to speak. But there are two very interesting points hidden within this set-up.

The first point regards the principles of mechanics, known as variation or "minimal principles" such as Gauss's principle of "least action" or Hertz's principle of "the straightest path". In the first part of his lectures, Hilbert gives a detailed mathematical analysis of the different principles and compares them with respect to their logical form and deductive strength. First at all, he distinguishes two types of principles: those consisting of a time integral, and those of a time-differential form. Next he inquires which principle implies which, if one takes into account different external forces and various constraints between the particles and the initial conditions. This part resembles to an astonishing degree Hilbert's procedure in geometry, but with two fundamental differences. In physics, Hilbert does not vary the axioms arbitrarily, at will so to speak, but takes them, as they already exist in the literature. Furthermore, he does not construct special number fields in order to prove the consistency of physical theories. In spite of the analytical form in which the physical concepts, functions and theories in general are presented, this is not necessary. It's sufficient for the consistency to show that the physical equations (or sets of equations) have "non-trivial" numerical solutions. It is beyond the scope of this essay to go into any details; but let me mention at least one result. It turns out that Hertz's principle of the "straightest path" is not only the "strongest" principle in the sense of having the widest range of possible applications, but also from a physical perspective the simplest and most distinct one, because it needs only local constraints between adjacent particles, contact forces, so to speak, instead of the dubious notion of "instant forces".

The second point is, perhaps, even more interesting. Hilbert tries to unite all physical theories which are known around 1900 into one fundamental theory, namely, that branch of mechanics which is called "continuum mechanics". This means that he tries to construct solid state physics, hydrodynamics, thermodynamics, statistical mechanics and even electrodynamics as particular *branches* of continuum mechanics. But here a serious question arises: How can he do this? Was it not already obvious that such a program was doomed to fail? In particular with respect to Maxwell's equations for the electromagnetic phenomena, because these equations were not Galilei-invariant? The answer to the last question is a definite no – at least at the turn of the 19th century – because there were still five years to come, before the "new" principle of relativity – which sheds light onto the relativity of rest and motion – was formulated by Einstein, and at least five years more, before it became accepted by the majority of physicists.

But, to capture all physical phenomena (known at the time) in one fundamental theory was an enormous task and required apart from great mathematical skills, a leading idea of how to achieve this, not to speak of the many minor ideas, which were necessary to incorporate the different branches of physics into the one fundamental theory. Nonetheless, the basic idea which Hilbert used is very simple; it consists of two steps: First, he introduces a (three-dimensional) continuum of mass points. Second, he divides the possible relative motions of the mass points into several types or classes. The simplest case is that, in which the (spatial) distance between the mass points in a certain volume is fixed. This leads to "solid-state physics" of finite bodies. The next simple case is that, in which neighboring mass-points can have a small constant velocity relative to each other. This leads to hydrodynamics with its different branches. The next complicated case is that, in which the velocities of the considered mass points respectively the molecules as ensembles of mass points are not correlated in time and space. This leads to statistical mechanics respectively thermodynamics in the limit of infinitely many molecules. Then also accelerated motions, in particular circular motions are admitted. This leads, or better should lead, to a conception of electrons as stable *vortexes* in an electromagnetic continuum. I stop here because now the guiding idea is obvious. The whole of physics shall be set up on the basis of *one* fundamental principle – for example the principle of Hamilton-Jacobi or Gauss or of Hertz – by combining two modes of division or classification of material systems. (1) The number of mass points respectively particles to be considered is exactly one, many or infinite. (2) Different types of relative motions of mass points are to be distinguished: rest, constant relative velocities (all the same or different), accelerations of different types, linear, circular, etc., and their combinations.

Of course, most of these ideas were *not new*! One can find them scattered over the works of many physicists of the 19th century, such as Maxwell, Kelvin, Helmholtz, Hertz, Boltzmann and Kirchhoff, to name only the best known. But, as far as I can see, it was only Hilbert and his circle, who really tried to put them together into one encompassing mathematical theory. I say "really", because Hilbert indeed worked very hard to come closer to the goal of a *unification* of all branches of physics in one fundamental theory – a theory built on three mathematical pillars: the calculus of variation and invariance theory, differential geometry, and the theory of probability. He was not content just with the proclamation of such a possibility or – as philosophers like to say – with its possibility *in principle*, but he used his huge mathematical knowledge and analytical talent to solve the concrete problems and obstacles which resist the idea of a unification of all the diverse physical phenomena in one universal mechanical theory. And, of course, as we all know with the wisdom of hindsight a century later, this ingenious idea of mechanics as the foundation of all physics could not withstand neither the experimental nor the theoretical development over the next two decades. Hilbert had to change, even to *exchange*, the guiding idea of mechanics as *the* fundamental theory of all physics and to adapt his conception of the foundations of physics to the new ideas about space, time and motion, which were forced upon physicists by experimental findings like the "negative" result of the Michelson experiment. And this is precisely, what we see Hilbert doing in the years from 1905 to 1925, while he struggled towards a new foundation of physics. In the beginning, in collaboration with his friend Minkowski, later with his students Born, Herglotz, Hecke and – last but not least – Nordheim, Courant and von Neumann.

It is far beyond the scope of this essay to unfold the whole process of remodeling the foundations of physics, but I will at least outline two aspects of this process, which throw an interesting light on the interrelation between axiomatics on the one hand and the search for a new foundation of physics on the other. The first aspect concerns the competition between two conflicting ideas with respect to a firm foundation for all physics, namely, atomism and the continuum. The second aspect touches the role of phenomenological theories as an intermediate step towards a solution of the foundational problem. In order to make the remodeling a bit more vivid and concrete, let me go back to Hilbert's remark about Boltzmann's book that it "stimulates us to investigate the limit processes, which lead from the atomistic conception (of matter) to the laws of motion of continua (and inversely)". Since Hilbert had made that remark, the question which of the two ideas is the more fundamental one had become extremely acute for several reasons. On the one hand, the enormous success of Maxwell's theory in handling electromagnetic phenomena such as radio waves and light waves, had made it pretty clear that physics had to use, in one way or the other, the notion of a "field" and thereby also, and quite inevitably, the idea of a continuum of the field quantities. On the other hand, the idea of atoms had gained much ground not only in chemistry but also in physics, because the atomic hypothesis could explain many phenomena, such as the different specific heats of molecules, which otherwise were incomprehensible.

In his second lecture on continuum mechanics in the winter term 1906/7, Hilbert still tries reconcile both competing ideas by reducing all physical phenomena of heat, electricity, chemistry, and so on, to the theory of point mechanics (of finite or infinite many mass points). Having reminded the audience of Hamilton's equations of point mechanics, he continues:

As goal of *mathematical physics* we can perhaps describe, to treat also all not purely mechanical phenomena according to the model of point mechanics; hence, one tries to gain detailed mechanical ideas of the singular phenomena in order to predict the whole process in a detailed form – on the basis of Hamilton's principle, perhaps after appropriately generalizing it. Physics has ever since likewise directed attention to this and already gained brilliant successes in this direction; one has only to recall the kinetic theory of gases, the theory of ions, the theory of electrons and chemical atomism. In any case, one is dealing with ideas, which go quite into the details of the motions of the gas molecules, the electric particles etc., from which the whole process is then inferred on the basis of analogous minimal principles. The deep and definitive recognition of the structure of matter, which we try to achieve and treat in this way, seems also to be the sole modality to offer a full satisfaction to our quest for physical knowledge.[24]

So much for the familiar view of mechanics as the foundation of all physics. But then, in a surprising turn, Hilbert confesses that the atomic hypothesis and the program to reduce all physics to point mechanics has not been as successful as physicists had hoped for, and for this reason he proposes a quite different procedure as an *intermediate* step:

Even if the keen hypotheses, which have been made in the realm of molecular physics, sometimes certainly come close to the truth because the predictions are often confirmed in a surprising manner, one has to characterize the achievements still as small and often as rather insecure, because the hypotheses are in many cases still in need of supplementation and they sometimes fail completely. ... Such considerations recommend it as advisable to take *meanwhile* a completely different, yet a *directly opposite* path in the treatment of physics – as it indeed has happened. Namely, one tries from the start to produce as little detailed ideas as possible of the physical process, but fixes instead only its general parameters, which determine its external development; then one can by *axiomatic physical assumptions* determine the form of the Lagrangian function L as function of the parameters and their differential quotients. If the development is given by the minimal principle $\int_{t_1}^{t_2} L \, dt = \text{Min.}$, then we can infer general properties of the state of motion solely from the assumptions with respect to the form of L, without any closer knowledge of the processes.[25]

In order to grasp the intention and significance of this statement correctly one has to remember the time and state of physics in which it was uttered. During the first decade of the 19th century it became gradually clear – mainly through the works of Poincaré, Fitzgerald and Lorentz, which were intensively studied in Göttingen – that the electron could not simply be treated as a *rigid body*, as

Abraham[26] had supposed. Also the "reduction" of thermodynamics to statistical mechanics via the kinetic theory of gases raised serious problems regarding the logical strength of the intended deduction. For these and similar reasons Hilbert postponed his program to formulate the entire physics within continuum mechanics and proposed instead a more *modest* procedure, a procedure which only wants to capture the macroscopic physical phenomena of a certain type without striving for an explanation of the phenomena by microscopic (atomic) assumptions. This phenomenological approach, as it was called by Hilbert, is especially well-suited for an axiomatic treatment because one can study various assumptions regarding the relevant parameters and see what consequences they have with respect to the phenomena. Hilbert gives the following instructive example.

In order to give ... an intuition of the new approach I refer to the theory of elasticity, which treats the deformations of solid bodies caused by the mutual influence and displacements of the molecules; we will have to renounce a detailed description of the molecular processes and instead only look for the parameters from which the measurable state of deformation of the body depends at each place. Then one has to determine the form of dependence of the Lagrangian function from these parameters which is, properly speaking, composed of the kinetic and potential energy of the single molecules. Similarly, in thermodynamics one will not deal with the oscillations of the molecules, but instead introduce the temperature as a general parameter and inquire the dependence of the energy from the latter. The presentation of physics just indicated, ... which permits the deduction of essential statements from formal assumptions about L, shall be the core of my lecture.[27]

I should emphasize that this phenomenological approach is only an intermediate step – an auxiliary device from a foundational point of view – which should be superceded in the long run by an explanation of the phenomena in terms of the respective microscopic structure. From an axiomatic point of view, however, the macroscopic-phenomenological approach is logically just as suited for an axiomatic investigation as the microscopic-molecular one. This shows that both points of views, the axiomatic and the foundational one, are not identical and can be pursued quite independently. Hilbert, however, for the most time in his life had both aims in mind. Hence, one should expect that he would later return to the foundational point of view. And this was in fact the case, but not yet in 1911, when he lectured for the last time on "continuum mechanics". Although the lecture begins precisely with the promise to reduce all known physical disciplines such as hydrodynamics, thermodynamics, optics and primarily electrodynamics to continuum mechanics, a short glimpse into the table of contents reveals that this is not the case, at least not in the same way, as one would expect from his earlier lectures. The lecture begins quite similar to the earlier lectures, but after chapter 7 on "thermodynamics" (which, by the way, is not reduced to statistical mechanics but treated as a phenomenological theory) a completely new and unexpected chapter entitled "The Principle of Relativity" follows. The rest of the lecture is then concerned with the significance and consequences of this principle with respect to the classical disciplines of mechanics, electrodynamics and thermodynamics. In other words, Hilbert tried to remodel the house of physics on the basis of the new principle of relativity. This task occupied his interest during the next years and came to a preliminary conclusion with the first and second communication of "Die Grundlagen der Physik" to the Göttingen Academy of Sciences in 1915/16, in which he presented his version of the "general theory of relativity". Yet, contrary to what Hilbert might have expected, the foundation of physics was not completed. With the occurrence of quantum phenomena a new chapter had to be opened, and Hilbert was quite aware of this.

Coming to the end of the paper let me briefly touch point 5) of the above list of topics: the "Foundations of the Kinetic Theory of Gases". This theme is insofar of considerable interest as many prejudices exist about what Hilbert has achieved in this work. The principle prejudice is presumably best expressed by Steven Brush, a well-known historian of statistical mechanics and the kinetic theory of gases:

When Hilbert decided to include a chapter on kinetic theory in his treatise on integral equations, it does not appear that he had any particular interest in the physical problems associated with gases. He did not try to make any detailed calculations of gas properties, and did not discuss the basic issues such as the nature of irreversibility and the validity of mechanical explanations, which had exercised the mathematician Ernst Zermelo in his debate with Boltzmann in 1896-97. A few years later, when Hilbert presented his views on the contemporary problems of physics, he did not even mention kinetic theory. [Brush refers here misleadingly to Hilbert's "Grundlagen der Physik", Math. Ann. 92,1 (1924)] We must therefore conclude that he was simply looking for another possible application of his mathematical theories, and when he had succeeded in finding and characterizing a special class of solutions, later called "normal", which are determined by a sequence of linear integral equations with symmetric kernels, his interest in the Boltzmann equation and in kinetic theory was exhausted. (Brush, 1976, Book 2, p. 448)

It is superfluous to say that I disagree with almost every sentence in this statement. Although this is not the place to disprove Brush's claim point by point, I will at least refute the basic error in this charge. It is simply not true that Hilbert was not concerned about the foundations of statistical mechanics and the question of irreversibility. On the contrary, all the documents on statistical mechanics and related topics we know show that these questions troubled him for many years. Let me just mention two of them, which Brush apparently is not familiar with. In 1912 Hilbert gave a talk entitled "Über meine Gasvorlesung", which refers back to his lecture on the kinetic theory of gases in the winter term 1911/12. In this talk he criticizes Boltzmann for his alleged deduction of the irreversible master equation from reversible mechanical assumptions with the following words:

You will convince yourself that in Boltzmann's expositions regarding the foundation of Maxwell's formula not one stone can remain on the other. [28]

In his lecture Hilbert had pointed out that Boltzmann's use of the notion of probability in his famous "Stoßzahlansatz" in order to deduce Maxwell's formula for the velocity distribution of the molecules was entirely futile and had to be corrected. Unfortunately, Hilbert did not publish this part of his lecture, but only the solution of Maxwell's equation for a simple *mono-atomic* gas (as chapter 22 of his book on the theory of linear integral-equations). This could (and obviously did) raise the impression that Hilbert was only interested in the mathematical problems of the kinetic theory of gases. That this is not the case is not only documented by several lectures, which Hilbert gave on the "Foundations of statistical mechanics" and the "Molecular theory of matter" over the next ten years, but also by a talk with the programmatic title "Natur und mathematisches Erkennen", which he gave in Copenhagen in 1921. The second half of this talk is exclusively devoted to the question: What is the *true* nature of irreversibility? Hilbert's answer is very astonishing and really intriguing: It is only an *anthropocentric* phenomenon and not an objective property of nature, because it is logically impossible to deduce an irreversible equation as the master equation from reversible ones and, as a matter of fact, the fundamental equations of physics are all reversible. And because there is no reason to change this, Hilbert arrives at the conclusion that irreversibility is a phenomenon rooted exclusively in our human perspective. Although it's tempting to discuss this solution from today's point of view, I have to come to an end. My personal conclusion with respect to Hilbert's achievement regarding mathematical physics is this:

Although Hilbert was in physical matters – that is, regarding the creation of new ideas and the integration of experimental results into new theories – certainly not as innovative and active as his contemporaries Planck, Lorentz, Minkowski and Sommerfeld and in particular the generation of younger physicists like Einstein, Bohr and Schrödinger, he was by no means the "formalist mathematician", as he is often portrayed, who did not understand the physical issues. On the contrary, he was an extremely gifted analytical thinker, who did not stop logical thinking shortly before physics. Instead, he was convinced that the main tool of analytical thinking, the axiomatic method, could also be used in physics and thus created "mathematical physics" as an autonomous research discipline.

NOTES

1. *6. Mathematische Behandlung der Axiome der Physik:*
    Durch die Untersuchungen über die Grundlagen der Geometrie wird uns die Aufgabe nahe gelegt, *nach diesem Vorbilde diejenigen physikalischen Disziplinen axiomatisch zu behandeln, in denen schon heute die Mathematik eine hervorragende Rolle spielt: dies sind in erster Linie die Wahrscheinlichkeitsrechnung und die Mechanik.* (Hilbert, 1900, p. 272, my translation)
2. But in any case, I confess that to date the birth of mathematical physics with such a precision, as I have just done, is a wild exaggeration, and has primarily a *symbolic* meaning.
3. For example by the two leading "logical empiricists", H. Reichenbach and R. Carnap, but also by G. Ludwig in (1978).

4. For the sake of clarity I even risk some exaggerations in the subsequent formulations, which I am prepared to temper, once the message has been received.
5. All lectures on geometry begin with the remark that geometry is a natural science – "die einfachste und vollkommenste Naturwissenschaft". See (Hilbert 1891, 1893/4 1898/99, 1904 and 1927). Of course, with the invention of general relativity this view had become a fundamental principle in the form that the *metric* structure of space is dependent on the "mass distribution". But before 1915 this was not so obvious and Poincaré, the arch conventionalist, would never have accepted this. Hilbert's insisting from the beginning that geometry is a natural science therefore shows that he entertained a conception of geometry different from the Helmholtz-Lie-Poincaré tradition; he belongs to the tradition of Gauß, Riemann and Pasch.
6. Whether they have also a "meaning" in Frege's sense, I will leave open at the moment.
7. Die Geometrie ist die Lehre von den Eigenschaften des Raumes. Sie unterscheidet sich wesentlich von den rein mathematischen Wissensgebieten wie z. B. Zahlentheorie, Algebra, Funktionentheorie. Die Resultate dieser Gebiete können durch reines Denken gewonnen werden, indem man durch klare logische Schlüsse die behaupteten Thatsachen auf immer einfachere zurückführt, bis man schließlich nur noch den Begriff der ganzen Zahl nöthig hat.

    Ganz anders verhält es sich mit der Geometrie. Ich kann die Eigenschaften des Raumes nimmer durch bloßes Nachdenken ergründen, so wenig wie ich die Grundgesetze der Mechanik, das Gravitationsgesetz oder irgend ein anderes physikalisches Gesetz so erkennen kann. Es ist ja der Raum nicht ein Produkt meines Nachdenkens, sondern er ist mir durch meine Sinne gegeben. Ich brauche daher zur Ergründung seiner Eigenschaften meine Sinne. Ich brauche die Anschauung und das Experiment, wie bei der Ergründung physikalischer Gesetze, wo auch noch die Materie als gegeben durch die Sinne hinzu kommt. (Hilbert, Projektive Geometrie, 1891, p. 5-7)
8. This sense is compatible with a certain imprecision or fuzziness of our observations and measurements. Points and lines on a blackboard still have a certain extension or, to take another example, the planets are, of course, not points in the geometrical sense although they can be treated as such in point mechanics because their diameter is extremely small in comparison to their distance from the sun. Only in pure (Euclidean) geometry we take these notions in their idealised (normative) sense, as we know it from our pure intuition.
9. A similar point is made by A. S. Wightman in his profound review of the development of "mathematical physics" after Hilbert had stated the sixth problem. It seems, however, that Wightman is more liberal regarding the application of the axiomatic method to a discipline in an unfinished state like quantum field theory than myself. (Wightman, 1976, p. 158-159)
10. The spatial intuition, of which Hilbert writes in the *Foundations of Geometry*, should not be equated with the space of visual perception of the psycho-physiologist. The latter has been investigated by psycho-physical means; its structure is non-Euclidean.
11. Such means were first used by E. Beltrami and they were later improved by F. Klein. Afterwards they became elaborated and refined mainly by Hilbert and eventually by A. Tarski.
12. To give just an idea how difficult this is, take the surface of a sphere as a two-dimensional model for the non-Euclidean geometry with more than one parallel through a point outside a given straight line. An appropriate interpretation of this geometry seems easily to construct: Primarily, one has to interpret the term "straight line" as meaning the "great circles" on a sphere and to adjust the other geometrical notions like "point, plane, between, etc.", correspondingly. Hence, there seems to be no problem at this point. A real difficulty first occurs, if we ask further, whether the interpretation of the term "straight line" as "great circles on a sphere" is consistent with the other properties of straight lines, such as the "infinite extension" of straight lines, which are asserted by the remaining unchanged axioms of Euclidean geometry. This is obviously not the case, and therefore we have also to change these axioms. This may go on and on indefinitely, and we are not sure, whether no contradiction will arise at a later stage in the process of further and further adaptation.
13. If either B or its negation $\neg B$ could be deduced from A, then necessarily a contradiction would arise in one of the two sets of sentences: either in A_1 to A_n and $\neg B$, if B is implied by A_1 to A_n, or in A_1 to A_n and B, if $\neg B$ is implied by A_1 to A_n.
14. Appropriateness is, of course, only a relative concept; it depends on the goal which one has in mind. The goal of the logician may be different from that of the physicist. Hilbert is completely aware of this. Indeed as a logician he uses quite another kind of axiomatization of geometry

than later as a physicist. From the physical perspective the Helmholtz-Lie approach to geometry may be the most appropriate one. From the logical perspective it is definitely not, because this approach presupposes the "differentiability" of the transformation functions of rigid bodies. See appendix IV of (Hilbert, 1987).

15. It is funny that even according to the Christian Religion there is no exception from this rule, because the New Testament testifies that Jesus, although his mother was virginal, had a father. The biological sense of this, however, is not clear.

16. See (Sauer, 1999).

17. There is, however, a minor difference: The logical empiricists mainly talked about logical reconstruction of theories in rather general terms, without presenting concrete examples, whereas Hilbert at least elaborated some concrete examples in a detailed mathematical form.

18. Über die Grundlagen der Mechanik liegen von physikalischer Seite bedeutende Untersuchungen vor. Ich weise hin auf die Schriften von *Mach, Hertz, Boltzmann* und *Volkmann*; es ist daher sehr wünschenswert, wenn auch von den Mathematikern die Erörterung der Grundlagen der Mechanik aufgenommen würde. (Hilbert, 1900, p. 272)

19. Similar things could be said of the two other treatises, yet I restrict my remarks to Hertz and Mach, because their books are the most extreme examples of the different styles regarding the presentation and arrangement of the principles and the content of mechanics.

20. The book of Volkmann is not mentioned any more in Hilbert's later work on mechanics.

21. So regt uns beispielsweise das *Boltzmann*sche Buch über die Principe der Mechanik an, die dort angedeuteten Grenzprozesse, die von der atomistischen Auffassung zu den Gesetzen über die Bewegung der Kontinua führen, streng mathematisch zu begründen und durchzuführen. Umgekehrt könnte man die Gesetze über die Bewegung starrer Körper durch Grenzprozesse aus einem System von Axiomen abzuleiten suchen, die auf der Vorstellung von stetig veränderlichen, durch Parameter zu definierenden Zuständen eines den ganzen Raum stetig erfüllenden Stoffes beruhen – ist doch die Frage nach der Gleichberechtigung verschiedener Axiomensysteme stets von hohem prinzipiellen Interesse. (Hilbert, 1900, p. 272)

22. After 1911 Hilbert again lectured on mechanics in 1913/14, 1919 and 1924. However, these later lectures have a different perspective than the first: Mechanics is no longer considered as the foundation of all branches of physics. Instead, mechanics itself becomes downgraded to a branch of the new much more encompassing physics. For this reason, the later lectures will be not considered in this essay.

23. Sauer (1999) argues convincingly that the short dissonance between Hilbert and Einstein in autumn 1915 had little to do with a "priority dispute" but with a disagreement about competing research strategies.

24. Als Ziel der *mathematischen Physik* kann man es nun vielleicht bezeichnen, auch alle nicht rein mechanischen Erscheinungen nach diesem Vorbilde der Punktmechanik zu behandeln; man wird sich also detaillierte mechanische Vorstellungen der einzelnen Erscheinungen zu verschaffen versuchen, um sodann auf Grund des Hamiltonschen, vielleicht in geeigneter Weise verallgemeinerten Prinzipes den ganzen Vorgang in detailliertester Weise vorauszusagen. Die Physik hat nun auch von jeher hierauf ihr Augenmerk gerichtet, und schon glänzende Erfolge in dieser Richtung erzielt; es sei nur an die kinetische Gastheorie, die Jonentheorie, die Elektronentheorie und die chemische Atomistik erinnert. Allemal handelt es sich hier um ganz ins einzelne gehende Vorstellungen über die Bewegungen der Gasmoleküle, der elektrischen Teilchen etc., aus denen dann auf Grund analoger Minimalprinzipien der ganze Vorgang erschlossen wird. Eine tiefe und definitive Erkenntnis der Struktur der Materie, die wir so anstreben und so behandeln, scheint auch allein eine volle Befriedigung unseres physikalischen Erkenntnisdranges bieten zu können. (Hilbert, 1906, p. 6/7)

25. Wenn nun auch die kühnen Hypothesen, die man auf diesem Gebiete der Molekularphysik gemacht hat, gewiß manchmal der Wahrheit nahe kommen, da die Voraussagen sich häufig in überraschender Weise bestätigen, so muß man doch andrerseits das Geleistete noch als gering und meist recht unsicher bezeichnen, da die Hypothesen vielfach noch der Ergänzung bedürfen, und mitunter auch ganz versagen. ... Solche Überlegungen lassen es gut erscheinen, *einstweilen* einen ganz andern, ja geradezu *entgegengesetzten* Weg in der Behandlung der Physik einzuschlagen, wie es auch tatsächlich geschehen ist. Man sucht sich nämlich vornherein möglichst wenig detaillierte Vorstellungen des physikalischen Processes zu machen, sondern legt zunächst

nur einmal die allgemeinen Parameter, die seinen äußeren Verlauf bestimmen, fest; alsdann kann man durch *axiomatische* physikalische *Annahmen* die Form der Lagrangeschen Funktion L als Funktion dieser Parameter und ihrer Differentialquotienten bestimmen. Wird dann der Vorgang durch das Minimalprinzip $\int_{t_1}^{t_2} L\, dt = $ Min. gegeben, so kann man alleine aus Annahmen über die Form von L allgemeine Eigenschaften des Bewegungszustandes herleiten, ohne eine nähere Kenntnis der Vorgänge zu besitzen.(Hilbert, 1906, p. 7/8)

26. Max Abraham was *Privatdozent* in Göttingen from 1900 to 1909 and, of course, took part in the famous seminar of Hilbert and Minkowski about the theory of the electron in 1905/6.

27. Um eine Anschauung dieser neuen Auffassung zu ermöglichen, verweise ich auf die Elasticitätstheorie, die die durch gegenseitige Einwirkung und Lageverschiebung der Moleküle entstehenden Deformationen fester Körper behandelt; wir werden hier auf eine eingehende Beschreibung dieser molekularen Vorgänge zu verzichten haben und dafür nur die Parameter aufsuchen, von denen der meßbare Verzerrungszustand der Körper an jeder Stelle abhängt. Alsdann wird festzustellen sein, wie die Form der Abhängigkeit der Lagrangeschen Funktion von diesen Parametern ist, die sich ja eigentlich aus kinetischer und potentieller Energie der einzelnen Molekel zusammensetzen wird. Ähnlich wird man in der Thermodynamik nicht auf die Schwingungen der Molekel eingehen, sondern die Temperatur selbst als allgemeinen Parameter einführen, und die Abhängigkeit der Energie von ihr untersuchen. Die hier angedeutete Darstellungsart der Physik, ... [welche] die Ableitung wesenticher Sätze aus formalen Annahmen über L gestattet, soll den Kern meiner Vorlesung bilden. (Hilbert, 1906, p. 8/9)

28. Sie werden sich überzeugen, daß an Boltzmanns Ausführungen zur Begründung der Maxwellschen Formel nicht ein Stein auf dem andern bleiben kann. (Hilbert, 1912, p. 1)

REFERENCES

The main bulk of the *Nachlass* of David Hilbert is preserved in the "Niedersächsische Staats- und Universitätsbibliothek" at the University of Göttingen; it is registered under Cod. MS. D. Hilbert, register number; this will be abreviated as (SUB, register number). More than seventy scripts of lectures given by Hilbert between 1895 and 1930 are in the Mathematical Institute of the University of Göttingen, they will be designated as (MI, register number).

1. Published writings of Hilbert

Hilbert, D. (1899), *Grundlagen der Geometrie. Festschrift zur Feier der Enthüllung des Gauss-Weber-Denkmals in Göttingen.* Teubner, Leipzig 1899.

Hilbert, D. (1900), "Mathematische Probleme", lecture held at the International Congress of Mathematics in Paris 1900; first published in *Nachrichten von der königlichen Gesellschaft der Wissenschaften zu Göttingen*, pp. 253-297.

Hilbert, D. (1912), Grundzüge einer allgemeinen Theorie der linearen Integralgleichungen. Teubner, Leipzig, Berlin, 1912.

Hilbert, D. (1915), "Die Grundlagen der Physik", *Nachrichten von der Königl. Gesellschaft der Wissenschaften zu Göttingen*, pp. 395-407.

Hilbert, D. (1932), *Gesammelte Abhandlungen;* 3 Volumes, Springer, Berlin.

Hilbert, D. (1987), *Grundlagen der Geometrie*, mit Anhängen und Supplementen, ed. by Paul Bernays, 13. Auflage, Teubner, Stuttgart.

2. Unpublished manuscripts from the Nachlass

Hilbert, D. (1891), Projektive Geometrie, SUB 535.

Hilbert, D. (1893/4), Die Grundlagen der Geometrie, SUB 541.

Hilbert, D. (1898/9), Euklidische Geometrie, SUB 551.

Hilbert, D. (1904), Grundlagen der Geometrie, MI 10.

Hilbert, D. (1906), Mechanik der Kontinua, MI 25,26.

Hilbert, D. (1927), Grundlagen der Geometrie, MI 75.

Hilbert, D. (1898), Mechanik, SUB 553.
Hilbert, D. (1902/3), Mechanik der Continua, MI 11.
Hilbert, D. (1911), Mechanik der Contiua, MI 41.
Hilbert, D. (1911/2), Kinetische Gastheorie, MI 42.
Hilbert, D. (1912), Strahlungstheorie, MI 43.
Hilbert, D. (1926/7), Mathematische Methoden der Quantentheorie, MI 73.

3. Other References

Boltzmann, L. (1897), *Vorlesungen über die Principe der Mechanik*; Theil I und II, Johann Ambrosi-
    us Barth, Leipzig, 1897 and 1904.
Brush, S.G. (1976), *The Kind of Motion we Call Heat – A History of the Kinetic Theory of Gases in
    the 19th Century*, Amsterdam - New York - Oxford, North Holland Publishing House. 1976.
Hertz, H. (1894), *Die Prinzipien der Mechanik*. (Gesammelte Werke. Band III.) Mit einem Vorworte
    von H. von Helmholtz. Johann Ambrosius Barth, Leipzig, 1894.
Ludwig, G. (1978), *Die Grundstrukturen einer physikalischen Theorie*, Springer, Berlin, 1990².
Mach, E. (1899), *Die Mechanik in ihrer Entwickelung historisch-kritisch dargestellt*. F.A. Brock-
    haus, Leipzig, 1889.
Neumann, J. von (1932), *Mathematische Grundlagen der Quantenmechanik*, Springer, Berlin 1932.
Sauer, T. (1999), "The Relativity of Discovery: Hilbert's First Note on the Foundations of Physics".
    Archive for the History of Exact Sciences **53**, pp. 529-575.
Wightman, A.S. (1976), "Hilbert's Sixth Problem: Mathematical Treatment of the Axioms of Phys-
    ics", in *Proceedings of the Symposium in pure Mathematics* of the American Mathematical So-
    ciety, ed. by F. E. Bowder.

Institut für Wissenschaftsgeschichte
Humboldtallee 11
D-37073 Göttingen
Germany

MICHAEL STÖLTZNER

OPPORTUNISTIC AXIOMATICS – VON NEUMANN ON THE METHODOLOGY OF MATHEMATICAL PHYSICS*

On December 10th, 1947, John von Neumann wrote to the Spanish translator of his *Mathematical Foundations of Quantum Mechanics:*[1]

Your questions on the nature of mathematical physics and theoretical physics are interesting but a little difficult to answer with precision in my own mind. I have always drawn a somewhat vague line of demarcation between the two subjects, but it was really more a difference in distribution of emphases. I think that in theoretical physics the main emphasis is on the connection with experimental physics and those methodological processes which lead to new theories and new formulations, whereas mathematical physics deals with the actual solution and mathematical execution of a theory which is assumed to be correct *per se*, or assumed to be correct for the sake of the discussion. In other words, I would say that theoretical physics deals rather with the formation and mathematical physics rather with the exploitation of physical theories. However, when a new theory has to be evaluated and compared with experience, both aspects mix.

By the time that these lines were written mathematics and theoretical physics had substantially moved away from each other leaving only few individuals appertaining to both cultures, such as von Neumann and his Hungarian born friend Eugene P. Wigner who once called 'The Unreasonable Effectiveness of Mathematics in the Natural Sciences' "a wonderful gift which we neither understand nor deserve" (Wigner, 1960, p.14).

Yet, already in the late 1930s and 1940s this gift did not satisfy many quantum field theorists who found themselves amidst a set of spectacularly successful rules of computation which, however, involved blatantly inconsistent mathematical objects. Richard P. Feynman whose famous rules for perturbation theory are still most startling to mathematically-minded physicists, accordingly, considered disciplinary separation as quite natural:

The mathematical rigor of great precision is not very useful in physics. But one should not criticize the mathematicians on this score ... They are doing their own job. If you want something else, then you work it out for yourself. (Feynman, 1965, p. 56f.)

Some day, when physics is complete and we know all the laws, we may be able to start with some axioms ... so that everything can be deduced. But while we do not know all the laws, we can use some to make guesses at theorems which extend beyond the proof (Ibid., p. 49f.)

35

M. Rédei and M. Stöltzner (eds.),
John von Neumann and the Foundations of Quantum Physics, 35–62.
© 2001 *Kluwer Academic Publishers. Printed in the Netherlands.*

– a feature alien to mathematics. Instead of following the Euclidean ideal, the theoretical physicist avails himself of the 'Babylonian method' according to which mathematics is mainly governed by rules that suit to calculate interesting examples.[2]

The present paper argues, contrary to Feynman, that von Neumann had good reasons to insist that progress in theoretical science does not require neatly separating it from mathematics. Rigorous formalization and axiomatization can prove fertile even in cases where the basic concepts of the science are not yet fully clarified and empirical evidence is still poor; the reason is that mathematization permits great flexibility and opportunism in concept formation. After all, method in the physical sciences "is primarily opportunistic – also that outside the sciences, few people appreciate how utterly opportunistic it is" (Neumann, 1955, p. 492). To von Neumann, mathematics is not a merely abstract tool of working scientists; instead, it is itself capable of heuristic development relevant to the sciences because its best inspirations stem from empirical problems. Thus, mathematics and the theoretical branches of the sciences share some common criteria of success, such as unification and simplicity. Mathematics proper also depends upon a further aesthetic criterion of success concerning the structure of concepts and proofs that often makes it drift away from the empirical motivations.

1. METHODOLOGICAL CONTINUITY AND ITS POSSIBLE EXTRAPOLATION

There is a remarkable continuity in von Neumann's methodological convictions throughout the years and the various topics which he devoted attention to. In 1932 he laid the mathematical foundations of a physical theory that had already attained conceptual maturity although it was – and still is – notoriously plagued with interpretational problems. In accordance with Hilbert's program of axiomatization of the sciences, he hoped to formulate these remaining questions with sufficient precision to become decidable. Von Neumann's book was indeed a major point of reference for the many interpretational discussions to come. Yet only a few contemporaries noticed that he quickly moved on towards infinite-dimensional systems – a field today called quantum statistical mechanics – because he was both mathematically and methodologically dissatisfied with the Hilbert space formulation.[3]

When in 1947 von Neumann and Oskar Morgenstern published their trailblazing *Theory of Games and Economic Behavior* they neither could build upon well-entrenched scientific concepts and models nor were they equipped with a ready-to-use mathematical theory that was as well elaborated as Hilbert's theory of integral equations had been in 1932. Accordingly, the Introduction calls for extensive empirical investigations and surmises that in economics "mathematical discoveries of a stature comparable to that of calculus will be needed in order to produce decisive success in this field" (Neumann/Morgenstern, 1947, p. 6). For the time being, scientists should strive at a mathematically precise formulation of elementary facts about simple games and try to extend them to more realistic situations. It makes little sense to prematurely formalize and abundantly apply an unsatisfactory theory of economic equilibrium that is formulated in terms of mathematical concepts which have proven useful in physics. Only a combination of exact mathematical and empirical studies will tell whether principles and concepts are sound. Already Newton's success was based on Tycho Brahe's vast empirical data. On the other hand, "before the development of the mathematical theory the possibilities of quantitative measurements [of heat] were less favorable than they are now in economics" (Ibid., p. 3).

According to his close friend Stanislaw Ulam, "beginning about 1938, von Neumann felt that the new facts and problems of nuclear physics gave rise to problems of an entirely different type and made it less important to insist on mathematically flawless formulation of a quantum theory of atomic phenomena alone" (Ulam, 1958, p. 20). Nonetheless, he clung "to the hope that the mathematical method would remain for a long time in conceptual control of the exact sciences!" (ibid.) Since in these fields one could no longer expect comparatively simple frameworks like those that 19th century physics had provided, numerical computations became increasingly important even for basic concepts. Apart from his numerous special contributions von Neumann was planning, by the time of his untimely death, to approach the very concepts of computation and automata in a rather general fashion.[4] Unfortunately, he also could not take part in the revival of mathematical physics that started about 1955 or 1960 involving the first proposals of an axiomatic quantum field theory and which subsequently led to an elaboration of several of von Neumann's ideas. As a matter of fact, both Arthur Wightman's and Rudolf Haag's first papers were written still during von Neumann's lifetime.[5]

2. MATHEMATIZATION AND SCIENTIFIC UNIVERSALITY

The 1980s and 1990s have come up with an entirely new feature: "not only is mathematics the language of physics, but ... in quite a large area of mathematical research today, *theoretical physics* has become the language of mathematics" (Jaffe, 1997, p. 138). Motivated by physical insight, string theorists have contributed immensely to the creation of new concepts in geometry. Fields medalist and physicist Edward Witten, for instance, succeeded in bringing together two whole mathematical subjects hitherto unrelated. But, appraising *"the unreasonable effectiveness of theoretical physics in mathematics"* (Ibid.), mathematicians face a severe problem. If theoretical physicists achieve genuinely mathematical results, their arguments usually do not fulfill those severe standards of rigor which many mathematicians consider as the core tenet of their discipline. But, typically string theorists' results have afterwards been proven rigorously to a higher percentage than it is common for heuristic reasoning or conjectural results.

In 1993 two eminent mathematical physicists, Arthur Jaffe and Frank Quinn, proposed to establish speculative, non-rigorous mathematics as a separate branch of mathematics in which scientists should obey special rules of conduct in the intercourse with rigorous mathematicians. Baptizing this field 'theoretical mathematics' the authors emphasize the analogy between rigorous proofs in mathematics and experimental verification in the sciences. Moreover, as string theory involves particles of energies that vastly exceed the possibilities of earthbound accelerators and deals with experimentally inaccessible events, for instance, parallel universes, rigorous mathematicians are the *only* partners which can provide 'theoretical mathematicians' with independent evidence and justification concerning their results. In this situation – which is entirely different from all other branches of applied mathematics – consistency plays a major role as a criterion for truth claims because string theory is often proposed as a candidate for the final and universal theory of physics. "In a logically isolated theory [a necessary property of the Theory of Everything] every constant of nature could be calculated from first principles; a small change in the value of any constant would destroy the consistency of the theory" (Weinberg, 1993, p. 189). Von Neumann to whom consistency was, of course, a key requirement for any theory, did not cherish such hopes. "It happens occasionally that a particular physical theory appears to provide the basis for a universal system, but in all such instances up to the present time this appearance has not lasted more than a decade at best" (Neumann/Morgenstern, 1947, p. 2).

David Hilbert who fathered the use of the axiomatic method in the 20th century had initially been more optimistic. When coupling his formulation of general relativity to Mie's electrodynamics, he believed that he had thus already accomplished the "newly emerged 'ideal of unity of field theory'" (Hilbert, 1915, p. 47) and published the results under the rather exalted title "The Foundations of Physics". When outlining his conception of axiomatic thinking in the sciences, Hilbert later concluded:

Once it has become sufficiently mature for the formation of a theory, anything which can at all be the object of scientific thinking succumbs to the axiomatic method and consequentially to mathematics. By penetrating into deeper levels of axioms ... we also gain deeper insight into the essence of scientific thinking and become more and more conscious of the unity of our knowledge. Under the banner of the axiomatic method, mathematics appears to be designated to a leading role in all science (Hilbert, 1918, p. 415).

In 1918, Hilbert's program of seeking successively deeper axiomatizations and his idea of the unity of science did not require a universal theory. Instead he emphasized that in physics apart from the internal consistency of the propositions of a theory one had also to assure "that they never contradict the propositions of neighboring domains of knowledge" (Ibid., p. 410). Again, in his famous 1930 Königsberg lecture Hilbert stated: "We do not master a scientific theory until we have scraped out and disclosed its mathematical core" (Hilbert, 1930, p. 962). Or in short, "physics is much too hard for physicists" (Reid, 1993, p. 127).

Recalling Hilbert's polemic with the experimentalist Ernst Pringsheim, Max Born in 1922 stressed the difference between deepening the foundations *(Tieferlegung)* and universality even more pointedly.

[B]eing conscious of the infinite complexity he faces in every experiment [the physicist] refuses to consider any theory as final. Therefore – in the healthy feeling that dogmatism is the worst enemy of science – he abhors the word 'axiom' to which common use clings the sense of final truth. Yet the mathematician does not deal with the factual happenings, but with logical relations; and in *Hilbert's* terms the axiomatic treatment of a discipline does not signify the final assertion of certain axioms as eternal truths, but the methodological requirement: state your assumptions at the beginning of your considerations, stick to them and investigate whether these assumptions are not partially superfluous or even mutually inconsistent (Born, 1922, p. 90f.).

These remarks indicate that during his Göttingen period – when he was most active in the foundations of quantum mechanics – von Neumann was able to obtain a more refined picture of the relation between mathematization and scientific ultimacy than the one which guides present-day string theorists' quest for structures that are both mathematically fundamental and physically final.

Von Neumann's commitment to Hilbert's program – in the way Born conceived of it – constitutes also the reference-point of his methodological papers written in the 1950s. Surprisingly, in the broad discussion about the goals and the standards of rigor in mathematical physics that ensued Jaffe and Quinn's proposal (Atiyah et al., 1994, and the special issue of *Synthese* in 1997) von Neumann is listed only as an example of a rigorous mathematical physicist, while his ideas on methodology are ignored.[6] However, von Neumann's ideas concern the core of the debate for three reasons. Firstly, the methodological value of axiomatization does not directly depend upon the state of empirical research, so that the peculiar situation of string theory is tractable. Secondly, having taken part in the foundational debate of the 1920s and 1930s, von Neumann concluded "that the very concept of 'absolute' mathematical rigor is not immutable" (Neumann, 1947, p. 4). Although mathematical theorems might accordingly be no more reliable than well-established scientific facts, mathematics is distinguished from the empirical sciences by its peculiar criteria of success. Thirdly, although von Neumann's ideas about axiomatization were inspired by set theory, he did not exaggerate foundational matters – which only played a marginal role in the Jaffe-Quinn debate.

The next section, which is devoted to von Neumann's ideas about rigor in mathematics and the significance of empirical science – theoretical physics in particular – for mathematics, argues that his position would be a good compromise to settle upon in the Jaffe-Quinn debate because it permits a proper dosage of a clearly formulated concept of rigor. The remainder of the paper is organized as follows. Section 4 compares the criteria of success in theoretical science and mathematics. Section 5 outlines how and to what extent mathematization supports theorizing in the sciences. In both the examples given by von Neumann, a mathematical isomorphism between two formulations contributes to physical insight. This motivates studying how von Neumann conceived the role of different representations in a logical foundationalist perspective (Section 6). In the final section I will return to his 1932 book and argue that his theory of measurement contradicts his own methodology of mathematization insofar as it introduces a (non-formalizable) ontological dualism that does not go along well with his emphasis on statistical causality.

3. THE CONCEPT OF RIGOR AND THE ROLE OF PHYSICS IN MATHEMATICS

In their 1993 paper, Jaffe and Quinn issued several prescriptions in order to ensure that fully rigorous proofs and incomplete theoretical arguments are clearly distinguished – above all, by adopting a standard nomenclature or flagging. These rules are intended to guarantee a reliable literature to build on, and to assure that due credit is given for making theoretical results rigorous. Otherwise mathematicians, instead of being receptive to physicists, would be discouraged. For, "[m]odern mathematics is nearly characterized by the use of rigorous proofs. This practice, the result of literally thousands of years of refinement, has brought to mathematics a clarity and reliability unmatched by any other science" (Jaffe/Quinn, 1993, p. 1). Although brilliant conjectures have inspired large research programs, weak standards of proof occasionally led astray or terminated a whole field. For instance, "in the eighteenth century, casual reasoning led to a plague of problems in analysis concerning issues like convergence of series and uniform convergence of functions" (Ibid., p.7).

The Jaffe-Quinn theses rest upon three presuppositions which are not shared by von Neumann. First, in the course of history the standard of mathematical rigor always increases. Second, there is a sound metamathematical basis of this rigor. Third, although 'theoretical mathematics' is sharply distinguished from rigorous mathematics, their ontologies must intersect. "For if we don't assume that mathematical speculations are about 'reality' then the analogy with physics is greatly weakened – and there is no reason to suggest that a speculative mathematical argument is a theory of anything, any more than a poem or novel is 'theoretical'" (Atiyah et al., 1994, p. 186) – thus Morris W. Hirsch. I shall address the first two topics concurrently in this section while the third will lead us to Section 4.

In "The Mathematician", von Neumann remembers: "I know myself how humiliatingly easily my own views regarding the absolute mathematical truth changed ... three times in succession" (Neumann, 1947, p. 6). After he had learned about Gödel's Incompleteness Theorem in 1930, von Neumann readily admitted that Hilbert's program was – though not based on wrong intentions – unfeasible.[7] In 1947, he continues:

The main hope for justification of classical mathematics – in the sense of Hilbert or of Brouwer and Weyl – being gone, most mathematicians decided to use that system anyway. After all, classical mathematics ... stood on at least as sound a foundation as, for example, the existence of the electron. Hence, if one is willing to accept the sciences, one might as well accept the classical system of mathematics (Ibid., p. 6).

Yet such a view does not boil down to holding "that rigor can be no more than a local and sociological criterion" (Atiyah et al., 1994, p. 203) – thus René Thom attempts to counter Jaffe and Quinn. To von Neumann's mind, mathematics – although any particular set of basic propositions can be doubted – "establishes certain standards of objectivity, certain standards of truth ... rather independently of everything else" (Neumann, 1954a, p. 478). This objectivity does not contradict the historical fact that many non-rigorous arguments were accepted – either with a certain sense of guilt or due to *bona fide* disagreements as to whether a particular proof was really a proof.[8] Already fluctuations in the style of proofs can come close to differences in rigor. "[I]n some respects the difference between the present and certain authors of the eighteenth or of the nineteenth centuries is greater than between the present and Euclid." (Neumann, 1947, p. 4)

"The variability of the concept of rigor shows that something else besides mathematical abstraction must enter into the makeup of mathematics" (Ibid., p. 4). Here empirical sciences cut in. "The most vitally characteristic fact about mathematics is ... its quite peculiar relationship ... to any science which interprets experience on a higher than purely descriptive level" (Ibid., p. 1). This relationship has two sides: On the one side,

[i]n modern empirical sciences it has become a major criterion of success whether they have become accessible to the mathematical method or to the near-mathematical methods of physics. Indeed, throughout the natural sciences an unbroken chain of pseudomorphoses, all of them pressing toward mathematics, and almost identified with the idea of scientific progress, has become more and more evident. (Ibid., p. 2)

On the other side, "[s]ome of the best inspirations of modern mathematics (I believe, the best ones) clearly originated in the natural sciences" (Ibid.). Von Neumann provides two examples which show that history was richer than a linear increase of abstractness and rigor and that an adequate appraisal of inexact results is not reached by merely stressing that they satisfied the standards of the day and that the gaps were filled in later years.

(i) The origin of *geometry* in antiquity was empirical and "it began as a discipline not unlike theoretical physics today" (Ibid.). Euclid's ensuing postulational treatment even served as a model for Newton's *Principia.* The 'de-empirization' of Euclidean geometry was never quite completed until with Hilbert the axiomatic method itself obtained a new abstract meaning and was extended to non-Euclidean geometries. Yet, in the form of general relativity empiry has not only the final say, but also initial doubt stems from there: "The prime reason, why, of all Euclid's postulates, the fifth was questioned, was clearly the unempirical character of the concept of the entire infinite plane which intervenes there, and there only." (Ibid., p. 3)

(ii) *Calculus*, Newton's fluxions in particular, was explicitly created for the purpose of celestial mechanics. "An inexact, semiphysical formulation was the only one available for over a hundred and fifty years after Newton!" (Ibid.) Despite major advances, "[t]he development was as confused and ambiguous as can be ... And even after the reign of rigor was essentially re-established with Cauchy, a very peculiar relapse into semiphysical methods took place with Riemann." (Ibid., p. 3f.)

Recalling his prognosis concerning game theory too, von Neumann apparently took as a generic feature of those scientific theories which cannot avail themselves of previously created mathematical structures that they are likely to incite their own mathematics that sets out in a rather informal way. Although quantum mechanics had been developed from Hilbert's theory of integral equations, in his 1954 address to the International Congress of Mathematicians – which the organizers had initially intended to provide a new version of Hilbert's famous problem list – he demanded a unification of probability theory and logics that would go beyond ordinary quantum logic (cf. Neumann, 1954b).

Yet, "[t]here are various important parts of modern mathematics in which the empirical origin is untraceable" (Neumann, 1947, p. 6) or very remote, such as topology or abstract algebra. The possibility of such genuinely mathematical transformations of a research field demonstrates that mathematics is not a purely empirical science and nor do virtually all mathematical ideas originate in the sciences. Moreover, "[t]wo strange examples are given by differential geometry and by group theory: they were certainly conceived as abstract, nonapplied disciplines ... After a decade in one case, and a century in the other, they turned out to be very useful in physics. And they are still mostly performed in the indicated, abstract, nonapplied spirit." (Ibid., p. 7) Hence, in order to cover all the examples given, there must be specific and self-contained mathematical criteria of success which, on the other hand, permit a rather smooth transition from empirical science to mathematics. To these criteria I shall now turn, taking as the example the science which has undergone the most pseudomorphoses towards mathematization – viz. theoretical physics.

4. CRITERIA OF SUCCESS IN THEORETICAL PHYSICS AND MATHEMATICS

To von Neumann, the sciences "mainly make models" (Neumann, 1955, p. 492) which are usually valid over limited scales only. "The justification of such a mathematical construct is solely and precisely that it is expected to work" (Ibid.), to wit, its empirical adequacy. Conceived as such constructs, mathematics and theoretical physics share many – mainly aesthetic – criteria of success. Apart from the fact that Euclid's system of geometry and modern axiomatizations in Hilbert's style have become – where feasible – the prototype of presentation,

the attitude that theoretical physics does not explain phenomena, but only classifies and correlates, is today accepted by most theoretical physicists. This means that the criterion of success of such a theory is simply whether it can, by a simple and elegant classifying and correlating scheme, cover very many phenomena, which without this scheme would seem complicated and heterogeneous, and whether this scheme covers phenomena which were not considered at the time when the scheme was evolved. (Neumann, 1947, p. 7)

In these lines, von Neumann expresses his adherence to the tradition of Kirchhoff, Mach and Boltzmann who considered equations or functional dependences as the basic entities of scientific theories. Apart from the phenomena, there existed no 'things' to be causally explained in the sense of classical metaphysics. Mach's redefinition of causality, however, made it possible to define as real the basic entities of a physical theory – for instance, collectives in statistical mechanics – without having to prove their ultimacy, as classical atomism had claimed. In this way, the negative aspects of Mach's phenomenalism could be avoided by insisting with Boltzmann that a major feature of theory was its excess content or predictive power and emphasizing that the unification and simplification had more that just economical value.[9]

"Simplicity is largely a matter of historical background ... and it is very much a function of what is explained by it" (Neumann, 1955, p. 492),[10] to wit, how heterogenous the material covered by the explanation is. Accordingly, simplicity and unificationary power have to be constantly equilibrated. Von Neumann attributes surprisingly little weight to whether prediction occurs before or after the fact. Heterogeneity ranks higher, in particular "confirmations in areas which were not in the mind of anyone who invented the theory" (Ibid., p. 493). Von Neumann emphasizes that both these criteria are "clearly to a great extent of an aesthetical nature" (Ibid.) which brings them rather close to the mathematical criteria of success.

But mathematics possesses a genuine criterion: "One expects a mathematical theorem or a mathematical theory not only to describe and to classify in a simple and elegant way ... One also expects 'elegance' in its 'architectural', structural makeup" (Neumann, 1947, p. 9), e.g., a surprising twist in the argument which immediately makes a point very easy, or some general principle which explains why difficulties crop up and which reduces the apparent arbitrariness. "These criteria are clearly those of creative art" (Ibid.), so that

the subject begins to live a particular life of its own and is better compared to a creative one, governed by almost entirely aesthetical motivations, than to anything else and in particular, to an empirical science ... As a mathematical discipline travels far from its empirical source ... it is beset with very grave dangers. It becomes more and more purely aestheticizing, more and more purely *l'art pour l'art* (Ibid.).

The field is then in danger of developing along the line of least resistance and will "separate into a multitude of insignificant branches" (Ibid.). "[W]henever this stage is reached, the only remedy seems … to be a rejuvenating return to the source: the reinjection of more or less directly empirical ideas" (Ibid.). Notice that the diagnosis of degeneration is reached by internal criteria of mathematical style or elegance, only. Thus, one cannot derive from these remarks a methodological restriction of the independence of mathematics. After all, "the principle of *laissez faire* has led to strange and wonderful results" (Neumann, 1954a, p. 490) once one had completely forgotten "about what one ultimately wanted" (Ibid., p. 489).

Von Neumann's idea that mathematics has to be rejuvenated every now and then appears also in Michael Atiyah's criticism of 'theoretical mathematics'.

Jaffe and Quinn present a sanitized view of mathematics which condemns the subject to an arthritic old age … But if mathematics is to rejuvenate itself and break new ground it will have to allow for the exploration of new ideas and techniques which, in their creative phase, are likely to be dubious as in some of the great eras of the past. Perhaps we now have high standards of proof to aim at but, in the early stages of new developments, we must be prepared to act in more buccaneering style (Atiyah et al., 1994, p. 178).

While the aesthetic criteria of success in mathematics and theoretical physics are quite similar, von Neumann locates major differences regarding their aims and their actual *modus procedendi*. Even without signs of degeneration, mathematics is more finely subdivided into subdisciplines because often the selection of problems itself is aesthetically oriented. Theoretical physics, on the contrary, is typically highly focused to resolve an internal difficulty or to solve a problem that was posed by experimental results. Once a break-through is reached, "the predictive and unifying achievements usually come afterward" (Neumann, 1947, p. 8).

From this diagnosis von Neumann draws ontological conclusions:

[T]he problems of theoretical physics are objectively given; and, while the criteria which govern the exploitation of a success are … mainly aesthetical, yet the portion of the problem, and that which I called above the original 'break-through', are hard, objective facts (Ibid., p. 8).

As typically theoretical physics is already a mathematized or even – to various degrees – an axiomatized empirical science, it seems that the ontology of mathematics somehow supervenes over the ontology of physical theory. In this way, however, one could neither assess the above-mentioned 'strange' examples of differential geometry and topology nor appreciate numerous other cases in which similar equations describe entirely different fields of physical reality. Thus attributing some genuine ontology to mathematics cannot result in so hermetic a separation of the ontologies of mathematics and physics that all 'best inspirations' are downgraded to mere heuristics.

To a certain extent, Imre Lakatos's proposal of a quasi-empirical ontology promises a way out. In this framework, merely the flow of truth is at stake: "*a theory which is quasi-empirical in my sense may be either empirical or non-empirical in the usual sense*" (Lakatos, 1967, p. 29). While Euclidean theories rest upon indubitable axioms from which truth flows downward through valid inferences, in quasi-empirical theories truth is injected at the bottom by virtue of a set of accepted basic statements, such that falsity is retransmitted upward. "[I]n a quasi-empirical theory the (true) basic statements are *explained* by the rest of the system" (Ibid., p. 28f.). Theoretical physics is, of course, quasi-empirical in this sense; it is also empirical in the usual sense.

In a Lakatosian perspective, physical problems provide the informal ancestors of formal mathematical theorems and proof quasi-experiments. Concerning the question as to when to axiomatize an informal theory, Lakatos and von Neumann substantially disagree. The former emphasizes that "we have no guarantee that our formal system contains the full empirical or quasi-empirical stuff in which we are really interested and with which we dealt in the informal theory. There is no formal criterion as to the correctness of formalization" (Lakatos, 1961, p.67). Agreed, probability theory without the Lebesgue integral or algebra without complex numbers would be much poorer theories and lack key theorems. But, modern mathematics contains many abstract concepts and axiomatized theories for which no informal ancestors can be imagined.[11] Von Neumann's conception of empirical injection avoids this major shortcoming of Lakatos's philosophy of mathematics because he considers axiomatization as much more flexible an enterprise than does Lakatos, who surprisingly neglects all relevant remarks in von Neumann's "The Mathematician" which he extensively quotes as a piece of evidence for "A renaissance of empiricism in the recent philosophy of mathematics?".

In contrast to Lakatos who denies than any proof could ever be final, von Neumann's considers his position on the empirical origin of mathematics as rather independent of the controversy over foundations because mathematics' "non-empirical character could only be maintained if one assumed that philosophy (or more specifically epistemology) can exist independent of experience. (And this assumption is only necessary but not in itself sufficient)" (Neumann, 1947, p. 4). Hence, von Neumann criticizes the view entertained by the early Carnap that mathematics is syntax of language. At the time, his Princeton colleague Kurt Gödel entertained similar views:

If it is argued that mathematical propositions have no content because, by themselves they imply nothing about experiences, the answer is that the same is true of laws of nature. For laws of nature without mathematics or logic imply as little about experiences as mathematics without laws of nature. That mathematics, at least in most applications, does add something to the content of the laws of nature is at best seen from examples where one has very simple laws about certain elements, e.g., those about the reactions of electronic tubes. Here mathematics clearly adds the general laws as to how systems of tubes connected in a certain manner will react. (Gödel, 1953/9, p. 360)

On the other hand, Gödel sustains an ontological difference between mathematics and physics. Mathematical propositions "are true in virtue of *the concepts* occurring in them*" (Ibid., p. 356), while "space-time reality ... is completely determined by the totality of particularities without any reference to the formal concepts" (Ibid., p. 354). As a member of the empiricist tradition, von Neumann certainly would have objected to Gödel's metaphysical realist determination of space-time reality. And in contrast to the post-formalist von Neumann, Gödel was a staunch platonist on foundational matters in mathematics. There, according to Saunders MacLane, lies the source that nourishes all non-rigorous mathematical buccaneers. "They tend to use set theory as THE foundation of mathematics, and so sometimes [philosophers of mathematics] eagerly spread the gospel that mathematics is the study of an ideal realm of sets – set theoretic platonism." (Mac Lane, 1997, p. 151) Instead, Atiyah's buccaneers should stand trial under a structuralist law of austerity: "If a result has not yet been given valid proof, it isn't yet mathematics: we should strive to make it such" (Ibid.). Von Neumann, to my mind, could stand trial with a clear conscience.

To summarize, von Neumann's recognition that – apart from empirical adequacy and logical correctness under an accepted metatheory – most criteria of scientific success are of an aesthetic nature, is neither tied to a platonist nor to a structuralist philosophy; rather it could be combined with an empiricist or even a pragmatist point of view. To my mind, this position is a reasonable compromise in assessing what the ontology of 'theoretical mathematics' should be like especially because the benefits of early axiomatization – or the search for sound basic concepts for axiomatization in cases like game theory – are not indissolubly bound to foundational issues. In the next section I shall discuss the list of pragmatic virtues of mathematization for scientific theorizing which von Neumann provides in his methodological papers.

5. MATHEMATIZATION AS THEORIZING

I feel that one of the most important contributions of mathematics to our thinking is, that it has demonstrated an enormous flexibility in the formation of concepts, a degree of flexibility to which it is very difficult to arrive in a non-mathematical mode (Neumann, 1954a, p. 482).

Here mathematization converges with the opportunism that, to von Neumann's mind, is basic to the scientific method. In two examples discussed in (Neumann, 1954a) and (Neumann, 1955) three further aspects emerge. First, after mathematization has revealed formal equivalences or isomorphisms between two competing approaches, certain philosophical problems connected to such a duplicity become simply meaningless. Second, mathematization makes it possible to formulate some sophisticated 'logical cycles' within and to find the absolute limitations of a theory. Third, beyond the models on which a particular axiomatization is based, mathematization may provide a certain excess content that can become heuristically fertile for scientific insight. Let me now turn to the examples.

(a) Causal versus teleological explanation in Newtonian mechanics

Since the times of Euler and Lagrange there exist two formulations of problems in classical mechanics. One can either set up a second-order differential equation which locally describes the dynamical evolution or apply the Principle of Least Action (or one of its relatives) over a finite time interval, i.e. globally. Since their inception in the 18th century action principles and their mathematical basis, variational calculus, have suffered from their vicinity to the philosophical debates on natural teleology.[12] Still, often the teleological connotation of these principles is erroneously equated with backward causation. As such a feature is manifestly absent in the time-reversal invariant Newtonian mechanics, some interpreters have considered this question of teleology in physics simply as a pseudo-problem.[13]

Von Neumann, however, takes a more sophisticated approach by analyzing what is required for a complete description of the state of the system. In the Newtonian framework "[i]t is just this amount of information, position plus velocity, which is hereditary in that theory and which can be propagated into the future by unambiguous calculations" (Neumann, 1955, p. 495) from one moment in time to the one immediately afterwards. In the teleological formulation "two moments which are definitely apart in time" (Ibid.) are taken into consideration and the dynamical evolution is not obtained stepwise, but in a single act, the variation of the action integral. Hence,

a teleological theory asserts that this entire historical process must satisfy certain criteria which are usually stated in terms of optimizing (maximizing) a suitable function of the process. The use of the word optimizing again illustrates the opportunism that even re-flects itself in the terminology. By optimizing one only means that one makes some quantity as large as possible. Whether that quantity is particularly desirable or not does not matter. By changing its sign one could transform the criterion in making it as small as possible. Thus, optimizing, maximizing, and minimizing are all neutral mathematical terms, to be substituted for each other on the basis of mathematical convenience and taste (Ibid.).

This neutrality reached through mathematization indeed rules out all types of *material* teleology in Newtonian mechanics. Yet one could still claim that the Principle of Least Action which – after due definition of the quantity to be mini-mized – is applicable in so many fields of modern physics represents some type of *formal* teleology in the sense described by Leibniz: "There is evidently in all things a principle of determination which is derived from a maximum or a mini-mum, such that without doubt the maximal effect is achieved at the least expense, so to speak" (Leibniz, 1890, p. 303). But mathematization makes it possible to exclude also this possibility:

Newton's description is causal and d'Alembert's description is teleological ... All the difference between the two is a purely mathematical transformation ... Thus whether one chooses to say that classical mechanics is causal or teleological is purely a matter of literary inclination at the moment of talking. This is very important, since it proves, that if one has really technically penetrated a subject, things that previously seemed in complete contrast, might be purely mathematical transformations of each other. Things which appear to represent deep differences of principle and of interpretation, in this way may turn out not to affect any significant statements and any predictions. They mean nothing to the content of the theory (Neumann, 1955, p. 496).

Von Neumann even contemplates whether this possibility of rendering philosophical questions meaningless by reducing them to mathematical isomorphisms could be generic in science.

I'm not trying to be facetious about the importance of keeping teleological principles in mind when dealing with biology; but I think one hasn't started to understand the problem of their role in biology, until one realizes that in mechanics, if you are just a little bit clever mathematically, your problem disappears and becomes meaningless. And it is perfectly possible that if one understood another area the same might happen. This is an insight which would probably never have been obtained without the purely mathematical trickery of transforming the equations of mechanics (Neumann, 1954a, p. 484).

Accordingly, this insight reached by mathematics proper represents an objective gain for physical research. It is surprising, however, that von Neumann does not mention the heuristic value which action principles have enjoyed in so many fields of physics. Moreover, although the strict equivalence between the causal and the teleological approach holds for the standard examples of classical mechanics, there exist problems which fulfill Newton's axioms, but in which the Principle of Least Action does not yield the proper solution, or does so only under further restrictions. On the other hand, the same differential equation may stem from several action principles which have quite different symmetry groups[14] – a feature which one could consider as the most important global characteristic of a dynamical evolution.

This deficiency of von Neumann's account combines with two terminological errors which are quite amazing for a mathematician who worked in the field of variational calculus even in later years. Calling the teleological formulation of Newtonian mechanics "the principle of minimum effect" (Neumann, 1954a, p. 483), he gives an equally reasonable translation from German as Principle of Least Action, but unfortunately not the common one. Moreover, considering D'Alembert's Principle teleological spoils, if taken literally, his whole point because this is a differential principle involving infinitesimal variations (virtual displacements), which does not express a global property. Admittedly, under rather mild conditions this principle is tantamount to the Principle of Least Action.

(b) Schrödinger versus Heisenberg Representation

In 1926, quantum mechanics entered the scene in two different versions that were proven mathematically equivalent shortly thereafter.[15] While the Schrödinger representation centers around the wave function and appears, at first glance, akin to classical optics, the Heisenberg interpretation is entirely probabilistic in spirit. Nevertheless, to von Neumann's mind "the difference in this case is not as striking and profound as" (Neumann, 1955, p. 497) above. On the other hand, equivalence holds only up to a certain point which is best described by a theorem of von Neumann that ensures the uniqueness of the representation (up to isomorphism) in quantum *mechanics*, but fails in quantum field theory.

The reason for preferring one version of quantum theory over the other has usually been the intuitive hope that one or the other would give better heuristic guidance in extending the theory into those areas which are not yet properly explained ... Questions of form, even when the mathematical contexts are equivalent, can therefore have great heuristic and guiding importance (Ibid.).

Fertility for empirically fruitful generalizations indeed constitutes a pragmatic criterion of theory choice commonly accepted among philosophers of science. But – in accordance with his above-mentioned criteria of scientific success – von Neumann even claims that ultimately fertility will even outscore the notorious interpretational problems of quantum mechanics. After expressing his preference for the statistical interpretation but acknowledging "that there have been in the last few years some interesting attempts to revive the other interpretation" (Ibid.), he concludes:

while there appears to be a serious philosophical controversy between the interpretations of Schrödinger and Heisenberg, it is quite likely that the controversy will be settled in quite an unphilosophical way ... It must be emphasized that this is not a question of accepting the correct theory and rejecting the false one. It is a matter of accepting that theory which shows greater formal adaptability for a correct extension. This is a formalistic, esthetic criterion, with a highly opportunistic flavor (Ibid., p. 498).

But it requires already a good deal of mathematical sophistication to properly formulate a single interpretation. While in Newtonian mechanics a complete specification of the state allowed causal predictions, in quantum theory one is left with probabilities.

There is, however, something else which is causally predictable, namely the so-called wave-function. The evolution of the wave-function can be calculated from one moment to the next, but the effect of the wave-function on observed reality is only probability ... And again an enormous contribution of the mathematical method to the evolution of our real thinking is, that it has made such logical cycles possible, and has made them quite specific. It has made possible to do these things in complete reliability and with complete technical smoothness. (Neumann, 1954a, p. 486)

Moreover, the mathematical language not only opens up a sophisticated way of reasoning, it also can describe its own domain of validity. Accordingly, von Neumann argues that in relativity and quantum theory,

> by the best descriptions we can give today, there are absolute limitations to what is knowable. However, they can be expressed mathematically very precisely, by concepts which would be very puzzling when attempted to be expressed by any other means. Thus, both in relativity and in quantum mechanics the things which cannot be known always exist; but you have a considerable latitude in controlling which ones they are ... This is certainly a situation of a degree of sophistication which it would be completely hopeless to develop or to handle by other than mathematical methods (Ibid., p. 487).

More concretely, Heisenberg's uncertainty relations represent such an absolute limitation to knowing simultaneously position and momentum of a quantum particle.

In treating the issue of a theory's limitation mainly on the formal level, von Neumann comes close to Logical Empiricists' rejection of metaphysical realism that would back claims that there are knowable things beyond the theory. Furthermore, his intention to resolve classical philosophical problems in an unphilosophical way – or even to render them meaningless – reminds one of their understanding of the role of philosophy. Yet vicinity is not identity. Although like Carnap, von Neumann emphasizes that choosing a conceptual (or linguistic) framework is largely a pragmatic or aesthetic matter[16], he did not consider mathematics as merely syntactic and conventional. If he additionally assumed that strict separability between internal and external questions hold throughout mathematical physics, there would be no genuine place for the empirical inspirations of mathematics. Moreover, von Neumann's strong belief that finding the mathematical core of a theory chiefly contributes to scientific understanding is hardly reconcilable with the strict verificationism prevailing among Logical Empiricists.

There is still another background that Logical Empiricists and von Neumann shared, to wit, Hilbert's program. Although after Gödel's results von Neumann quickly admitted that this could not lay an absolute foundation of mathematics, he nevertheless took it as a guideline in the axiomatization of physics. This can be best seen by his constant emphasis that mathematization renders scientific questions empirically or logically decidable. The foundational debates among logicians also concerned the definability of rigorous mathematical truth. Moreover, these debates taught that axiom systems might suffer from the existence of unintended models, to wit, the failure of categoricity. The next section investigates the role these primarily logical features played within von Neumann's understanding of mathematical physics.

6. DECIDABILITY, DEFINABILITY AND CATEGORICITY

It is almost a commonplace that every scientist hopes to answer or decide open questions as precisely as possible and to obtain unique solutions to his problems. Yet, in von Neumann – so I will argue – one can discern quite a definite use of this motive of decidability because the scientific matters to be decided have been turned into clearly phrased questions by setting up an axiomatic framework.

In both the examples of the previous section, mathematization decided – at least partially – a philosophical alternative linked to two formulations of a physical theory in an unphilosophical way by establishing an isomorphism between them. Hilbert's program of the axiomatization of science is, however, not exhausted by unraveling partial identities. On this basis one can also decide scientific matters empirically, i.e. by testing the predictions of the theory, or logically, i.e. by proving certain properties given the axioms of the theory. Both motives figure prominently in the *Mathematical Foundations of Quantum Mechanics* the main intention of which was to lay the foundations of quantum mechanics as a statistical theory. By adopting a certain definition of classical causality the factual question as to whether the statistical features of quantum theory were objective or merely due to a lack of knowledge on part of the observer, became mathematically decidable. Consequently, already in the 1932 book, von Neumann tried to cast the concept of measurement in logical terms. I shall approach these issues from a broader philosophical perspective.

In 1955, von Neumann recalled three serious scientific crises that had occurred during his lifetime: "the conceptual soul-searching connected with the discovery of relativity and the conceptual difficulties connected with discoveries in quantum mechanics ... [plus one in mathematics] dealing with rigor and the proper way to carry out a correct mathematical proof" (Neumann, 1955, p.491). All three crises concerned definability. This is most noticeable in the third case where Tarski proved on the basis of Gödel's Incompleteness Theorem that one cannot define a truth predicate within an axiom system that is sufficiently rich to comprise arithmetic. The *Mathematical Foundations* offer an interesting comparison between the other two.

The indeterminacy relations have at first glance a certain similarity to the basic postulates of relativity theory. There it is maintained that it is impossible in principle to determine the simultaneity of two events occurring at points a distance r apart, more precisely than within a time interval of magnitude r/c (c is the velocity of light), while the indeterminacy relations predict that it is impossible in principle to give the position of a material point in phase space more precisely than within a region of volume $(h/4\pi)^3$. Nevertheless there exists a fundamental difference. The relativity theory ... [makes] possible, by the introduction of a Galilean frame of reference, to put a coordinate system in the world which makes a simultaneity definition possible that is in reasonable agreement with our normal concepts on this subject. An objective meaning will not be attributed to such a definition of distant simultaneity only because such a coordinate system can be chosen in an infinite number of different ways ... That is, behind the impossibility of measurement we find an infinite multiplicity of possible theoretical definitions. It is otherwise in quantum mechanics, where it is in general not possible to describe a system with the wave function φ by points in phase space, not even if we introduce new (hypothetical, unobserved) coordinates, the 'hidden parameters', – since this would lead to dispersion-free ensembles [which contradict the axioms] ... The principle of impossibility of the measurement thus arises in one case from the fact that there is an infinite number of ways in which the relevant concepts can be defined without conflicting directly with experience (or, with the general, basic assumptions of the theory) – while in the other case no such way exists at all (Neumann, 1932, p. 171f./325f.).

To give a more detailed account, von Neumann's argument commences with the attempt to define causality as an expression of the principle of sufficient reason in such a way that nature could never violate it.

That is, the theorem that two identical objects S_1, S_2 – i.e. two replicas of the system S which are in the same state – will remain identical in all conceivable interactions is true because it is tautological. For if S_1, S_2 could react differently to the same intervention in their interaction (i.e., if they gave different values in the measurement of a quantity $\mathfrak{R}$), then we would not call them identical. Therefore, in an ensemble $[S_1,...,S_N]$ which has dispersion relative to a quantity $\mathfrak{R}$ the individual systems $S_1,...,S_N$ cannot (by definition) all be in the same state (Ibid., p. 160/302f.).

This line of reasoning leads into postulating 'hidden parameters' which would permit a further division of dispersing ensembles. Thus, "the attempt to interpret causality as an equality definition led to a question of fact ... : is it really possible to represent each ensemble $[S_1,...,S_N]$, in which there is a quantity $\mathfrak{R}$ with dispersion, by the superposition of two (or more) ensembles different from one another and from it?" (Ibid., p. 162/305) Ultimately, this question is answered negatively in a purely mathematical manner by von Neumann's no-hidden variable proof. This conclusion rests upon two principles:

I. If the quantity $\mathfrak{R}$ has the operator R, then the quantity $f(\mathfrak{R})$ has the operator $f(R)$.
II. If the quantities $\mathfrak{R}, \mathfrak{S},...$ have the operators $R, S,...$, then the quantity $\mathfrak{R}+\mathfrak{S}+...$ has the operator $R+S+...$ (The simultaneous measurability of $\mathfrak{R}, \mathfrak{S},...$ is not assumed ...) (Ibid., p. 167/313f.)

Later debates on the interpretation of quantum theory revealed that condition II. is too restrictive because it excludes modifications of the theory which allow one to recover all predictions of ordinary quantum theory. The first of these extensions was proposed by David Bohm, still during von Neumann's lifetime in 1952. This highly non-local theory which represents a modernized version of de Broglie's pilot-wave theory of 1926 supplements the Schrödinger equation with a guidance equation that determines how the (unobservable) positions of the quantum particles evolve in time. According to Bohm, "[i]t appears that von Neumann has agreed that my interpretation is logically consistent and leads to all results of the usual interpretation. (This I am told by some people.)" (from a letter to Wolfgang Pauli, in von Meyenn, 1996, p. 392) As a matter of fact, already in 1932 von Neumann wrote: "certainly quantum mechanics has, in its present form, several lacunae, and it may even be that it is false, although this latter possibility is highly unlikely" (Neumann, 1932, p. 173/327). Axiomatization here permits a large degree of openness because one can easily devise alternative theories just by dropping or modifying a single axiom. Yet, the success of these alternatives will be decided by means of the above-mentioned criteria. At present the odds are strongly against the Bohmian interpretation mainly due to its bad performance in simplicity and fertility.[17]

Shortly after 1932, von Neumann was himself seeking modifications of his axiomatic framework in order to account for certain conceptual lacunae. Motivated by problems of the interpretation of quantum probabilities, he became dissatisfied with the Hilbert space framework and investigated what are today called type II von Neumann algebras. Physically this is tantamount to considering the quantum theory of systems with infinitely many particles as more fundamental than simple quantum mechanics. Already in the *Mathematical Foundations*, he had made the first steps towards quantum logic which he regarded as conceptually sharper than the ordinary theory: "the calculus of these propositions, based on projections, has the advantage over the calculus of quantities, which is based on the totality of (hypermaximal) Hermitian operators, that the concept of 'simultaneous decidability' represents a refinement of the concept of 'simultaneous measurability'" (Neumann, 1932, p. 134/254). By the time of his untimely death, the reconciliation between logic and probability was, to his mind, still a – widely neglected – open problem although mathematically both were based on the concept of orthogonality. In his 1954 address to the International Congress of Mathematicians, he surmises:

one has a formal mechanism, in which logics and probability arise simultaneously and are derived simultaneously. I think that it is quite important and will ... probably alter the whole formal structure of logics considerably, if one succeeds in deriving this system from first principles, in other words from a suitable set of axioms. All the existing axiomatisations of this system are unsatisfactory in this sense, that they bring in quite arbitrarily algebraical laws which are not related to anything that one believes to be true or that one has observed in quantum theory to be true (Neumann, 1954b, p. 245).

Two other key topics of the foundational debates in logics were – on the part of Hilbert's Program – the problems of categoricity and impredicativity. The former property concerns whether the axioms uniquely determine the system they describe. In his 1925 paper on the axiomatization of set theory, von Neumann affirms that

[proving the categoricity of a set of axioms is] highly important. For we only know that the propositions already proven follow from it. The unproven ones, however, such as the continuum problem could (if categoricity fails) be true in one system satisfying them [the axioms], but false in another. Thus, it would be at all uncertain whether these axioms suffice for the solution of, e.g., the continuum problem (Neumann, 1925, p. 54).

Categoricity is explicitly established if one finds an isomorphism between each two systems fulfilling the same set of axioms, as was the case – at least under suitable conditions – between the causal and teleological formulation of classical mechanics. As to quantum mechanics, von Neumann even succeeded in proving that *all* representations of the canonical commutation relations are equivalent up to isomorphism.

If categoricity fails, unintended interpretations can pop up, such as the Skolem functions in ordinary number theory. This feature also "leaves the mark of irreality (or, in a widely used word: 'impredicativity') on each axiomatic set theory" (Ibid., p. 51). Interestingly, as early as in 1925, von Neumann considered the prospects of finding a categoric axiomatization of set theory as rather dim. In mathematical physics such a scepticism is even more appropriate – already without having recourse to Gödel's Incompleteness Theorems.

Here a problem is lurking that, to my mind, cannot be shifted onto the aesthetic criteria of success, only. What happens if no partial isomorphisms of the type discussed in the preceding section can be established or, even worse, if empirical inspiration leads mathematical physicists to two entirely different sets of axioms? In general, there are three possibilities. First, both sets of axioms do not match each other on the level of the quasi-empirical stuff, to wit, the theorems or models that describe a given set of physical phenomena. In this case, scientific theorizing and empirical investigations are still wanted in order to find an appropriate set of axioms. The balance between simplicity and breadth of the explanation and the heterogeneity of the explanandum is the most important pragmatic criterion in this case. Second, two axiomatizations can be isomorphic on the level of the axioms. Formalistic fictions might still occur, but they could be excluded by preferring certain theories for physical reasons and comparing those judged reasonable in their aesthetic value, fertility and heuristic power. Third, both sets of axioms yield all known theorems, models or interpretations of the other. But, in this case they might differ in infinitely many theorems unknown so far, even if all known theories are isomorphic. These isomorphisms might even tempt us to erroneously consider a philosophical issue as meaningless within a wider area than it actually is. For instance, the Principle of Least Action is also applicable in general relativity where Newton's axioms cease to hold true and where both the notion of causality and the relation between local and global solutions are quite involved. Some of these features, such as the question of foci or conjugate points, can already be illustrated in classical mechanics. Hence, even isomorphisms cannot always eliminate differences in the heuristic background. Moreover, the aesthetic criteria cannot fully thrive because the appearance of unintended interpretations quickly spoils any structural beauty – at least, if this class cannot be easily separated out. On the other hand, the known theorems might still not encompass the quasi-empirical stuff of a scientific field. At this point empirical investigations and the study of models will regain importance.

This third case represents, to my mind, a certain limitation to the opportunism of axiomatics as conceived by von Neumann. While in the second case the deficiencies of the axiom systems are manifest, here one might prematurely dispense with the informal heuristic power present in both formulations by prematurely equating certain parts by means of isomorphisms. Yet, in stressing thus the importance of informal reasoning I do not want to go as far as Lakatos because, as stated above, some informal heuristics will appear only *after* formalization.

There is still another problem connected with the plurality of axiom systems. Von Neumann's above-mentioned emphasis on decidability is focused on single problems; he does not require a decision between the systems themselves. On the contrary, being able to choose between two approaches serves practical needs and reduces computational expenditure. A Lakatosian conception of mathematics, for instance, is unable to accommodate any relation other than competition on the level of the entire research program. From this one can draw an important lesson for the Jaffe-Quinn debate. As the question of mathematical success chiefly depends upon the standards of proof it, is very hard to compare two programs that are at a different stage of formal development or follow different standards of rigor, such as string theory and geometry. The Lakatosian has to make a choice and he or she certainly will be most impressed by the heuristic excess content of informal string theory or 'theoretical mathematics'. The lesson of von Neumann's thinking is that there cannot be a uniform approach. While in the intersection in question one might – at least temporarily – allow for a 'more buccaneering style', this permission should not be extended to well-entrenched fields, such as quantum mechanics of atomic systems or the mathematical formulation of classical dynamics.

7. STATISTICAL CAUSALITY VERSUS DUALISM

Despite the openness in principle towards future modifications which any axiomatic approach has to concede, von Neumann left little doubt concerning his opinion on the issue of causality.

In the macroscopic case there is no experience which supports it, and none can be devised because the apparent causal order of the world in the large (i.e., for objects visible to the naked eye) has certainly no other cause than the 'law of large numbers'; and this is completely independent of whether the natural laws governing the elementary processes (i.e. the actual laws) are causal or not. (Neumann, 1932, p. 172/326)

At this point the author adds footnote that appears at another location too: "Cf. the extremely lucid discussions of Schrödinger on this subject: Naturwiss. *17* (1929), number 37" (Ibid., p. 258/327). This refers to a slightly abbreviated version of Schrödinger's inaugural address to the Prussian Academy of Sciences in which he paid tribute to his teachers Fritz Hasenöhrl and Franz Serafin Exner. The former had conveyed to the young Schrödinger Boltzmann's indeterminism which just results from the practical unfeasibility of determining the initial states of all particles of a macroscopic body – a view that, to Schrödinger's mind, did not contradict the causal postulate. "Franz Exner, to whom I am personally indebted for unusually great encouragement, was the first to mention the possibility and the advisability of an acausal concept of nature" (Schrödinger, 1929b, p. 732/xvii). With a reference to Poincaré's remarks concerning the choice between Euclidean and non-Euclidean geometry, Schrödinger concludes that at present no final decision on causality is in sight. "The most that can be decided is whether the one or the other leads to the simpler and clearer survey of all observed facts" (Ibid., p.732/xviii). As noted above, this was a position also shared by von Neumann. Several other features of his account are, however, absent in Schrödinger's short speech. As I shall argue, they can be found in an earlier paper in *Die Naturwissenschaften* of the same year, to wit, in Schrödinger's famous paper "What is a law of nature?" which was originally his inaugural speech delivered at the University of Zurich already on December 9th, 1922. It is highly probable that von Neumann met Schrödinger at least after 1923 when he spent most of his time at the ETH in order to obtain a degree in chemistry. Not only were there close contacts between the University and the ETH, but also were both Schrödinger and von Neumann upon very good terms with the mathematician Hermann Weyl. It is, moreover, not too bold to conjecture that von Neumann had already been directly acquainted with Exner's views because, at the time, the *Lectures on the Physical Foundations of the Natural Sciences* (1922) were a rather well-received introductory textbook.

In the fourth part of his *Lectures,* Exner, in my interpretation[18], contrived a synthesis between a Machian methodology and Boltzmann's atomism which became the basis of a far-reaching and comprehensive indeterminism prevailing in the sciences and the humanities. Similarly to Exner, Schrödinger's Zurich lecture takes a quite determined stance.

Within the past four of five decades physical research has clearly and definitely shown – strange discovery – that *chance* is the common root of all the rigid conformity to Law that has been observed, at least in the overwhelming majority of natural processes, the regularity and invariability of which have led to the establishment of the postulate of universal causality (Schrödinger, 1929a, p. 9/136).

Thus, macroscopic causality appears in virtue of the law of large numbers, only. Also the second element of von Neumann's observation reaches back to Schrödinger and Exner. Whether microscopic events are governed by deterministic laws or not remains an open question because the microscopic events must show just the regularity on the average required for the observed macroscopic laws. An empiricist of a Machian brand cannot prefer either alternative in an absolute sense. Viewing causality as *a priori* condition of any possible experience, i.e., as a Kantian category, merely represents a habit of thought that has been shaped by millennia of experiences, but which with the advent of quantum mechanics loses its rational foundation. As postulating causality transcends experience, "*[t]he burden of proof falls on those who champion absolute causality, and not on those who question it.*" (Schrödinger, 1929a, p. 11/147) Maintaining, nonetheless, the idea of absolute causality would result in entertaining simultaneously two types of laws, statistical and causal ones, a duplication of natural law that "so closely resembles the animistic duplication of natural *objects,* that I cannot regard it as at all tenable" (Ibid.). Von Neumann concludes in rather similar terms,

there is at present no occasion and no reason to speak of causality in nature – because no experience supports its presence ... To be sure, we are dealing with an age old way of thinking of all mankind, but not with a necessity of thought ... , and anyone who enters upon the subject without preconceived opinions has no reason to adhere to it. Under such circumstances, are there any motives to sacrifice a reasonable physical theory for its sake? (Neumann, 1932, p. 173/328)

Yet there are two important differences between Schrödinger and von Neumann: the first of which, interestingly, brings the latter even closer to the former's most revered teacher. Exner's fundamental indeterminism was grounded on drawing ontological conclusions from the relative frequency interpretation of probability, by considering collectives as possible basic entities of physical theory. Similarly, von Neumann was sticking so closely to the frequency interpretation that he was even willing to sacrifice his own brainchild, the Hilbert space formalism because, within this framework, one could not define a meaningful *a priori* probability that was acceptable to the frequentist.[19] Von Neumann's subsequent achievements in developing an algebraic theory, assigned to infinite quantum systems a conceptual priority over single particles. Schrödinger, on the other hand, – albeit an anti-realist on the metaphysical level – initially intended to interpret his wave function as the basic entity of quantum mechanics. But on this track there was no easy cure in sight for his qualms about the quantum ontology and its relation to the macroworld. His dissatisfaction found its pictorial expression in the famous cat-paradox. Nevertheless, he was not in the least longing for a return to classical causality which the Bohm interpretation seemed to offer. Instead, he came to expect a resolution of the interpretational problems from a unified field theory. Schrödinger's dissatisfaction with the Copenhagen interpretation will indicate a certain fracture in von Neumann's philosophy of quantum mechanics that, to my mind, has led some readers of the *Mathematical Foundations* into misconceiving the virtues of the axiomatic method.

On the metaphysical level, Schrödinger's problems with the Born-Heisenberg interpretation resulted from his ontological monism. Within Mach's neutral monism, functional dependences between internal and external elements represented the ontological basis of all scientific experiences. According to Bitbol (1996), Schrödinger's approach to quantum mechanical reality was relational too – an attitude that was hardly reconcilable with the Heisenberg cut. Although von

Neumann endorsed all the main theses of the tradition of Vienna Indeterminism concerning the issue of causality, in his analysis of the measuring process – which actually constitutes one of the major accomplishments of the *Mathematical Foundations* – a dualistic ontology becomes manifest. I shall not provide here a detailed analysis of this very subtle topic, but focus only on the question of the ontological status of the observer which, to my mind, overtly violates von Neumann's program of mathematization.

Like Logical Empiricists, von Neumann repeatedly emphasized that the problem of measurement had been posed already by classical electrodynamics.[20] In quantum mechanics the core of the problem is the difference between two processes a system undergoes. On the one hand, the time evolution of the system is uniquely hence causally determined by the Schrödinger equation. This reversible process does not yield any increase of uncertainty. On the other hand, every measurement leads to a discontinuity in the wave function and turns pure into mixed states, hence it yields an irreversible increase of the uncertainty. When interpreting this distinction, von Neumann unexpectedly resorts to classical metaphysical dualism.

Let us now compare these circumstances with those which actually exist in nature or in its observation. First, it is in general entirely correct that the measurement or the related process of the subjective perception [German: *Apperzeption*] is a new entity relative to the physical environment which is not reducible to the latter. For, subjective perception leads us out of the latter, or more precisely: it leads into the intellectual inner life of the individual, which is uncontrollable since it must be taken for granted by any attempt of [empirical] control ... Nevertheless, it is a fundamental requirement of the scientific world view – the so-called principle of the psycho-physical parallelism – that it must be possible so to describe the (in reality) extra-physical process of the subjective perception as if it would occur in the physical world, i.e., to assign to its parts equivalent physical processes in the objective environment, in ordinary space ... But in any case, no matter how far we calculate – ... to the retina, or into the brain, at some time we must say: and this is perceived by the observer. That is, we must always divide the world into two parts, the one being the observed system, the other the observer. In the former, we can follow up all physical processes (in principle at least) with arbitrary precision. In the latter this is meaningless. The boundary between the two can be chosen at random to a very large extent ... That this boundary can be pushed arbitrarily deeply into the observer is the content of the principle of psycho-physical parallelism (Neumann, 1932, p. 223f./418f.).

More concretely, the world is divided into three parts: *I* is the system actually observed, *II* the measuring instrument, *III* the actual observer. In the sequel von Neumann proves that in quantum mechanics the boundary can be drawn either between *I+II* and *III* or between *I* and *II+III*. When he emphasizes that "in each case *III* itself remains outside of the calculation" (Ibid., p. 224/421), this violates the main methodological principles he had subscribed to, because for the most inner part of the world any formalization is in principle excluded. Rather than being open to scientific investigation, this inner part appears more akin to a pre-reflexive Cartesian *cogito*. From this dualism many metaphysical questions emerged that represented typical pseudo-problems according to the Logical Empiricists: such as in how far a conscious observer actually constitutes reality; whether there is an irreducible voluntary element in the quantum world and, if so, does this explain consciousness or free will; and so forth.

I shall conclude this paper with a historical conjecture concerning the reception of von Neumann's ideas. Both his methodological inconsistency concerning the formalization of the observer and the ontological distinction of the observer that was at odds with a fundamental indeterminism – at least if one wanted to circumvent certain notorious metaphysical questions –, prevented his position from being considered on its own merits. Instead, the *Mathematical Foundations* were conceived as the mathematical codification of an orthodox Copenhagen interpretation – perhaps a historical artifact in itself – and the no-hidden variable proof, in particular, became the point of attack for generations of alternative interpretations: which, pointing at the metaphysically questionable status of the observer took recourse to common-sense realism, to wit, to 'things' or 'beables'. In the debates waged on the foundation of quantum mechanics, von Neumann's mathematical accomplishments of the late 1930s were neglected and instead of observing the openness to modification that is part and parcel of the axiomatic method, axiomatization was equated to the infamous finality claim of Copenhagen orthodoxy. Thus it should not surprise us that also von Neumann's posthumously published Silliman lectures *The Computer and the Brain* did not lead to a modification of this view. There von Neumann compares the architecture of computers with the structural properties of our brain "from the mathematician's point of view ... [which carries a different distribution of emphasis:] the logical and the statistical aspects will be in the foreground" (Neumann, 1958, p. 1), viewed primarily as the basic tools of information theory. Again as in the *Theory of Games*, the author emphasizes that these studies will in turn modify our understanding of mathematics. The lectures end with a comparison that teaches that the later Neumann, unlike the young one, was pondering a formalization of the observer, although he was most aware of the basic difficulties:

it is only reasonable to assume that logics and mathematics are similarly historical, accidental forms of expression. They may have essential variants, i.e. they may exist in other forms than the ones to which we are accustomed. Indeed, the nature of the central nervous system and of the message systems that it transmits indicate positively that this is so. We have now accumulated sufficient evidence to see that whatever language the central nervous system is using, it is characterized by less logical and arithmetical depth than what we are normally used to ... It ought to be noted that the language here involved may well correspond to a short code in the sense described earlier, rather than to a complete code: when we talk mathematics, we may be discussing a *secondary* language, built on the *primary* language *truly* used by the central nervous system. However, the above remarks about reliability and logical and arithmetical depth prove that whatever the system is, it cannot fail to differ considerably from what we consciously and explicitly consider as mathematics (Ibid., p. 81f.).

NOTES

*   The author wants to express his thanks to Jeremy Butterfield, Hannes Leitgeb, Ulrich Majer, and Miklós Rédei for inspiring suggestions and critical comments.

1.   Letter to Ramon Ortis Fornaguera, Library of Congress, Manuscript Division. I thank Marina von Neumann-Whitman for the permission to quote from this letter.
2.   Through his widely-read introductory course (Feynman, Leighton, Sands, 1964) Feynman's way of thinking has molded an entire generation of physicist.
3.   See (Rédei, 1996).
4.   See (Neumann, 1958).
5.   See (Haag, 1992), §II, for the respective references.
6.   However, it deserves mention that Jaffe's first remarks on "The bad influence of physics on mathematics" (Jaffe, 1990) appeared in a volume commemorating John von Neumann.
7.   See the letter to Carnap, June 7th, 1931; Carnap Papers, University of Pittsburgh, Hillman Library.
8.   The latter also happened in the case of Hilbert's solution of Gordan's Problem (see Reid, 1996) which marked the first break-through reached by means the new non-constructive method.
9.   I have recently proposed to call this tradition 'Vienna Indeterminism' (Stöltzner, 1999a). As will be mentioned in Section 7, von Neumann could be comprised under this heading up to one important qualification, to wit, his notion of the observer.
10.   In view of von Neumann's rejection of the term "explanation" in the preceding quote of 1947 this signifies an interesting shift because it coincides which the reinstatement of the term in philosophy that began with the Hempel-Oppenheim paper of 1948. As Salmon (1989, p.11ff.) relates, in spite of the large impact of this paper on our present notion of explanation it remained virtually unknown for about a decade. The only book that had a substantial influence on philosophers was Braithwaite's *Scientific Explanation* of 1953. I thank Wesley Salmon for these hints.
11.   See (Corfield, 1997) and (Stöltzner, 1999b).
12.   For the early history, see (Schramm, 1985).
13.   See (Yourgrau/Mandelstam, 1960).
14.   See (Stöltzner, 1994, §§ 4.1-4.4) for further details.
15.   This is presumably not the whole story. F.A. Muller (1997) has recently argued that the equivalence proof taken literally did not concern the theories actually proposed in 1926 but the modified versions current as of 1932. Yet this diagnosis only strengthens von Neumann's point that by finding isomorphisms mathematization renders interpretational problems meaningless without destroying the respective heuristic fertility of the single theories.
16.   See (Carnap, 1950).
17.   See (Stöltzner, 1999c) for a reconstruction of von Neumann's position.
18.   See (Stöltzner, 1999a).
19.   See, once again, (Rédei, 1996).
20.   Cf. (Neumann, 1932, p. 126, 159).

REFERENCES

If translations of German texts exist, their page numbers are given after those of German originals. In the case of (Neumann, 1932) I have made some modifications to the translation cited here because it is inaccurate in some philosophical passages.

Michael Atiyah et al. (1994): "Responses to 'Theoretical Mathematics: Toward a Cultural Synthesis of Mathematics and Theoretical Physics'", *Bulletin of the American Mathematical Society* **30**, pp. 178-211.
Marcel Bitbol (1996): *Schrödinger's Philosophy of Quantum Mechanics*, Dordrecht: Kluwer.
Max Born (1922): "Hilbert und die Physik", *Die Naturwissenschaften* **10**, pp. 88-93.
Rudolf Carnap (1950): "Empricism, Semantics and Ontology", *Revue internationale de philosophie* **4**, pp. 20-40.

David Corfield (1997): "Assaying Lakatos's Philosophy of Mathematics", *Studies in the History and Philosophy of Science* **28**, pp. 99-121.

Franz Serafin Exner (1922): *Vorlesungen über die physikalischen Grundlagen der Naturwissenschaften*, Leipzig-Wien: Franz Deuticke.

Richard P. Feynman, Robert B. Leighton, Matthew Sands (1964): *The Feynman Lectures on Physics.* Addison-Wesley, Reading, Mass. vol. 1 and 2.

Richard Feynman (1965): *The Character of Physical Law,* Cambridge, MA: M.I.T. Press.

James Glimm, John Impagliazzo, and Isadore Singer (eds.) (1990): *The Legacy of John von Neumann* (Proceedings of Symposia in Pure Mathematics vol. 50), Providence, RI: American Mathematical Society.

Kurt Gödel (1953/9): "Is mathematics syntax of language?", in *Collected Works*, vol. III, ed. by Solomon Feferman et.al., New York, Oxford: Oxford University Press, 1995.

Rudolf Haag (1992): *Local Quantum Physics*, Berlin-Heidelberg-New York: Springer.

David Hilbert (1915): "Die Grundlagen der Physik", in: *Hilbertiana – Fünf Aufsätze von David Hilbert.* Darmstadt: wbg, 1964, pp. 47-78.

David Hilbert (1918): "Axiomatisches Denken", *Mathematische Annalen* **78**, pp. 405-415.

David Hilbert (1930): "Naturerkennen und Logik", *Die Naturwissenschaften* **18**, pp. 959-963.

Arthur Jaffe (1990): "Mathematics motivated by physics", in Glimm et al. (1990), pp. 137-150.

Arthur Jaffe, Frank Quinn (1993): "'Theoretical Mathematics': Toward a Cultural Synthesis of Mathematics and Theoretical Physics", *Bulletin of the American Math. Society* **29**, pp. 1-13.

Arthur Jaffe (1997): "Proof and the Evolution of Mathematics", *Synthese* **111**, pp. 133-146.

Imre Lakatos (1961): " What does a mathematical proof prove?", in (Lakatos, 1978), pp. 61-69.

Imre Lakatos (1967): "A renaissance of empiricism in the recent philosophy of mathematics?", in: (Lakatos, 1978), pp. 24-42.

Imre Lakatos (1976): *Proofs and Refutations. The Logic of Mathematical Discovery*, edited by John Worrall and Elie Zahar, Cambridge: Cambridge University Press.

Imre Lakatos (1978): *Mathematics, science and epistemology* (Philosophical Papers Volume 2), edited by John Worrall and Gregory Currie, Cambridge: Cambridge University Press.

Gottfried Wilhelm Leibniz (1890): *Die philosophischen Schriften.* Ed. by G.J. Gerhardt. Berlin. 1890 (reprint: Olms, Hildesheim, 1961), vol. VII.

Saunders MacLane (1997): "Despite Physicists, Proof is Essential in Mathematics", *Synthese* **111**, pp. 147-154.

Karl von Meyenn (ed.) (1996): *Wolfgang Pauli – Scientific Correspondence with Bohr, Einstein, Heisenberg, a.o.*, Volume IV, Part I: 1950-1952, Berlin-New York: Springer.

Frederick A. Muller (1997): "The Equivalence Myth of Quantum Mechanics", *Studies in the History and Philosophy of Modern Physics* **28**, pp. 35-61 & 219-247.

Johann von Neumann (1925): "Eine Axiomatisierung der Mengenlehre", in *Collected Works,* ed. by A.H. Taub, Oxford: Pergamon Press, 1961, vol. 1, pp. 35-56; originally in *Journal für die reine und angewandte Mathematik* **154** (1925), pp. 219-240.

Johann von Neumann (1932): *Mathematische Grundlagen der Quantenmechanik,* Berlin: Julius Springer. English translation by Robert T. Beyer *Mathematical Foundations of Quantum Mechanics*, Princeton: Princeton University Press, 1955.

John von Neumann (1947): "The Mathematician", *Collected Works,* vol. 1, pp. 1-9; originally in R.B. Heywood (ed.): *The Works of the Mind,* Chicago: University of Chicago Press, 1947, pp.180-196.

John von Neumann (1954a): "The Role of Mathematics in the Sciences and in Society", *Collected Works*, vol. 6, pp. 477-490; originally in Princeton Graduate Alumni, June 1954, pp. 16-29.

John von Neumann (1954b): "Unsolved Problems in Mathematics", Typescript, first published in this volume.

John von Neumann (1955): "Method in the Physical Sciences", *Collected Works,* vol. 6, pp. 491-498; originally in L. Leary (Ed.): *The Unity of Knowledge*, New York: Doubleday, pp. 157-164.

John von Neumann (1958): *The Computer and the Brain*, New Haven: Yale University Press.

John von Neumann and Oskar Morgenstern (1947): *Theory of Games and Economic Behavior,* Princeton: Princeton University Press.

Miklós Rédei (1996): "Why von Neumann did not like the Hilbert space formalism of quantum mechanics (and what he liked instead)", *Studies in the History and Philosophy of Modern Physics* **27**, pp. 493-510.

Constance Reid (1996): *Hilbert,* New York: Copernicus.

Matthias Schramm (1985): *Natur ohne Sinn ? – Das Ende des teleologischen Weltbildes,* Graz–Wien–Köln: Styria.

Wesley C. Salmon (1989): *Four Decades of Scientific Explanation,* Minneapolis: University of Minnesota Press.

Erwin Schrödinger (1929a): "Was ist ein Naturgesetz?", *Die Naturwissenschaften* **17**, pp. 9-11; English translation by James Murphy and W.H. Johnston in *Science and the Human Temperament,* New York: W.W. Norton & Co., pp. 133-147.

Erwin Schrödinger (1929b): "Aus der Antrittsrede des neu in die Akademie eintretenden Herrn Schrödinger", *Die Naturwissenschaften* **17**, p. 732; English translation of the unabbreviated text in the Introduction to *Science and the Human Temperament,* pp. xiii-xviii.

Michael Stöltzner (1994): "Action Principles and Teleology", in: Harald Atmanspacher, Gerhard Dalenoort (Eds.): *Inside Versus Outside,* Berlin-New York: Springer, pp. 33–62.

Michael Stöltzner (1999a): "Vienna Indeterminism: Mach, Boltzmann, Exner", *Synthese* **119**, pp. 85-111.

Michael Stöltzner (1999b): "What Lakatos Could Teach the Mathematical Physicist", in: George Kampis, Ladislav Kvasz, Michael Stöltzner (Eds.): *Appraising Lakatos – Mathematics, Methodology and the Man,* Dordrecht: Kluwer, to appear.

Michael Stöltzner (1999c): "What John von Neumann Thought of the Bohm Interpretation", in: Daniel Greenberger, Wolfgang Reiter, Anton Zeilinger (Eds.): *Epistemological and Experimental Perspectives on Quantum Physics,* Dordrecht: Kluwer, pp. 257-262.

Stanislaw Ulam (1958): "John von Neumann, 1903-1957", *Bulletin of the American Mathematical Society,* Special Issue May 1958, edited by J.C. Oxtoby, B.J. Pettis, G.B. Price, pp. 1-49.

Steven Weinberg (1993): *Dreams of a Final Theory,* London: Vintage.

Eugene P. Wigner (1960): "The Unreasonable Effectiveness of Mathematics in the Natural Sciences", *Communications in Pure and Applied Mathematics* **13**, pp. 1-14.

Wolfgang Yourgrau and Stanley Mandelstam (1960): *Variational Principles in Dynamics and Quantum Theory,* London: Pitman.

Institut Wiener Kreis
Museumstraße 5/2/19
A-1070 Vienna
Austria

JEFFREY BUB

VON NEUMANN'S THEORY OF QUANTUM MEASUREMENT

In a series of lectures written around 1952, Schrödinger refers to von Neumann's account of measurement in quantum mechanics as follows:

I said quantum physicists bother very little about accounting, according to the accepted law, for the supposed change of the wave-function by measurement. I know of only one attempt in this direction, to which Dr. Balazs recently directed my attention. You find it in John von Neumann's well-known book. With great acuity he constructs one analytical example. It does not refer to any actual experiment, it is purely analytical. He indicates in a simple case a supplementary operator which, when added to the internal wave operator, would *with any desired approximation* turn the wave function as time goes on into an eigenfunction of the observable that is measured. He found it necessary to show that such a mechanism is *analytically possible.* The idea has not been taken up and worked out since – in about twenty years or more. Indeed I do not think it would pay. I do not believe any real measuring device is of this kind. ([1], p. 83)

This characterization of von Neumann's analysis of the measurement process in his *Mathematische Grundlagen der Quantenmechanik* [2] is rather odd. Schrödinger undoubtedly has in mind the discussion in the final chapter of the book ('The Measuring Process'), but the 'analytical example' developed there is not designed to show how the wave function can evolve over time into an eigenfunction of the measured observable.

In the following I shall outline von Neumann's theory and show what it does achieve, and why a fundamental issue remains unresolved.

1. VON NEUMANN'S TWO PROCESSES

Von Neumann proposed two modes of evolution for a quantum system S that he called 'process 1' and 'process 2' (see Chapter III of [3], p. 351 and pp. 417–18)[1] : a non-unitary, stochastic, entropy-increasing evolution induced by measurements, and a unitary evolution described by Schrödinger's time-dependent equation of motion between measurements.

For a (pure or mixed) state of S represented by a density operator W, the evolution induced by the measurement of a maximal (non-degenerate) observable A with eigenvectors α_k (process 1) is:

$$W \rightarrow W' = \sum tr(W P_{\alpha_k}) P_{\alpha_k} \qquad (1)$$

63

M. Rédei and M. Stöltzner (eds.),
John von Neumann and the Foundations of Quantum Physics, 63–74.
© 2001 Kluwer Academic Publishers. Printed in the Netherlands.

where P_{α_k} is the projection operator onto the eigenvector α_k and tr represents the trace function. The density operator W' represents a mixture of pure states α_k, with weights equal to the probabilities assigned by W to the corresponding eigenvalues a_i of A. So if the state of the system immediately before the measurement of A is a pure state ψ associated with the density matrix $W = P_\psi$, the measurement results in the projection of ψ onto one of the eigenvectors α_k with probability $tr(W P_{\alpha_k}) = |(\alpha_k, \psi)|^2$.

The unitary evolution (process 2) is:

$$W \to W(t) = U(t) W U^{-1}(t) \tag{2}$$

where $U(t) = \exp -iHt/\hbar$.

The initial argument for process 1 is formulated on pp. 212–18. Von Neumann considers three 'degrees of causality or non-causality' that might occur in nature (p. 213):

(i) the result of a measurement of an observable A might be predictable only statistically with a certain dispersion, and a subsequent measurement of A performed immediately after the first measurement might also be predictable only statistically, possibly with the same dispersion;

(ii) the result of a measurement of A might be predictable only statistically, but an immediate subsequent measurement might be constrained to yield the same result as the first measurement, with no dispersion;

(iii) the result of the measurement might be determined causally from the outset. Von Neumann observes that option (i) characterizes the Bohr-Kramers-Slater theory and argues that this theory and option (i) can be considered refuted by the Compton-Simon experiment (pp. 212–14). Option (iii) is not characteristic of quantum measurements, so the only remaining possibility is option (ii). It follows that a quantum system initially in a state in which the result of a measurement of an observable A cannot be predicted with certainty must be transformed by a measurement of A into a state in which the value of A is predictable with certainty. Von Neumann argues that this state must be an eigenstate of A corresponding to the eigenvalue obtained in the measurement.

The claim that option (ii) is excluded by the Compton-Simon experiment is somewhat problematic, but I shall not pursue this here. I note only that von Neumann believed there was experimental evidence for process 1, or what has since become known as the 'projection postulate.' Dirac comes to a similar conclusion on the basis of a continuity argument:

When we measure a real dynamical variable ξ, the disturbance involved in the act of measurement causes a jump in the state of the dynamical system. From physical continuity, if we make a second measurement of the same dynamical variable ξ immediately after the first, the result of the second measurement must be the same as that of the first. Thus after the first measurement has been made, there is no indeterminacy in the result of the second. Hence, after the first measurement has been made, the system is in an eigenstate of the dynamical variable ξ, the eigenvalue it belongs to being equal to the result of the first measurement.

This conclusion must still hold if the second measurement is not actually made. In this way we see that a measurement always causes the system to jump into an eigenstate of the dynamical variable that is being measured, the eigenvalue this eigenstate belongs to being equal to the result of the measurement. ([5], p. 36)

The argument for the applicability of process 1 to measurements does not resolve the issue of how process 1 and process 2 are related. Von Neumann takes up this question in Chapters V and VI. Why should process 2 not suffice to describe measurement interactions? Von Neumann's answer defines the remaining problem as he sees it (p. 232):

Indeed a physical intervention can be nothing else than the temporary insertion of a certain energy coupling into the observed system, i.e., the introduction of an appropriate time dependency of H (prescribed by the observer). Why then do we need the special process 1 for the measurement? The reason is this: in the measurement we cannot observe the system S by itself, but must rather investigate the system $S + M$, in order to obtain (numerically) its interaction with the measuring apparatus M. The theory of the measurement is a statement concerning $S + M$, and should describe how the state of S is related to certain properties of the state of M (namely, the positions of a certain pointer, since the observer reads these). Moreover it is rather arbitrary whether or not one includes the observer in M, and replaces the relation between the S state and the pointer position in M by the relations of this state and the classical changes in the observer's eye or even in his brain (i.e., to that which he has 'seen' or 'perceived'). ... In any case, therefore, the application of 2 is of importance only for $S + M$. Of course, we must show that this gives the same result for S as the direct application of 1 to S. If this is successful, then we have achieved a unified way of looking at the physical world on a quantum mechanical basis.

The remainder of Chapter V deals with the problem of defining the entropy of a quantum ensemble, in terms of which process 1 can be shown to be irreversible, and a treatment of macroscopic measurements.[2] The solution to the 'consistency problem' for process 1 and process 2 is formulated in the final Chapter VI.

2. THE CONSISTENCY PROBLEM

Von Neumann shows that if we divide the world into three parts – the observed system (S), the measuring instrument (M_1), and the observer (M_2) – then the boundary between the observer and the observed system can be drawn between S and $M_1 + M_2$ or, equivalently, between $S + M_1$ and M_2. So S could be taken as the system measured, M_1 as the pointer instrument, and M_2 as the light plus the observer. Or S could represent the system measured together with the pointer system, M_1 the light and the eye of the observer as a refracting device, and M_2 the observer interfacing with $S + M_1$ through the retina. Or S could represent everything up to the retina, M_1 the retina and optic nerve, and M_2 the observer's brain.

That this boundary can be shifted arbitrarily is shown by the equivalence *for* S of applying process 1 to S (in a measurement of S by $M_1 + M_2$) and applying

process 1 to $S + M_1$ (in a measurement of $S + M_1$ by M_2), after an appropriate unitary interaction described by process 2 between S and M_1 characterizing
the measurement. The same description is obtained for S whether the boundary
between the observed system and the observing system is placed between S and
$M_1 + M_2$ or between $S + M_1$ and M_2.

The demonstration depends on features of the description of composite systems
in tensor product Hilbert spaces. For a composite system $S + M$ in a pure or mixed
state represented by the density operator W on the Hilbert space $\mathcal{H}_S \otimes \mathcal{H}_M$, von
Neumann shows that the statistics of all S-observables $X \otimes I_M$, where I_M is the
unit operator on $\mathcal{H}_M$, is given by a density operator W_S on $\mathcal{H}_S$ that is the partial
trace[3] of W over $\mathcal{H}_M$:

$$W_S = tr_M W$$

Similarly, the statistics of all M-observables is given by $W_M = tr_S W$, where tr_S
is the partial trace over $\mathcal{H}_S$.

Now suppose a measuring instrument M_1 measures an observable A with
eigenstates α_k on a system S initially in the pure state

$$\psi = \sum_k (\alpha_k, \psi)\alpha_k$$

If M_1 is in some zero state ρ_0 of the 'pointer' observable R with eigenstates ρ_k,
we expect a measurement interaction to correlate A-eigenstates with R-eigenstates,
resulting in the state

$$\Psi = \sum_k (\alpha_k, \psi)\alpha_k \otimes \rho_k$$

Von Neumann shows that there is a unitary transformation that will generate the
required transformation:

$$\sum_k (\alpha_k, \psi)\alpha_k \otimes \rho_0 \overset{U}{\to} \Psi = \sum_k (\alpha_k, \psi)\alpha_k \otimes \rho_k \tag{3}$$

This is a description of the measurement as an instance of process 2.

The solution to the consistency problem is obtained as follows: Applying process 1, the projection postulate, to the measurement of the observable A of S by
the instrument $M_1 + M_2$, we obtain

$$W_S = \sum_k |(\alpha_k, \psi)|^2 P_{\alpha_k}$$

for the density matrix representing the statistics of S-observables after the measurement. Applying process 2 to the interaction between S and M_1, we obtain the
state

$$\Psi = \sum_k (\alpha_k, \psi)\alpha_k \otimes \rho_k$$

for $S + M_1$. Finally, the application of process 1 to a further measurement of the pointer observable R of M_1 by a second instrument M_2 (which, von Neumann remarks on p. 440, is simultaneously a measurement of the value A of S, because the eigenvalues of A and R are correlated in the state Ψ) yields the density operator

$$W = \sum_k |(\alpha_k, \psi)|^2 P_{\alpha_k} \otimes P_{\rho_k}$$

for the state of $S + M_1$ after the measurement. Taking the partial trace of W over the Hilbert space $\mathcal{H}_M$ yields

$$W_S = tr_M W = \sum_k |(\alpha_k, \psi)|^2 P_{\alpha_k}$$

for the density operator of S, as before.

So, applying process 1 directly to S yields the same density operator for the statistics of all S-observables after the measurement as we obtain by considering the measurement as a process 2 interaction between S and M_1, and then applying process 1 to the measurement of $S + M_1$ by a second instrument M_2 and taking the partial trace of the resulting density operator over the Hilbert space $\mathcal{H}_{M_1}$.

The solution to the consistency problem, which Schrödinger seems to have misconstrued as a 'purely analytical' example of a wave function evolving to an eigenfunction of the measured observable, depends simply on the existence of a certain unitary transformation on the tensor product of the Hilbert spaces of the measured system S and the measuring instrument M. As von Neumann remarks (p. 441):

In this case only the principle is of importance to us, i.e., the existence of any such [U]. The further question, whether the [U] corresponding to simple and plausible measuring arrangements also have this property, shall not concern us.

Von Neumann's analysis concerns only ideal measurements, in which eigenstates of the measured observable are strictly correlated with eigenstates of a pointer observable of the measuring instrument. In quantum mechanics measurements are typically non-ideal. For example, consider the unitary tranformation generated by the interaction Hamiltonian

$$H_{int} = g(t) A \otimes V$$

on the Hilbert space $\mathcal{H}_S \otimes \mathcal{H}_M$, where $RV - VR = iI$ (taking $\hbar = 1$) and $g(t)$ is a coupling constant that is nonzero only during the time interval from $t = 0$ to $t = T$, when the interaction is 'switched on.' If $g(t)$ is sufficiently large so that the total Hamiltonian can be approximated by H_{int} during this time interval, Schrödinger's time-dependent equation reduces to

$$i\frac{\partial \Psi}{\partial t} = H_{int} \Psi = g(t) A \otimes V \Psi \tag{4}$$

during the interaction and has the solution

$$\Psi(t) = e^{\int_0^t -ig(t')A\otimes V dt'}\Psi_0 \tag{5}$$

If $g(t) = g$ during the interaction and $\Psi_0 = \psi \otimes \rho_0 = \sum_k(\alpha_k,\psi)\alpha_k \otimes \rho_0$ then

$$\Psi(t) = e^{-igtA\otimes V}\Psi_0 = \sum_k(\alpha_k,\psi)\alpha_k \otimes e^{-igta_k V}\rho_0 \tag{6}$$

for $0 \leq t \leq T$. (Here a_k is the eigenvalue of A corresponding to the eigenstate α_k.) So, after the measurement at time T:

$$\Psi(T) = \sum_k(\alpha_k,\psi)\alpha_k \otimes \rho'_k \tag{7}$$

where

$$\rho'_k = e^{-igTa_k V}\rho_0$$

The shifted states ρ'_k will generally not be mutually orthogonal, and hence not eigenstates of R. For example, suppose R represents position and V momentum. Then in the position representation the operator V is the differential operator $-i\frac{\partial}{\partial q}$ and

$$\Psi(T) = \sum_k(\alpha_k,\psi)\alpha_k \otimes e^{-gTa_k\frac{\partial}{\partial q}}\rho_0(q) = \sum_k(\alpha_k,\psi)\alpha_k \otimes \rho(q - gTa_k) \tag{8}$$

If $\rho_0(q)$ is a narrow Gaussian wave packet symmetric about the origin, then the wave functions $\rho'_k = \rho(q - gTa_k)$ are narrow Gaussians, symmetric about gTa_k, $k = 1, 2, \ldots$ They are relatively localized if gT is sufficiently large, so that the separation between peaks is very much greater than the width of the Gaussians. But since their tails always overlap, the wave functions are not strictly orthogonal, even if they can be considered to be 'effectively orthogonal.'

A Stern-Gerlach measurement of the spin of a spin-$\frac{1}{2}$ particle, in which the coarse-grained position of the particle itself acts as the pointer R, is non-ideal in this sense. In a measurement of spin in the z-direction, the particle, represented initially by a relatively localized symmetric wave packet moving in the x-direction, enters a region between two magnets and is subjected to the influence of an inhomogeneous magnetic field that gives the particle a momentum in the positive or negative z-direction, depending on whether the spin of the particle in the z-direction, σ_z, is 'up' ($+$) or 'down' ($-$). So the position of the particle in the z-direction, after the particle has left the magnetic field, indicates the value of the spin component:

$$(\frac{1}{\sqrt{2}}\alpha_+ + \frac{1}{\sqrt{2}}\alpha_-) \otimes \rho_0(q) \rightarrow \frac{1}{\sqrt{2}}\alpha_+ \otimes \rho(q - gt) + \frac{1}{\sqrt{2}}\alpha_- \otimes \rho(q + gt) \tag{9}$$

where α_+ and α_- represent the 'up' and 'down' spin states of the particle, respectively.

Here $gt = \mu t(\frac{\partial H_z}{\partial z})_{z=0}$ to a first order approximation, where μ is the magnetic moment of the particle and H_z is the component of the magnetic field in the z-direction. The two wave packets $\rho_+ = \rho(q - gt)$ and $\rho_- = \rho(q + gt)$, centred at gT and $-gT$, respectively, at time T, begin to spread out after time T but continue to move in the positive and negative z-directions. If the separation between the peaks is much greater than the width of the packets (which depends essentially on the time t and the gradient of the magnetic field in the z-direction), there is, for all practical purposes, a correlation between the packet moving up and the spin eigenstate α_+ and the packet moving down and α_-, but there is no strict correlation between z-position and the eigenvalues, ± 1, of σ_z. The separating packets will always have infinitely long tails, so they never *completely* separate in the configuration space of the particle. They are not orthogonal wave functions, and hence not 'position eigenfunctions.'

Nevertheless, we can see that the structure of the consistency problem is similar for the non-ideal case and the ideal case considered by von Neumann. The application of process 1 to a non-ideal measurement of the spin observable will yield a density matrix for the spin that is 'close to' the density matrix $\frac{1}{2}P_{\alpha_+} + \frac{1}{2}P_{\alpha_-}$. We should obtain the same density matrix by applying process 1 to a coarse-grained measurement of position on the system in the state $\frac{1}{\sqrt{2}}\alpha_+ \otimes \rho(q - gt) + \frac{1}{\sqrt{2}}\alpha_- \otimes \rho(q + gt)$ and then taking the partial trace over the Hilbert space of the position observable.

3. SCHRÖDINGER'S PROBLEM

Consistency aside, there is a remaining problem that von Neumann does not address. Observables in quantum mechanics are represented by self-adjoint Hilbert space operators. Properties, or 'yes-no' observables, are represented by projection operators, or their corresponding subspaces. On pp. 252–3, von Neumann formulates certain principles for the relations between properties, projections, and subspaces. Principle β on p. 253 states that a property P is 'certainly present or certainly absent' for the states in the subspace that is the range of P and the states in the orthogonal complement of this subspace, respectively. Dirac formulates a similar principle for observables:

The expression that an observable 'has a particular value' for a particular state is permissible in quantum mechanics in the special case when a measurement of the observable is certain to lead to the particular value, so that the state is an eigenstate of the observable .... In the general case we cannot speak of an observable having a value for a particular state, but we can speak of its having an average value for the state. We can go further and speak of the probability of its having any specified value for the state, meaning the probability of this specified value being obtained when one makes a measurement of the observable.
([5], p. 253)

There is a clear statement of the principle in the Einstein-Podolsky-Rosen argument [6] for the incompleteness of quantum mechanics. Indeed, the argument is formulated as a *reductio* for the principle: Einstein, Podolsky, and Rosen show that it follows from this principle, together with certain realist assumptions, that quantum mechanics is incomplete. Following Fine ([7], p. 20), the principle is often referred to as the 'eigenvalue-eigenstate link.'

An ideal measurement interaction, according to von Neumann, results in the transition 3:

$$\sum_k (\alpha_k, \psi) \alpha_k \otimes \rho_0 \to \Psi = \sum_k (\alpha_k, \psi) \alpha_k \otimes \rho_k$$

After the measurement interaction (represented by process 2), each subsystem, S and M, is no longer associated with its own pure state. The final state is the 'entangled' state $\Psi = \sum_k (\alpha_k, \psi) \alpha_k \otimes \rho_k$, in Schrödinger's terminology: a linear superposition of product sates. According to von Neumann's principle β, the only properties of $S+M$ that are 'certainly present or certainly absent' in the state Ψ are represented by subspaces of $\mathcal{H}_S \otimes \mathcal{H}_M$ that contain the state Ψ or are orthogonal to the state Ψ. But since such subspaces are not the range of any projection operator of the form $P \otimes I_M$ or $I_S \otimes P$, neither A-properties nor R-properties are definite after a process 2 measurement interaction!

As Schrödinger pointed out, if M represents a cat and the pointer R takes two possible values, associated with the cat being alive and the cat being dead, and the cat interacts with a microsystem S, such as an atom that can either decay or not decay in a certain time (where these events are associated with the two possible values of A), and the decay event triggers a device that kills the cat, then the cat will be neither alive nor dead after the measurement interaction, according to the principle β. Solving the consistency problem does not resolve the problem of Schrödinger's cat.

Evidently, the root of the problem is the interpretative principle β, coupled with the linearity of the unitary tranformation in process 2. If we want the zero pointer state to indicate the k'th eigenvalue of the measured observable when the measured system is in the k'th eigenstate

$$\psi_k \otimes \rho_0 \to \psi_k \otimes \rho_k$$

then linearity requires that

$$\sum_k (\alpha_k, \psi) \alpha_k \otimes \rho_0 \to \sum_k (\alpha_k, \psi) \alpha_k \otimes \rho_k$$

But what is the justification for accepting the principle β?

The transition from classical to quantum mechanics involves the transition from a commutative to a noncommutative algebra of dynamical variables. Equivalently, one can see this as the transition from a Boolean representation of the properties of a system, in which properties are associated with subsets of a set or elements of a Boolean algebra, to the representation of properties as a projective geometry, in which properties are represented as the subspaces of a linear space.

In a address on 'Unsolved Problems in Mathematics' to the International Mathematical Congress, Amsterdam, September 2, 1954, von Neumann presents the significance of this structural change as follows:

If you take a classical mechanism of logics, and if you exclude all those traits of logics which are difficult and where all the deep questions of the foundations come in, so if you limit yourself to logics referred to a finite set, it is perfectly clear that logics in that range is equivalent to the theory of all sub-sets of that finite set, and that probability means that you have attributed weights to single points, that you can attribute a probability to each event, which means essentially that the logical treatment corresponds to set theory in that domain and that a probabilistic treatment corresponds to introducing measures. I am, of course, taking both things now in the completely trivialized finite case.

But it is quite possible to extend this to the usual infinite sets. . . .

In the quantum mechanical machinery the situation is quite different. Namely instead of the sets use the linear sub-sets of a suitable space, say of a Hilbert space. The set theoretical situation of logics is replaced by the machinery of projective geometry, which in itself is quite simple.

However all quantum mechanical probabilities are defined by inner products of vectors. Essentially if a state of a system is given by one vector, the transition probability in another state is the inner product of the two which is the square of the cosine of the angle between them. In other words, probability corresponds precisely to introducing the angles geometrically. . . .

And therefore, as soon as you have introduced into the projective geometry the ordinary machinery of logics, you must have introduced the concept of orthogonality. . . . So in order to have logics you need in this set a projective geometry with a concept of orthogonality in it. In order to have probability all you need is a concept of all angles, I mean angles other than 90°. Now it is perfectly quite true that in a geometry, as soon as you can define the right angle, you can define all angles. Another way to put it is that, if you take the case of an orthogonal space, those mappings of this space on itself, which leave orthogonality intact, leave all angles intact, in other words, in those systems which can be used as models of the logical background for quantum theory, it is true that as soon as all the ordinary concepts of logics are fixed under some isomorphic transformation, all of probability theory is already fixed. ([4], pp. 21-22; p. 244f. in this volume)

The 'classical mechanism of logics' is the fixed Boolean or set-theoretical structure that represents the possible ways in which properties can 'fit together' in a classical world. In classical mechanics, the properties of a system are represented by the (Borel) subsets of phase space, and a classical state, represented by a point in phase space, defines a 2-valued homomorphism on the Boolean algebra of (Borel) subsets: an assignment of truth values (1 or 0, true or false) to the properties. Probabilities are represented as measures over 2-valued homomorphisms, that is, as measures over truth possibilities. Von Neumann's point is that, when you replace the classical 'possibility structure' or logic with the 'machinery of projective geometry' as a representation of the 'possibility structure' of a quantum world, probabilities emerge as a new structural feature of the way properties can 'fit together' in a quantum world. In this sense, quantum probabilities are 'logical.'

It is natural then to take the quantum analogue of the classical state (a minimal element in the Boolean possibility structure of a classical system) as a ray in Hilbert space (a minimal element in the projective geometry or non-Boolean possibility structure defined by the lattice of subspaces of the Hilbert space of a quantum system). The quantum state in this sense (a pure state) assigns truth values to some properties in the quantum possibility structure – those that are 'certainly present or certainly absent' in the state according to principle β – and probabilities to the remaining properties. There are no 2-valued homomorphisms on the quantum possibility structure (except in the case of a 2-valued Hilbert space), and the probabilities defined by the quantum state are not representable as measures over 2-valued homomorphisms: they represent structural relations between quantum propositions that have no classical counterpart.

Once we accept the quantum analogue of the classical state – the catalogue of properties that are 'certainly present or certainly absent' – as a ray in Hilbert space, Schrödinger's problem is unavoidable. Is there an alternative? It seems that von Neumann excluded any alternative in his negative answer to the question as to whether one can add 'other characteristics or coordinates' to the quantum state, which would 'give the values of all physical quantities exactly and with certainty' (p. 209). But is this the relevant question to ask?

I want to propose another way of looking at what has come to be called the problem of 'hidden variables' in quantum mechanics. The family of observables associated with a given basis in Hilbert space, or a given partition of Hilbert space into mutually orthogonal eigenspaces for a nonmaximal or degenerate observable, defines a Boolean algebra, and the probabilities specified by any quantum state can be represented as a classical (Kolmogorov) measure over the 2-valued homomorphisms or truth possibilities on this algebra. We know that this representation cannot be extended to all observables, or even to certain finite sets of observables (in virtue of various 'no go' theorems for hidden variables). Suppose we start with a given quantum state ψ and the Boolean algebra generated by a given 'preferred' observable R, associated with a partition of Hilbert space into mutually orthogonal eigenspaces. What is the maximal lattice extension $\mathcal{L}(\psi, R)$ of this Boolean algebra on which sufficiently many 2-valued homorphisms can be defined, so that the probabilities spcified by the quantum state for the properties in $\mathcal{L}(\psi, R)$ can be represented by a measure over the set of 2-valued homomorphisms on $\mathcal{L}(\psi, R)$?

It turns out that there is a unique answer to this question, if we demand that $\mathcal{L}(\psi, R)$ is invariant under lattice isomorphisms that preserve ψ and R. The lattice $\mathcal{L}(\psi, R)$ is generated by the non-zero projections of ψ onto the eigenspaces of R, together with all the rays in the orthocomplement of the subspace spanned by these projections. (See [9] for details.) Von Neumann's principle β then amounts to taking the 'preferred' observable R as the unit, I. The lattice $\mathcal{L}(\psi, I)$ contains the properties represented by subspaces that include the ray ψ and the properties represented by subspaces orthogonal to ψ. Since there is only one 2-valued homomorphism on $\mathcal{L}(\psi, I)$ – the homomorphism that assigns 1 to all properties represented by subspaces that include ψ and 0 to the remaining properties – $\mathcal{L}(\psi, I)$

indeed contains all the properties that are 'certainly present or certainly absent' in the state ψ.

This suggests an alternative interpretation of the transition from classical to quantum mechanics to that proposed by von Neumann. Instead of taking the lattice of subspaces of Hilbert space as the quantum analogue of the fixed Boolean possibility structure of a classical world, take $\mathcal{L}(\psi, R)$ as the quantum analogue of this structure. The significance of the transition from classical to quantum mechanics then lies in the replacement of a *fixed* (Boolean) possibility structure by a *dynamically evolving* (non-Boolean) possibility structure, because the lattice $\mathcal{L}(\psi, R)$ evolves with the dynamical evolution of the quantum state ψ. The analogue of the classical state is not ψ but a 2-valued homomorphism on the possibility structure $\mathcal{L}(\psi, R)$. What the dynamics of quantum mechanics describes, via Schrödinger's equation of motion, is not the evolution of what is *actual* over time, but what is *possible*. We need a new dynamics to describe the evolution of the states that select what is actually the case as 2-valued homomorphisms on the possibility structure $\mathcal{L}(\psi, R)$. This dynamics must 'mesh' with the evolution of possibilities defined by Schrödinger's equation of motion for ψ. This question has been investigated. (Again, see [9] for details.) It turns out that if we take the 'preferred' observable R as position in configuration space, the dynamics is just the dynamics of position in Bohm's hidden variable theory [8].

What of the measurement problem? If the observable R can be chosen so that the pointer observables of our measuring instruments are functions of R, then pointer properties will always belong to $\mathcal{L}(\psi, R)$. In an ideal measurement in which ψ correlates pointer readings with eigenvalues of the measured observable, both pointer readings and properties corresponding to ranges of values of the measured observable will belong to $\mathcal{L}(\psi, R)$, and the probabilities specified by ψ can be understood as measures over the 2-valued homomorphisms that catalogue the various possible outcomes of the measurement.

Evidently, this is a contingent matter, reflecting features of our universe. It seems, though, that there are good grounds for taking position, or a coarse-grained position, as 'preferred' in this sense. Then all measurements are non-ideal, but pointer readings are always definite, and we can still interpret the quantum probabilities as measures over 2-valued homomorphisms associated with different pointer readings. So Schrödinger's problem is avoided.

NOTES

1. All page number references in the following to von Neumann are to *Mathematical Foundations of Quantum Mechanics*, the English translation of [2].
2. The analysis associates an 'unsharp' measurement of an observable A with a 'sharp' measurement of an discrete observable that is a function of A. Von Neumann did not develop the notion of POV measures to handle 'unsharp' or macroscopic measurements.
3. The partial trace over $\mathcal{H}_M$, tr_M, is defined as follows: If the density matrix of W with respect to some basis in $\mathcal{H}_S \otimes \mathcal{H}_M$ is $W_{ij,i'j'}$ then $(tr_M W)_{ii'} = \sum_j W_{ij,i'j}$.

REFERENCES

[1] E. Schrödinger, *The Interpretation of Quantum Mechanics* (Woodbridge, CT: Ox Bow Press, 1995).

[2] J. von Neumann, *Mathematische Grundlagen der Quantenmechanik* (Berlin: Springer, 1932).

[3] J. von Neumann, *Mathematical Foundations of Quantum Mechanics* (Princeton: Princeton University Press, 1955).

[4] J. von Neumann, 'Unsolved Problems in Mathematics' Unpublished address to the International Mathematical Congress, Amsterdam, September 2, 1954. Typescript, von Neumann Archives, Library of Congress, Washington, DC; first published in this volume.

[5] P.A.M. Dirac, *Quantum Mechanics* (Oxford: Clarendon Press, 1958).

[6] A. Einstein, B. Podolsky, and N. Rosen, 'Can Quantum Mechanical Description of Physical Reality be Considered Complete?' *Physical Review* 47 (1935), 777–80.

[7] A. Fine, 'Probability and the Interpretation of Quantum Mechanics,' *British Journal for the Philosophy of Science* 24 (1973), 1–37.

[8] D. Bohm, 'A Suggested Interpretation of Quantum Theory in Terms of "Hidden Variables" ' Parts I and II, *Physical Review* 85, 166–79, 180–93.

[9] J. Bub, *Interpreting the Quantum World* (Cambridge: Cambridge University Press, 1997).

Philosophy Department
University of Maryland
College Park, MD, 20742
U.S.A.

THOMAS BREUER

VON NEUMANN, GÖDEL AND QUANTUM INCOMPLETENESS

John von Neumann was among the first to learn about Kurt Gödel's results on the incompleteness of formal systems. Did this shape his views on the completeness of quantum mechanics? I will investigate this question from two viewpoints: von Neumann's no-hidden-variables proof and his treatment of the quantum measurement problem.

1. VON NEUMANN AND GÖDEL IN KÖNIGSBERG

John von Neumann was an eminent logician. In his dissertation he developed a new axiomatisation of set theory, which became well known only after it was employed by Gödel and Bernays. He also was one of the outstanding protagonists of Hilbert's formalist programme.

It was as logician that he was invited to the second Conference on Epistemology of the Exact Sciences held in Königsberg on 5-7 September 1930, a meeting organised by the *Gesellschaft für empirische Philosophie*, which was closely allied with the Vienna Circle. I will briefly report on the conference, following John Dawson's account in Gödel's *Collected Works* (1986, pp. 196–199). Von Neumann gave one of the three main addresses of the first day, the others being delivered by Rudolf Carnap and Arend Heyting. His talk entitled "On the axiomatic grounding of mathematics" was intended to be a manifesto for formalism in the foundations of mathematics. It is reprinted in von Neumann (1931).

Being a formalist he advocated a concept of mathematics as a formal theory distinct from its interpretation. Problems with infinity and philosophical qualms may arise at the level of interpretation, but the formal theory should be kept separate from this. There are, however, parts of mathematics which can be interpreted in an unobjectionable way: the finitary parts. Von Neumann expected of a formal system that finitary formulas should be true whenever they are provable in the formal system (soundness) and provable whenever they are informally true (completeness). So in his opinion, the main unfinished task of proof theory is to prove that the finitary part of mathematics is correctly reproduced by the formalism.

Von Neumann presented an argument why for finitary formulas the criterion of soundness may be replaced by consistency: Assuming completeness, consistency implies soundness (and vice versa). And completeness he took for granted for finitary formulas, since these formulas are essentially arithmetical and therefore can be proved by a calculation.

75

M. Rédei and M. Stöltzner (eds.),
John von Neumann and the Foundations of Quantum Physics, 75–82.
© 2001 *Kluwer Academic Publishers. Printed in the Netherlands.*

So von Neumann expected completeness for the finitary part of mathematics, but not for the infinitary. Incompleteness for the infinitary part was not a problem for a formalist following Hilbert's strategy of considering as meaningless those parts of mathematics which cannot be interpreted in an unobjectionable way. To a formalist it does not matter if completeness fails for the infinitary part of mathematics, because for him this part of mathematics is meaningless anyway.

The conference continued with three main addresses by Hans Reichenbach, Werner Heisenberg and Otto Neugebauer, followed by three twenty-minute talks, including one by Kurt Gödel on his completeness proof (1930) for the restricted functional calculus. The conference concluded with a roundtable discussion on the foundations of mathematics, which was intended as an adjunct to the addresses of Carnap, Heyting, and von Neumann. It was during this discussion that Gödel for the first time mentioned in public his incompleteness results. He argued against Carnap's proposal to accept consistency as a criterion of adequacy for formal systems, because even in a consistent system there might be provable but false statements. As if to clarify Gödel's remarks, von Neumann interposed that it remained unsettled whether all intuitionistically acceptable means of proof were formally representable. Then Gödel forthrightly asserts that one can give examples of true but unprovable formulas in the formalised framework of classical mathematics. This is the first part of Gödel's incompleteness theorem (1931). Gödel did not mention the second part, which asserts that within a formal system the consistency of the system is not provable.

It is not clear whether everybody appreciated the significance of Gödel's announcement. For instance, there is evidence in Carnap's *Nachlaß* that several months after the conference he found the result difficult to understand, although Gödel explained it to him when they met in a Vienna coffee shop before the conference. But certainly von Neumann immediately understood the relevance of Gödel's announcement. He is reported to have pressed Gödel for further details after the discussion, and on 20 November he wrote to Gödel to announce his own discovery of the unprovability of consistency. Too late. Three days earlier the editors of the *Monatshefte für Mathematik und Physik* had received the manuscript of Gödel's (1931) proof of both incompleteness results.

Why was Gödel's argument so devastating to the formalist position? If we take 'finitary' in von Neumann's argument to mean 'strictly finite' or 'primitive recursive', then von Neumann's conclusions hold, as Gödel showed in Theorem V of his (1931): if formal mathematics is consistent, then a primitive recursive formula is true whenever it is provable and vice versa. (So completeness and consistency imply soundness.) However, a proposition of the form $\exists x F(x)$ with F primitive recursive, is not primitive recursive. Gödel pointed to a proposition of this kind which is not provable and not refutable, although in informal mathematics it is true. The shock for the formalists was that there are *meaningful* propositions which are undecidable by the means of the formal system.

After this von Neumann never again worked on the foundations of mathematics. His further work in logic was devoted to the development of quantum logic, the foundations of computer science and neural networks. He did, however, continue work in quantum mechanics, which reached a first climax in his (1932).

Was von Neumann's view on quantum mechanics in any way influenced by Gödel's incompleteness results? There are two parts of the 1932 book where the topic of completeness surfaces: in his no-go proof for hidden variable formulations of quantum mechanics (Chapter 4) and in his presentation of the quantum measurement problem (Chapter 6). To these two topics I will turn now.

2. THE NO-HIDDEN-VARIABLES PROOF

Did von Neumann intend his no-hidden-variables proof as an answer to the threat of incompleteness leaking from the foundations of mathematics into quantum mechanics? I think there is evidence that von Neumann in 1927 did not regard the question of completeness of quantum mechanics as important, but that he did so in 1932. Still, in the end he settled against the incompleteness of quantum mechanics.

In his 1927 papers on quantum mechanics there is no reference to the questions of completeness and hidden variables. These papers were published well before Gödel's incompleteness results. Interestingly enough, the second of the three 1927 papers contains all the technical ingredients of the no-hidden-variables proof: he showed that (1) the pure ensembles are represented by the one-dimensional projections on the vectors of the Hilbert space (1927, p. 257), and that (2) there are no ensembles which give dispersion-free expectation values on all quantum observables (1927, p. 259). But he did not relate these results to the question of completeness, as he did in (1932, p. 170).

The question of completeness according to von Neumann is: All wave functions yield expectation values with non-zero dispersion for some observables; can this be explained by assuming that the wave functions are not the true states of the system but mixtures of different "true" states? Then the true state would have to be specified not just by the wave function but additionally by the values of some 'hidden variables'. Von Neumann argued that this is impossible, because otherwise a wave function would represent a mixed ensemble, which contradicts (1); and furthermore by (2) the dispersion-free ensembles corresponding to the true states do not exist. (This proof of von Neumann later was regarded as not conclusive by many, most famously by Bell (1966) who criticised von Neumann's assumption that execpectation values should be additve not only for quantum states but also for the presumed true states.)

In 1927 von Neumann did not combine (1) and (2) in order to establish the completeness of quantum mechanics. He bothered to do this only after the event of Gödel's incompleteness results. This seems to be evidence that im 1927 he simply did not regard the question of completeness as important in quantum mechanics.

Still, I do not think that von Neumann seriously intended his proof to establish the irrelevance of Gödel's result for quantum mechanics. We should not allow ourselves to be fooled by the polysemy of the term 'completeness'. The term does not have the same meaning when applied to quantum mechanics and to formal systems. Von Neumann's exclusion of hidden variable formulations seeks to establish that the quantum mechanical description of reality is complete in the sense that the unpredictability of measurement outcomes is an objective property of nature, and not just due to subjective ignorance. Does this imply – or is it at least intended to imply – that there are no undecidable but true statements in quantum mechanics? It does not, and it is not intended to.

If von Neumann's views on quantum mechanics were influenced at all by Gödel's incompleteness results, the only place to look for it is his treatment of the quantum measurement problem. This is what I will do now.

3. THE MEASUREMENT PROBLEM

According to von Neumann quantum systems can undergo two different kinds of time evolutions: During a measurement the statistical state of the system evolves into a mixture of eigenstates of the measured observable (Process 1); systems which are not in interaction with an 'observing' system evolve according to the Schrödinger equation (Process 2). These two kinds of evolution are fundamentally different: Process 2 is causal (since it carries pure states into pure states; so given the initial state and the hamiltonian, and the time span, for every future time there is only one state possible) and reversible, Process 1 is non-causal (since the outcome in a single run of an experiment is not determined by the initial state) and irreversible. The duality of time evolutions threatens to violate the principle of psycho-physical parallelism, which requires that every mental process can be described as happening in the physical world.

Von Neumann argues that this principle is not violated in quantum mechanics because the border between observed and observing system can be set arbitrarily. What he shows is that a measurement on a system I performed by an observing system $II + III$ yields the same result about I as a measurement on $I + II$ performed by III: If I is in some unknown state ϕ and we measure an observable with eigenstates $\phi_1, \phi_2, \ldots$, Process 1 leads to outcome n with probability $|(\phi, \phi_n)|^2$; if $I + II$ is governed by Process 2 one can find an interaction hamiltonian, a pointer basis $\xi_1, \xi_2, \ldots$ in II, an initial state of II and a time t such that at t $I + II$ is in the state $\sum_n c_n \phi_n \otimes \xi_n$. If III measured in this final state of $I + II$ the original measured observable of I or the pointer observable of II, he would always get the same result, namely n with probability $|(\phi, \phi_n)|^2$. In this sense, and only in this sense, it does not matter whether Process 1 applies to I directly, or $I + II$ follows Process 2 first and then Process 1.

Note that the two scenarios lead to different statements about II. This von Neumann regards as irrelevant because only I belongs in both scenarios to the observed part of the world, and only there he requires agreement. The observed part of the world is what physics is about. The subjective perception of the observer is not the object of physics, rather it is presupposed in each experiment (1932,

p. 223). Wigner (1963), one year above von Neumann in the same high school, did not regard the state of II after the measurement as irrelevant. He emphasised that if II is a human observer $I + II$ must not be governed by Process 2. This is not quite the same position as von Neumann's: According to Wigner animate systems do not follow the same laws as inanimate systems; according to von Neumann observing systems do not follow the same laws as observed systems.

The measurement problem arises if one postulates universal validity of quantum mechanics and the Schrödinger equation. This postulate contradicts the possibility of measurements. Primas (1990) suggests to view the measurement problem as the problem of compatibility of *endo-* and *exophysics*. These terms are used by Rössler, Svozil (1993), and Finkelstein (1988) in a different way. To avoid these terminological discussions I will use the terms "from inside" and "from outside". As physics-from-inside I dub the observations and the formalism used by an observer contained in the observed or described system; as physics-from-outside the physics of an outside observer.

A description from outside presupposes a separation of observer and observed system. This is not deep philosophy or exciting physics, it is a logical necessity. When we speak about measurements from outside we have to specify outside what. So the observed system must be conceivable as something distinct from the observer. This does not mean that there may not be any interaction between the two, nor that they must not be correlated, nor that there is an absolute observer outside all systems described by physics-from-outside. But since measurements without interaction are impossible, systems which are observed from outside have to be *open* systems. Only open systems can be the subject of physics-from-outside. The universe cannot be the subject of physics-from-outside, so physics-from-outside cannot be universally valid.

Closed systems, as for example the universe, can be observed only from inside. Thus only physics-from-inside is able to deal with closed systems, and only physics-from-inside can be universally valid. The price to pay for potential universality is the failure of full measurability. Full measurability fails in physics-from-inside for problems of self-reference (Breuer, 1995).

This is true in general but there are some unexpected consequences for quantum mechanics. Quantum-mechanics-from-inside refers to closed systems and their time evolution is governed by the Schrödinger equation. They evolve linearly, deterministically, and reversibly. Measurability of such systems is problematic because they can be observed only from inside; firstly, this gives rise to problems of self-reference (Breuer, 1995); secondly, the measurement problem arises since an observer being part of the observed quantum mechanical system must be described quantum mechanically as well.

In contradistinction, quantum-mechanics-from-outside avoids problems of self-reference and the measurement problem. The measurement problem does not arise because quantum mechanics is only applied to the observed open system but not to the observer. Neither do problems of self-reference arise for observations from outside. But the universe cannot be described by quantum-mechanics-from-outside.

The Copenhagen interpretation and von Neumann's interpretation are paradigm examples of quantum-mechanics-from-outside. They require that quantum mechanics must not be applied to the observer.

The measurement problem can be viewed as a problem of compatibility of quantum-mechanics-from-inside with quantum-mechanics-from-outside. Is it possible for quantum mechanics both to be universally valid, as is quantum-mechanics-from-inside, and to allow for measurements, as does quantum-mechanics-from-outside in von Neumann's or the Copenhagen interpretation? The two can be compatible only if there is no contradiction between the validity of the Schrödinger equation for closed systems and the occurrence of measurement results. Is it consistent to apply the projection postulate to quantum systems observed from outside and to believe that the wave function of the universe follows the Schrödinger equation? This is the measurement problem.

The alleged incompatibility of quantum-mechanics-from-inside and quantum-mechanics-from-outside, and the measurement problem, arises not from the fact that the evolution during the collapse is non-linear and stochastic. This does not necessarily contradict the validity of the Schrödinger equation for closed systems since during the collapse the observed system is open. The problem rather is that if the Schrödinger equation holds for closed systems the evolution during the collapse cannot be explained by the observed system being open during the measurement.

In his theory of the quantum measurement process von Neumann (1932) introduced a chain of observers. If an observer A makes a measurement on a system O the values of the pointer observable will be related to the values of the measured observable. Knowing the value of the pointer we know the value of the measured observable. But if the Schrödinger equation holds for the measurement interaction and the measured observable did not have a well-defined value before the measurement, the pointer observable will not have a well-defined value afterwards. Now a second observer A_2 can try to measure the value of the pointer observable. But this measurement performed by A_2 on A faces the same problem: the value of the second pointer observable can be related to that of the first and thus to that of the measured observable, but if the first pointer did not have a well-defined value the second will not have one either. The same will be true if a third observer A_3 tries to determine the value of the second pointer, etc., etc.

What precisely is the analogy of this situation with Tarski's hierarchy of object theory, meta-theory, meta-theory etc.? This hierarchy is enforced by Tarski's (1935) result that a truth predicate applicable to all sentences of a language cannot be part of that language. It is intimately related to Gödel's theorem, of which it can be regarded as contraposition after replacing truth by provability. Put this way Gödel's theorem can be paraphrased as: A concept of provability which is formulated within the formal system cannot apply to all sentences of that system.

Let us return to the analogy of von Neumann's chain of observers to Tarski's hierarchy. In a formal system there are propositions like the Gödel formula which cannot be proved or refuted by the rules of the system. Correspondingly, a quantum measurement of A on O does in general not have a result if $A\&O$ is closed. In the meta-theory there are propositions which cannot be decided by the rules of the meta-theory. Correspondingly, a measurement of A_2 on A does not have a result if $O\&A\&A_2$ is closed. Is there an analogy to the decidability of the Gödel formula in the meta-theory? Yes, if the measurement problem could be solved by assuming $A\&O$ to be open. Then measurements of A on O would not have a result for A but they would have a result for an observer outside $A\&O$.

This is how far the analogy carries in a traditional analysis of the measure-ment problem. It was based on an analogy of "having a proof of a statement" with "having a result of a measurement". This is not entirely mistaken. After all, measurement and proof are both semantic concepts in that they establish a relation between a physical or mathematical formalism, and what is referred to by the for-malism. Still, if the occurrence of results cannot be explained by assuming $A\&O$ to be open, the analogy is vague and wanting: there is no analogy to the decidability of the Gödel formula on the meta-level .

It can be shown (Breuer, 1996) that A cannot distinguish between states s, s' of $A\&O$ for which $R(s) = R(s')$, where R denotes the partial trace over O. This is the case for the pure state $s_1 := \sum_n c_n \phi_n \otimes \xi_n$ and the decohered $s_2 := \sum_n |c_n|^2 |\phi_n \otimes \xi_n\rangle\langle\phi_n \otimes \xi_n|$: we have $R(s_1) = R(s_2) = \sum_n |c_n|^2 |\xi_n\rangle\langle\xi_n|$. Thus a statement like

$$\text{The state of } A\&O \text{ after the measurement is } s_1 \text{ and not } s_2. \tag{1}$$

or "In the final statistical state the interference terms between A and O vanish" is undecidable for A.

Primas (1990) formulates a different quantum mechanical undecidability theo-rem: *The statement 'The cat is either dead or alive' is endophysically undecidable (i.e. not provable and irrefutable), even it were true.* Unfortunately he does not give an argument for his undecidability theorem. If one translates "endophysically" with "for the cat" and does not refer to a single experiment but rather to a series then Primas' undecidability theorem reads: *The statement 'In the final statistical state the interference terms between cat and observed system vanish' is undecid-able for the cat.* This is roughly my (1). But it seems that Primas' statement is not undecidable if read as referring to single experiments. Definite life-state and superpositions of definite life states of $A\&O$ have different restrictions to A and therefore can be distinguished by A.

A proposition which is undecidable on the meta-level, i.e. for A_2, is: "The interference terms between A_2 and $A\&O$ vanish." The system $O\&A$ corresponds to the object theory, $O\&A\&A_2$ to the meta-theory etc. Similar to the language of the meta-theory being richer than that of the object theory and containing it, the physical system of the meta-level contains that of the object level.

Is (1) decidable for A_2? It seems that so, because an outside observer like A_2 can distinguish s_1 from s_2 simply by measuring the interference terms between A and O. This corresponds to the decidability of the Gödel-formula on the meta-level.

REFERENCES

Bell J. S. (1966), "The Problem of Hidden Variables in Quantum Mechanics", *Reviews of Modern Physics* **38**, 447–452.

Breuer T. (1995), "The Impossibility of Accurate State Self-Measurements", *Philosophy of Science*, **62**, 197.

Breuer T. (1996), "Subjective Decoherence in Quantum Measurements", *Synthese* **107**, 1.

Finkelstein D. (1988), "Finite Physics", in R.Herken (ed.): *The Universal Turing Machine. A Half-Century Survey*, Oxford: University Press.

Gödel K. (1930), "Die Vollständigkeit der Axiome des logischen Funktionenkalküls", *Monatshefte für Mathematik und Physik* **37**, 349–360.

Gödel K. (1931), "Über formal unentscheidbare Sätze der *Principia Mathematica* und verwandter Systeme",*Monatshefte für Mathematik und Physik* **38**, 173–198.

Gödel K. (1986), *Collected Works, Vol. I*, Oxford: University Press.

London F., Bauer E.(1939), *La théorie de l'observation en mécanique quantique*, Paris: Hermann.

von Neumann J. (1927), "Wahrscheinlichkeitstheoretischer Aufbau der Quantenmechanik", *Nachrichten von der Gesellschaft der Wissenschaften zu Göttingen, Mathematisch-physikalische Klasse* **1927**, 245-272.

von Neumann J. (1929), "Über eine Widerspruchsfreiheitsfrage in der axiomatischen Mengenlehre", *Journal für reine und angewandte Mathematik* **160**, 227–241.

von Neumann J. (1931), "Die formalistische Grundlegung der Mathematik", *Erkenntnis* **2**, 116–121.

von Neumann J. (1932), *Mathematische Grundlagen der Quantenmechanik*, Berlin: Springer.

Primas H. (1990) in J. Audretsch, K. Mainzer (eds): *Wieviele Leben hat Schrödingers Katze?*, Mannheim: BI-Wissenschaftsverlag, 209-243.

Rössler O.E. (1987), "Endophysics", in J.L.Casti and A. Karlqvist (eds.): *Real Brains – Artificial Minds*, New York: North-Holland, 25-46.

Svozil K. (1993), *Randomness and Undecidability in Physics*, Singapore: World Scientific.

Tarski A. (1935), "Der Wahrheitsbegriff in den formalisierten Sprachen", *Studia Phiosophica* **1**, 261. Übersetzung von "Pojęcie prawdy w językach nauk dedukcyjnych", *Prace Towardzystwa Naukowego Warszawskiego, wydział* **34** (1933).

Wigner E. P. (1963), "The problem of measurement", *American Journal of Physics* **31**, 6-15.

Fachhochschule Vorarlberg
Achstraße 1
A-6850 Dornbirn
Austria

DÉNES PETZ

ENTROPY, VON NEUMANN
AND THE VON NEUMANN ENTROPY[*]

Dedicated to the memory of Alfred Wehrl

The highway of the development of entropy is marked by many great names, for example, Clausius, Gibbs, Boltzmann, Szilárd, von Neumann, Shannon, Jaynes, and several others. In this article the emphasis is put on von Neumann and on quantum mechanics. The selection of the subjects reflects the taste (and the knowledge) of the author and it must be rather restrictive. In the past 50 years entropy has broken out of thermodynamics and statistical mechanics and invaded communication theory, ergodic theory and shown up in mathematical statistics, social and life sciences. It is practically impossible to present all of its features. The favourite subjects of entropy is about macroscopic phenomena, irreversibility and incomplete knowledge. In the strictly mathematical sense entropy is related to the asymptotics of probabilities or it is a kind of asymptotic behaviour of probabilities.

This paper is organized as follows. After a short introduction to entropy, von Neumann's gedanken experiment is repeated, which led him to the formula of thermodynamic entropy of a statistical operator. In the analysis of his ideas we stress the role of (the lack of) superselection sectors and summarize von Neumann's knowledge about quantum mechanical entropy. The final part of this article is devoted to some important developments of the von Neumann entropy which were discovered long after von Neumann's work. Subadditivity and interpretation of the von Neumann entropy as the capacity of a communication channel are among those.

1. GENERAL INTRODUCTION TO ENTROPY

The word "entropy" was created by Rudolf Clausius and it appeared in his work "Abhandlungen über die mechanische Wärmetheorie" published in 1864. The word has a Greek origin, its first part reminds us of "energy" and the second part is from "tropos" which means turning point. Clausius' work is the foundation stone of classical thermodynamics. According to Clausius, the change of entropy of a system is obtained by adding the small portions of heat quantity received by the system divided by the absolute temperature during the heat absorption. This definition is satisfactory from a mathematical point of view and gives nothing other than an integral in precise mathematical terms. Clausius postulated that the entropy of a

83

M. Rédei and M. Stöltzner (eds.),
John von Neumann and the Foundations of Quantum Physics, 83–96.

closed system cannot decrease, which is generally referred to as the second law of thermodynamics. On the other hand, he did not provide any heuristic argument to support the law. This fact might partly be responsible for the mystery surrounding entropy for a long time. As an extreme position, we can cite Alfred Wehrl who had the opinion in 1978 that "the second law of thermodynamics does not appear to be fully understood yet" [13].

The concept of entropy was really clarified by Ludwig Boltzmann. His scientific program was to deal with the mechanical theory of heat in connection with probabilities. Assume that a macroscopic system consists of a large number of microscopic ones, we simply call them particles. Since we have ideas of quantum mechanics in mind, we assume that each of the particles is in one of the energy levels $E_1 < E_2 < \ldots < E_m$. The number of particles in the level E_i is N_i, so $\sum_i N_i = N$ is the total number of particles. A macrostate of our system is given by the occupation numbers $N_1, N_2, \ldots, N_m$. The energy of a macrostate is $E = \sum_i N_i E_i$. A given macrostate can be realized by many configurations of the N particles, each of them at a certain energy level E_i. Those configurations are called microstates. Many microstates realize the same macrostate. We count the number of ways of arranging N particles in m boxes (i.e., energy levels) such that each box has $N_1, N_2, \ldots, N_m$ particles. There are

$$\binom{N}{N_1, N_2, \cdots, N_m} := \frac{N!}{N_1! \, N_2! \ldots N_m!} \tag{1}$$

such ways. This multinomial coefficient is the number of microstates realizing the macrostate $(N_1, N_2, \ldots, N_m)$ and it is proportional to the probability of the macrostate if all configurations are assumed to be equally likely. Boltzmann called (1) the thermodynamical probability of the macrostate, in German "thermodynamische Wahrscheinlichkeit", hence the letter W was used. Of course, Boltzmann argued in the framework of classical mechanics and the discrete values of energy came from an approximation procedure with "energy cells".

If we are interested in the thermodynamic limit N increasing to infinity, we use the relative numbers $p_i := N_i/N$ to label a macrostate and, instead of the total energy $E = \sum_i N_i E_i$, we consider the average energy per particle $E/N = \sum_i p_i E_i$. To find the most probable macrostate, we wish to maximize (1) under a certain constraint. The Stirling approximation of the factorials gives

$$\frac{1}{N} \log \binom{N}{N_1, N_2, \cdots, N_m} = H(p_1, p_2, \ldots, p_m) + O(N^{-1} \log N), \tag{2}$$

where

$$H(p_1, p_2, \ldots, p_m) := \sum_i -p_i \log p_i. \tag{3}$$

If N is large then the approximation (2) yields that instead of maximizing the quantity (1) we can maximize (3). For example, maximizing (3) under the constraint

$\sum_i p_i E_i = e$, we get

$$p_i = \frac{e^{-\lambda E_i}}{\sum_j e^{-\lambda E_j}}, \tag{4}$$

where the constant λ is the solution of the equation

$$\sum_i E_i \frac{e^{-\lambda E_i}}{\sum_j e^{-\lambda E_j}} = e.$$

Note that the last equation has a unique solution if $E_1 < e < E_m$, and the distribution (4) is known as the discrete Maxwell-Boltzmann law today.

Let $p_1, p_2, \ldots, p_n$ be the probabilities of different outcomes of a random experiment. According to Shannon, the expression (1) is a measure of our ignorance prior to the experiment. Hence it is also the amount of information gained by performing the experiment. (1) is maximum when all the p_i's are equal. In information theory logarithms with base 2 are used and the unit of information is called bit (from binary digit). As will be seen below, an extra factor equal to Boltzmann's constant is included in the physical definition of entropy.

2. VON NEUMANN'S CONTRIBUTION TO ENTROPY

The comprehensive mathematical formalism of quantum mechanics was first presented in the famous book *Mathematische Grundlagen der Quantenmechanik* published in 1932 by Johann von Neumann. In the traditional approach to quantum mechanics, a physical system is described in a Hilbert space: Observables correspond to self-adjoint operators and statistical operators are asssociated with the states. In fact, a statistical operator describes a mixture of pure states. Pure states are the really physical states and they are given by rank one statistical operators, or equivalently by rays of the Hilbert space.

Von Neumann associated an entropy quantity with a statistical operator in 1927 [5] and the discussion was extended in his book [6]. His argument was a gedanken experiment on the ground of phenomenological thermodynamics. Let us consider a gas of $N(\gg 1)$ molecules in a rectangular box K. Suppose that the gas behaves like a quantum system and is described by a statistical operator D which is a mixture $\lambda|\varphi_1\rangle\langle\varphi_1| + (1 - \lambda)|\varphi_1\rangle\langle\varphi_2|$, $|\varphi_i\rangle \equiv \varphi$ is a state vector $(i = 1, 2)$. We may take λN molecules in the pure state φ_1 and $(1 - \lambda)N$ molecules in the pure state φ_2. On the basis of phenomenological thermodymanics we assume that if φ_1 and φ_2 are orthogonal, then there is a wall which is completely permeable for the φ_1-molecules and isolating for the φ_2-molecules. (In fact, von Neumann supplied an argument that such a wall exists if and only if the state vectors are orthogonal.) We add an equally large empty rectangular box K' to the left of the box K and we replace the common wall with two new walls. Wall (a), the one to the left is impenetrable, whereas the one to the right, wall (b), lets through the φ_1-molecules but keeps back the φ_2-molecules. We add a third wall (c) opposite to (b) which is semipermeable, transparent for the φ_2-molecules and impenetrable for the φ_1-ones. Then we push slowly (a) and (c) to the left, maintaining their distance. During this process the φ_1-molecules are pressed through (b) into K' and the φ_2-molecules diffuse through wall (c) and remain in K. No work is done against the gas pressure, no heat is developed. Replacing the walls (b) and (c) with a rigid absolutely impenetrable wall and removing (a) we restore the boxes K and K' and succeed in the separation of the φ_1-molecules from the φ_2-ones without any work being done, without any temperature change and without evolution of heat. The entropy of the original D-gas (with density N/V) must be the sum of the entropies of the φ_1- and φ_2-gases (with densities $\lambda N/V$ and $(1 - \lambda)N/V$, respectively.) If we compress the gases in K and K' to the volumes λV and $(1 - \lambda)V$, respectively, keeping the temperature T constant by means of a heat reservoir, the entropy change amounts to $\kappa\lambda N \log \lambda$ and $\kappa(1 - \lambda)N \log(1 - \lambda)$, respectively. Indeed, we have to add heat in the amount of $\lambda_i N \kappa T \log \lambda_i$ (< 0) when the φ_i-gas is compressed, and dividing by the temperature T we get the change of entropy. Finally, mixing the φ_1- and φ_2-gases of identical density we obtain a D-gas of N molecules in a volume V at the original temperature. If $S_0(\psi, N)$ denotes the entropy of a ψ-gas of N molecules (in a volume V and at the given temperature), we conclude that

$$S_0(\varphi_1, \lambda N) + S_0(\varphi_2, (1 - \lambda)N)$$
$$= S_0(D, N) + \kappa\lambda N \log \lambda + \kappa(1 - \lambda)N \log(1 - \lambda)$$

must hold, where κ is Boltzmann's constant. Assuming that $S_0(\psi, N)$ is proportional to N and dividing by N we have

$$\lambda S(\varphi_1) + (1 - \lambda)S(\varphi_2)$$
$$= S(D) + \kappa\lambda \log \lambda + \kappa(1 - \lambda)\log(1 - \lambda), \tag{5}$$

where S is certain thermodynamical entropy quantity (relative to the fixed temperature and molecule density). We arrived at the mixing property of entropy, but we should not forget about the initial assumption: φ_1 and φ_2 are supposed to be orthogonal. Instead of a two-component mixture, von Neumann operated by an infinite mixture, which does not make a big difference, and he concluded that

$$S\left(\sum_i \lambda_i |\varphi_i\rangle\langle\varphi_i|\right) = \sum_i \lambda_i S(|\varphi_i\rangle\langle\varphi_i|) - \kappa \sum_i \lambda_i \log \lambda_i. \tag{6}$$

Before we continue to follow his considerations, let us note that von Neumann's argument does not require that the statistical operator D is a mixture of pure states. What we really needed is the property $D = \lambda D_1 + (1 - \lambda)D_2$ in such a way that the possible mixed states D_1 and D_2 are disjoint. D_1 and D_2 are disjoint in the thermodynamical sense, when there is a wall which is completely permeable for the molecules of a D_1-gas and isolating for the molecules of a D_2-gas. In other words, if the mixed states D_1 and D_2 are disjoint, then this should be demonstrated by a certain filter. Mathematically, the disjointness of D_1 and D_2 is expressed in the orthogonality of the eigenvectors corresponding to nonzero eigenvalues of the two density matrices. The essential point is in the remark that equation (5) must hold also in a more general situation when possibly the states do not correspond to density matrices but orthogonality of the states makes sense:

$$\lambda S(D_1) + (1 - \lambda) S(D_2)$$
$$= S(D) + \kappa \lambda \log \lambda + \kappa (1 - \lambda) \log(1 - \lambda) \qquad (7)$$

Equation (6) reduces the determination of the (thermodynamical) entropy of a mixed state to that of pure states. The so-called Schatten decomposition $\sum_i \lambda_i |\varphi_i\rangle\langle\varphi_i|$ of a statistical operator is not unique even if $\langle\varphi_i, \varphi_j\rangle = 0$ is assumed for $i \neq j$. When λ_i is an eigenvalue with multiplicity, then the corresponding eigenvectors can be chosen in many ways. If we expect the entropy $S(D)$ to be independent of the Schatten decomposition, then we are led to the conclusion that $S(|\varphi\rangle\langle\varphi|)$ must be independent of the state vector $|\varphi\rangle$. This argument assumes that there are no superselection sectors, that is, any vector of the Hilbert space can be a state vector. On the other hand, von Neumann wanted to avoid degeneracy of the spectrum of a statistical operator (as well as the possible degeneracy of the spectrum of observables as we shall see below).

Von Neumann's proof of the property that $S(|\varphi\rangle\langle\varphi|)$ is independent of the state vector $|\varphi\rangle$ was different. He did not want to refer to a unitary time development sending one state vector to another, because that argument requires great freedom in choosing the energy operator H. Namely, for any $|\varphi_1\rangle$ and $|\varphi_2\rangle$ we would need an energy operator H such that

$$e^{itH}|\varphi_1\rangle = |\varphi_2\rangle.$$

This process would be reversible. (It is worthwhile to note that the problem of superselection sectors appears also here.)

Von Neumann proved that $S(|\varphi_1\rangle\langle\varphi_1|) \leq S(|\varphi_2\rangle\langle\varphi_2|)$ by constructing a great number of measurement processes sending the state $|\varphi_1\rangle$ into an ensemble, which differs from $|\varphi_2\rangle\langle\varphi_2|$ by an arbitrarily small amount. The measurement of an observable $A = \sum_i \lambda_i |\psi_i\rangle\langle\psi_i|$ in a state $|\varphi\rangle$ yields an ensemble of the pure states $|\psi_i\rangle\langle\psi_i|$ with weights $|\langle\varphi|\psi_i\rangle|^2$. This was a basic postulate in von Neumann's measurement theory when the eigenvalues of A are non-degenerate, that is, λ_i's are all different. In a modern language, von Neumann's measurement is a conditional expectation onto a maximal Abelian subalgebra of the algebra of all bounded operators acting on the given Hilbert space. Let $(|\psi_i\rangle)_i$ be an orthonormal basis consisting of eigenvectors of the observable under measurement. For any bounded operator T we set

$$E(T) = \sum_i \langle\psi_i|T|\psi_i\rangle |\psi_i\rangle\langle\psi_i|. \qquad (8)$$

The linear transformation E possesses the following properties:

(i) $E = E^2$.

(ii) If $T \geq 0$ then $E(T) \geq 0$.

(iii) $E(I) = I$.

(iv) $\operatorname{Tr}\big(E(T)\big) = \operatorname{Tr} T$.

In particular, for a statistical operator D its transform $E(D)$ is a statistical operator as well. It follows immediately from definition (8) that

$$E(|\varphi\rangle\langle\varphi|) = \sum_i |\langle\varphi|\psi_i\rangle|^2 |\psi_i\rangle\langle\psi_i|$$

and the conditional expectation E acts on the pure states exactly in the same way as it is described in the measurement procedure. It was natural for von Neumann to assume that

$$S(D) \leq S\big(E(D)\big), \tag{9}$$

at least if the statistical operator D corresponds to a pure state. Inequality (9) is nothing other than the manifestation of the second law for the measurement process.

In the proof of the inequality $S(|\varphi_1\rangle\langle\varphi_1|) \leq S(|\varphi_2\rangle\langle\varphi_2|)$ one can assume that the vectors $|\varphi_1\rangle$ and $|\varphi_2\rangle$ are orthogonal. The idea is to construct measurements $E_1, E_2, \ldots, E_k$ such that

$$E_k\Big(\ldots\big(E_1(|\varphi_1\rangle\langle\varphi_1|)\big)\ldots\Big) \tag{10}$$

is in a given small neighbourhood of $|\varphi_2\rangle\langle\varphi_2|$. The details are well-presented in von Neumann's original work, but we confine ourselves here to his definition for E_n. He set a unit vector

$$\psi^{(n)} = \cos\frac{\pi n}{2k}|\varphi_1\rangle + \sin\frac{\pi n}{2k}|\varphi_2\rangle$$

and extended it to a complete orthonormal system. The measurement conditional expectation E_n corresponds to this basis ($1 \leq n \leq k$). It is elementary to show that (10) tends to $|\varphi_2\rangle\langle\varphi_2|$ as $k \to \infty$. We stress again that the argument needs that $|\varphi_1\rangle$ and $|\varphi_2\rangle$ are in the same superselection sector, so that their linear combinations may be state vectors.

Let us summarize von Neumann's discussion of the thermodynamical entropy of a statistical operator D. First of all, he assumed that $S(D)$ is a continuous function of D. He carried out a reversible process to obtain the mixing property (5) for orthogonal pure states, and he concluded (6). He referred to the second law again when assuming (9) for pure states. Then he showed that $S(|\varphi\rangle\langle\varphi|)$ is independent of the state vector $|\varphi\rangle$ so that

$$S\left(\sum_i \lambda_i |\varphi_i\rangle\langle\varphi_i|\right) = -\kappa \sum_i \lambda_i \log \lambda_i \qquad (11)$$

up to an additive constant which could be chosen to be 0 as a matter of normalization. (11) is von Neumann's celebrated entropy formula; it has a more elegant form

$$S(D) = \kappa \operatorname{Tr} \eta(D), \qquad (12)$$

where $\eta : \mathbb{R}^+ \to \mathbb{R}$ is the continuous function $\eta(t) = -t \log t$. (The modern notation for $-t \log t$ comes from information theory which did not exist at that time.)

When von Neumann deduced (12), his natural intention was to make mild assumptions. For example, the monotonicity (9) was assumed only for pure states. If we already have (12) as a definition, then (9) can be proved for an arbitrary statistical operator D. The argument is based on the Jensen inequality, and von Neumann remarked that for

$$S_f(D) = \operatorname{Tr} f(D)$$

with a differentiable concave function $f : [0, 1] \to \mathbb{R}$,

$$S_f(D) \le S_f(E(D)) \qquad (13)$$

holds for every statistical operator D. His analysis also indicated that the measurement process is typically irreversible, the finite entropy of a statistical operator definitely increases if a state change occurs.

Von Neumann solved the maximization problem for $S(D)$ under the constraint $\operatorname{Tr} DH = e$. This means the determination of the ensemble of maximal entropy when the expectation of the energy operator H is a prescribed value e. It is convenient to rephrase his argument in terms of conditional expectations. $H = H^*$ is assumed to have a discrete spectrum and we have a conditional expectation E determined by the eigenbasis of H. If we pass from an arbitrary statistical operator D with $\operatorname{Tr} DH = e$ to $E(D)$, then the entropy is increasing on the one hand and the expectation of the energy does not change on the other hand, so the maximizer should be searched among the operators commuting with H. In this way we are (and von Neumann was) back to the classical problem of statistical mechanics treated at the beginning of this article. In terms of operators the solution is in the form

$$\frac{\exp(-\beta H)}{\operatorname{Tr} \exp(-\beta H)},$$

which is called the Gibbs state today.

3. SOME TOPICS ABOUT ENTROPY FROM VON NEUMANN
TO THE PRESENT

After Boltzmann and von Neumann, it was Shannon who initiated the interpretation of the quantity $-\sum_i p_i \log p_i$ as "uncertainty measure" or "information measure". The American electrical engineer/scientist Claude Shannon created communication theory in 1948. He posed a problem in the following way:

Suppose we have a set of possible events whose probabilities of occurence are $p_1, p_2, \ldots, p_n$. These probabilities are known but that is all we know concerning which event will occur. Can we find a measure of how much "choice" is involved in the selection of the event or how uncertain we are of the outcome?

Denoting such a measure by $H(p_1, p_2, \ldots, p_n)$ he listed three very reasonable requirements which should be satisfied. He concluded that the only H satisfying the three assumptions is of the form

$$H = -K \sum_{i=1}^{n} p_i \log p_i \,,$$

where K is a positive constant. For H he used different names such as information, uncertainty and entropy. Many years later Shannon said [12]:

My greatest concern was what to call it. I thought of calling it "information", but the word was overly used, so I decided to call it "uncertainty". When I discussed it with John von Neumann, he had a better idea. Von Neumann told me, "You should call it entropy, for two reasons. In the first place your uncertainty function has been used in statistical mechanics under that name, so it already has a name. In the second place, and more important, nobody knows what entropy really is, so in a debate you will always have the advantage."

Shannon's postulates were transformed later into the following axioms:

(a) Continuity: $H(p, 1 - p)$ is continuous function of p.

(b) Symmetry: $H(p_1, p_2, \ldots, p_n)$ is a symmetric function of its variables.

(c) Recursion: For every $0 \le \lambda < 1$ the recursion $H(p_1, \ldots, p_{n-1}, \lambda p_n, (1 - \lambda)p_n) = H(p_1, \ldots, p_n) + p_n H(\lambda, 1 - \lambda)$ holds.

These axioms determine a function H up to a positive constant factor. With the exceptional the above story about a conversation between Shannon and von Neumann, we do not know about any mutual influence. Shannon was interested in communication theory and von Neumann's thermodynamical entropy was in the formalism of quantum mechanics. Von Neumann himself never made any connection between his quantum mechanical entropy and information. Although von Neumann's entropy formula appeared in 1927, there was not much activity concerning it for several decades. At the end of the 1960s, the situation changed. Rigorous statistical mechanics came into being [10] and soon after that the needs of rigorous quantum statistical mechanics forced new developments concerning von Neumann's entropy formula.

Von Neumann was aware of the fact that statistical operators form a convex set whose extreme points are exactly the pure states. He also knew that entropy is a concave functional, so

$$S\left(\sum_i \lambda_i D_i\right) \geq \sum_i \lambda S(D_i) \qquad (14)$$

for any convex combination. To determine the entropy of a statistical operator, he used the Schatten decomposition, which is an orthogonal extremal decomposition in our present language. For a statistical operator D there are many ways to write it in the form

$$D = \sum_i \lambda_i |\psi_i\rangle\langle\psi_i|$$

if we do not require the state vectors to be orthogonal. The geometry of the statistical operators, that is the state space, allows many extremal decompositions and among them there is a unique orthogonal one if the spectrum of D is not degenerate. Non-orthogonal pure states are essentially nonclassical. They are between identical and completely different. Jaynes recognized in 1956 that from the point of view of information the Schatten decomposition is optimal. He proved that

$$S(D) \quad = \quad \sup\left\{-\sum_i \lambda_i \log\lambda_i : D = \sum_i \lambda_i D_i\right.$$

$$\left. \text{for some convex combination and statistical operators}\right\}.$$

This is Jaynes contribution to the von Neumann entropy [2]. (However, he became known for the very strong advocacy of the maximum entropy principle.)

Certainly the highlight of quantum entropy theory in the 1970s was the discovery of subadditivity. Before we state it in precise mathematical form, we describe the setting where this property is crucial. A one-dimensional quantum lattice system is a composite system of $2N + 1$ subsystems, indexed by the integers $-N \leq n \leq N$. Each of the subsystems is described by a Hilbert space $\mathcal{H}_n$; those Hilbert spaces are isomorphic if we assume that the subsystems are physically identical, and even the very finite dimensional case $\dim\mathcal{H}_n = 2$ can be interesting if the subsystem is a "spin 1/2" attached to the lattice site n. The finite chain of $2N + 1$ spins is described in the tensor product Hilbert space $\otimes_{n=-N}^{N}\mathcal{H}_n$, whose dimension is $(\dim\mathcal{H}_n)^{2N+1}$. For a given Hamiltonian H_N and inverse temperature β the equilibrium state maximizes the free energy functional

$$F_N(D_N) = \mathrm{Tr}_N H_N D_N - \frac{1}{\beta}S(D_N), \qquad (15)$$

and the actual maximizer is the Gibbs state

$$\frac{\exp(-\beta H_N)}{\text{Tr} \exp(-\beta H_N)}. \tag{16}$$

It seems that this was already known in von Neumann's time but not the thermodynamical limit, $N \to \infty$. Rigorous statistical mechanics of spin chains was created in the 1970s. Since entropy, energy, and free energy are extensive quantities, the infinite system should be handled by their normalized versions, called entropy density, energy density, etc. One possibility to describe the equilibrium of the infinite system is to carry out a limiting procedure from the finite volume equilibrium states, and another is to solve the variational principle for the free energy density on the state space of the infinite system. In a translation invariant theory the two approaches lead to the same conclusion, but many technicalities are involved. The infinite system is modeled by a C^*-algebra and their states are normalized linear functionals instead of statistical operators. The rigorous statistical mechanics of quantum spin systems was one of the successes of the operator algebraic approach. ([11] and Sec. 15 of [7] are suggested further readings about details of quantum spin systems.) One of the key points in this approach is the definition of entropy density of a state of the infinite system which goes back to the subadditivity of the von Neumann entropy. Let $\mathcal{H}_1$ and $\mathcal{H}_2$ be possibly finite dimensional Hilbert spaces corresponding to two quantum systems. A mixed state of the composite system is determined by a statistical operator D_{12} acting on the tensor product $\mathcal{H}_1 \otimes \mathcal{H}_2$. Assume that we are to measure observables on the first subsystem. What is the statistical operator we need? The statistical operator D_1 has to fulfill the condition

$$\text{Tr}_1 A D_1 = \text{Tr}_{12}(A \otimes I)D_{12} \tag{17}$$

for any observable A. Indeed, the left-hand side is the expectation of A in the subsystem and the right hand side is that in the total system. It is easy to see that condition

$$\langle \psi | D_1 | \psi \rangle = \sum_i \langle |\psi\rangle \otimes |\varphi_i\rangle, D_{12}|\psi\rangle \otimes |\varphi_i\rangle \rangle \tag{18}$$

gives the statistical operator D_1, where $|\psi\rangle \in \mathcal{H}_1$ and $|\varphi_i\rangle$ is an arbitrary orthonormal basis in $\mathcal{H}_2$. (In fact equation (18) is obtained from (17) by putting $|\psi\rangle\langle\psi|$ in place of A.) It is not difficult to state the subadditivity property now:

$$S(D_{12}) \leq S(D_1) + S(D_2). \tag{19}$$

This is a particular case of the strong subadditivity

$$S(D_{123}) \leq S(D_{12}) + S(D_{23}) - S(D_2) \tag{20}$$

for a system consisting of three subsystems. (We hope that the notation is self-explanatory, otherwise see [4], [13] or p. 23 in [7].) If the second subsystem is absent, (20) reduces to (19). (19) was proven first by Lieb and Ruskai in 1973 [4].

The measurement conditional expectation was introduced by von Neumann as the basic irreversible state change, and it is of the form

$$D \mapsto \sum_i P_i D P_i, \qquad (21)$$

where P_i are pairwise orthogonal projections and $\sum_i P_i = I$. (We are in the Schrödinger picture.) The measurement conditional expectation has the following generalization. Assume that our quantum system is described by an operator algebra $\mathcal{M}$ whose positive linear functionals correspond to the states. A functional $\tau : \mathcal{M} \to \mathbb{C}$ is a state if $\tau(A) \geq 0$ for any positive observable A and $\tau(I) = 1$. An operational partition of unity is a finite subset $\mathcal{W} = (V_1, V_2, \ldots, V_n)$ of $\mathcal{M}$ such that $\sum_i V_i^* V_i = I$. In the Heisenberg picture $\mathcal{W}$ acts on the observables as

$$A \mapsto \sum_i V_i^* A V_i$$

and the corresponding state change in the Schrödinger picture is

$$\tau(\cdot) \mapsto \tau\left(\sum_i V_i^* \cdot V_i\right).$$

Let us compare this with the traditional formalism of quantum mechanics. If $\tau(A) = \operatorname{Tr} D A$, then

$$\tau\left(\sum_i V_i^* A V_i\right) = \operatorname{Tr}\left(D \sum_i V_i^* A V_i\right) = \operatorname{Tr}\left(\sum_i V_i D V_i^*\right) A,$$

hence the transformation of the statistical operator is

$$D \mapsto \sum_i V_i D V_i^*$$

which is an extension of von Neumann's measurement conditional expectation (21). Given a state τ of the quantum system, the observed entropy of the operational process is defined to be the von Neumann entropy of the finite statistical operator

$$[\tau(V_i^* V_j)]_{i=j=1}^n,$$

which is an $n \times n$ positive semidefinite matrix of trace 1. If we are interested in the entropy of a state, we perform all operational processes and compute their entropy. If the operational process changes the state of our system, then the observed operational entropy includes the entropy of the state change. Hence we have to restrict ourself to state invariant operational processes when focusing on the entropy of the state. The formal definition

$$S^L(\tau) = \sup\left\{ S\left([\tau(V_i^* V_j)]_{i=j=1}^n\right)\right\}$$

is the operational (or Lindblad) entropy of the state τ if the least upper bound is taken over all operational partitions of unity $\mathcal{W} = (V_1, V_2, \ldots, V_n)$ such that

$$\tau(A) \mapsto \tau\left(\sum_i V_i^* A V_i\right),$$

for every observable A. For a statistical operator D we have

$$S^L(D) = 2S(D),$$

and we may imagine that the factor 2 is removable by appropriate normalization, so that we are back to the von Neumann entropy. The operational entropy satisfies von Neumann's mixing condition and is a concave functional on the states even in the presence of superselection rules. However, it has some new features. To see a concrete example, assume that there are two superselection sectors and the operator algebra is $M_2(\mathbb{C}) \oplus M_3(\mathbb{C})$, that is, the direct sum of two full matrix algebras. Let a state τ_0 be the mixture of the orthogonal pure states $|\psi_i\rangle$ with weights λ_i, where $|\psi_1\rangle, |\psi_2\rangle$ are in the first sector and $|\psi_3\rangle$ is in the second. This assumption implies that there is no dynamical change sending $|\psi_1\rangle$ into $|\psi_3\rangle$, and superpositions of those states are also prohibited. One computes

$$
\begin{aligned}
S^L(\tau_0) \;=\; & -2\sum_i \lambda_i \log \lambda_i \\
& -(\lambda_1 + \lambda_2)\log(\lambda_1 + \lambda_2) - (\lambda_1 + \lambda_2 + \lambda_3)\log(\lambda_1 + \lambda_2 + \lambda_3),
\end{aligned}
$$

which shows that this entropy is really sensitive to the superselection sectors. (For further properties on S^L we refer to pp. 121–124 of [7].)

Nowdays some devices are based on quantum mechanical phenomena, and this holds also for information transmission. For example, in optical communication a polarized photon can carry information. Although von Neumann apparently did not see an intimate connection between his entropy formula and the formally rather similar Shannon information measure, many years later an information theoretical reinterpretation of von Neumann's entropy is becoming common. Communication theory deals with the coding, transmission, and decoding of messages. Given a set $\{a_1, a_2, \ldots, a_n\}$ of messages, a coding procedure assigns to each a_i a physical state, say a quantum mechanical state $|\psi_i\rangle$. The states are transmitted and received. During the transmission some noise can enter. The receiver uses some observables to recover the transmitted message. Shannon's classical model is stochastic, so it is assumed that each message a_i should be teleported with some probability λ_i, $\sum_i \lambda_i = 1$. Hence in the quantum model the input state of the channeling transformation is a mixture; its statistical operator is $D_{\text{in}} = \sum_i p_i |\psi_i\rangle\langle\psi_i|$. This is the state we need to transmit, and after transmission it changes into $T(D_{\text{in}}) = D_{\text{out}}$ which is formally a statistical operator but may correspond to a state of a very different system. Input and output could be far away in space as well as in time. The observer measures the observable A_j and $p_i = \text{Tr}\, D_{\text{out}} A_i$ is the probability with which he concludes the message a_j was transmitted. Here we need $\sum_j A_j = I$ and $0 \leq A_i$. More generally, we assume that

$$p_{ji} = \operatorname{Tr} T(|\psi_i\rangle\langle\psi_i|)A_j$$

is the probability that the receiver deduces the transmission of the message a_j when actually the message a_i was transmitted. If we forget about the quantum mechanical coding, transmission and decoding (measurement), we see a classical information channel in Shannon's sense. According to Shannon, the amount of information going through the channel is

$$I = \sum_{ij} p_{ji} \log \frac{p_{ji}}{p_i}.$$

One of the basic problems of communication theory is to maximize this quantity subject to certain constraints. For the sake of simplicity, assume that there is no noise. This may happen when the channel is actually the memory of a computer; storage of information might be a noiseless channel in Shannon's sense. We have then $T =$ identity, $D_{\text{in}} = D_{\text{out}} = D$ and the inequality

$$I \leq S(D)$$

holds. If we fix the channel state D and optimize with respect to the probabilities λ_i, the states $|\psi_i\rangle$ and the observables A_j, then the maximum information transmittable through the channel is exactly the von Neumann entropy. What we are considering is a simple example, probably the simplest possible. However, it is well demonstrated that the von Neumann entropy is actually the capacity of a communication channel. Recently, there has been a lot of discussion about capacities of quantum communication channels, which is outside of the scope of the present article. However, the fact that von Neumann's entropy formula has much to do with Shannon theory and possesses an interpretation as measure of information must be conceptually clear without entering more sophisticated models and discussions. More details are in [8] and a mathematically full account is [1]. Further sources about quantum entropy and quantum information are [3], [9] and [7].

NOTES

*   Work supported by the Hungarian National Foundation for Scientific Research grant no. OTKA T 032662.

REFERENCES

[1] A.S. Holevo, "Quantum coding theorems", *Russian Math. Surveys*, **53** (1998), 1295–1331.

[2] E.T. Jaynes, "Information theory and statistical mechanics. II", *Phys. Rev.* **108** (1956), 171–190.

[3] R. Jozsa, "Quantum information and its properties", in *Introduction to Quantum Computation and Information*, H.-K. Lo, S. Popescu and T. Spiller (eds.), World Scientific, 1998.

[4] E.H. Lieb, M.B. Ruskai, "Proof of the strong subadditivity of quantum mechanical entropy", *J. Math. Phys.* **14** (1973), 1938–1941.

[5] J. von Neumann, "Thermodynamik quantummechanischer Gesamheiten", *Gött. Nach.* **1** (1927), 273-291.

[6] J. von Neumann, *Mathematische Grundlagen der Quantenmechanik*, Springer, Berlin, 1932.

[7] M. Ohya, D. Petz, *Quantum Entropy and its Use*, Springer, 1993.

[8] A. Peres, *Quantum Theory: Concepts and Methods*, Kluwer, 1993.

[9] D. Petz, "Properties of quantum entropy", in *Quantum Probability and Related Topics VII*, 275–297, World Sci. Publishing, 1992.

[10] D. Ruelle, *Statistical mechanics. Rigorous results*, Benjamin, New York-Amsterdam, 1969.

[11] G.L. Sewell, *Quantum theory of collective phenomena*, Clarendon Press, New York, 1986.

[12] M. Tribus, E.C. McIrvine, "Energy and information", *Scientific American* **224** (September 1971), 178–184.

[13] A. Wehrl, "General properties of entropy", *Rev. Mod. Phys.* **50** (1978), 221–260.

Mathematical Institute
Budapest University of Technology and Economics
H-1521 Budapest XI
Hungary

ECKEHART KÖHLER*

WHY VON NEUMANN REJECTED
CARNAP'S DUALISM OF INFORMATION CONCEPTS

> *Positivists ... have too narrow a notion of experience.*
> Kurt Gödel in Wang (1996, p. 173)

0. INTRODUCTION. TWO CONCEPTS OF INFORMATION?

In a private discussion in Princeton in 1952, von Neumann argued against Carnap's view that epistemological concepts like information should be treated separately from physics or from any natural science. For Carnap, such a separation had been taken so much for granted that he was very nonplussed and he consequently withheld from publication – out of apprehension it did not sufficiently clarify important points – a study on information.

The real underlying issue between von Neumann and Carnap has not been analyzed to date. It is my view that, although there *is* a difference between logical and physical (thermodynamic or engineering) theories of information and/or entropy, this is *not* because the *concepts* of information and entropy are inherently different. Rather, the difference is due to different *modalities* of the respective theories: the thermodynamic theory is *factual*, whereas the logical theory is *normative*. The distinction is best motivated and most commonly observed in decision theory, but also in probability theory, as will be shown. Even mathematical theories can be made to display the distinction, although it is not so well-known there.

My recent studies[1] on Gödel's philosophy of mathematics have led me to believe the distinction can be properly analyzed and drawn, in particular by paying attention to the special rôle of *mathematical intuition*. (For normative theories beyond mathematics, *e.g.* inductive probability theory or statistics, we extend the view to *rational intuition* in general.) The result is that logic, mathematics, and statistics, although certainly "formal", are in a special sense "real" sciences as well, insofar as they have specially identifiable *intuitable contents*. These can even be seen as *physical* insofar as logic and mathematics involve operations on symbols, either on paper, or in memories, or in processing registers, or in brains – as von Neumann correctly saw. Thus logic and mathematics indeed contain information of an empirical sort, even though they are normative. But on the other hand, logic and mathematics are still "formal" in the sense that

M. Rédei and M. Stöltzner (eds.),
John von Neumann and the Foundations of Quantum Physics, 97–134.
© 2001 *Kluwer Academic Publishers. Printed in the Netherlands.*

determining their validity is not dependent on empirical observation – in accordance with Carnap's analytic/synthetic (formal/real) distinction. This "formal" property of logic and mathematics is due to their being justified not by empirical data, but by *aesthetic criteria* as judged by logical and mathematical intuition.

1. CARNAP'S GOAL FOR HIS LOGICAL CONCEPTS OF INFORMATION AND ENTROPY

This section is a sketch of how Carnap came to hold his view that there are *two different*[2] concepts of probability and consequently of information. In his syntax-program, Carnap (1934) represented logic and mathematics as (sets of) sentences universally valid in every empirical theory.[3] In this scheme, logical and mathematical sentences (*qua* analytic) were to be distinguished from empirical sentences (*qua* synthetic[4]) by the latter's property of making essential use of empirical terms.[5] The possibility that the purely logical or mathematical terms (*syncategoremata*)[6] of the former could *also* be empirically interpreted was ignored – although the difficulty of a *general distinction between the two classes for all possible languages* was indeed noted. To be sure, naturalists like Quine (1951) made the point that, if no *general* distinction were possible (particularly, if *all* terms could be interpreted empirically), then there was no *general* way to distinguish logical (*i.e.* analytic) from empirical (synthetic) propositions.

Carnap (1936) subsequently accepted the approach of Tarski (1936) of representing formal (*i.e.* analytic) properties and relations of deductive logic and mathematics in *metatheoretical semantic* systems. Although Carnap's syntax program already provided for a *syntactical* metatheory in which validity of formal systems could be defined in terms of shapes of signs, the use of semantical metatheory first allowed explicit treatment of logical and metamathematical applications to various subject matters – rather than just relations between *signs*. The difference lay in the provision for explicit treatment of designation, satisfaction, and truth within semantics.

Extending his semantic work into *inductive* logic, Carnap (1945), popularized in Carnap (1953), made with "Two Concepts of Probability" what came to be a widely accepted, even orthodox, distinction between physical and logical probabilities, thus settling for many philosophers and some statisticians the dispute between two major schools of probability theory by granting legitimacy to *both*. The classical concept of probability,[7] as Carnap successfully demonstrated, in fact required a formal system such as the one he offerred in his logical semantics since Carnap (1942). The more recently developed statistical concept of probability championed by von Mises (1919) and Reichenbach (1920) is an empirical measure based on (limits of) relative frequencies of events. Thus, the distinction of "two concepts of probability" with different meanings in two different domains appeared well motivated by the idea that logic/mathematics and physics/natural science seem to be about *two different* domains. This immediately implies the existence of two *different concepts of information* (or entropy), each defined by one of the two different concepts of probability.

Such a dualism depended on Carnap's older "dogma" of a clean separation of analytic from synthetic propositions. This was rejected by Tarski (in private communication) and Quine (1951).[8] Quine's position is plausible since Carnap had not established any *separate and independent content* for logic and mathematics; indeed, Wittgenstein's influential view that logic and mathematics were tautological made them *empty of content.*[9] The *syncategorematic* logical/mathematical terms were otherwise *underdetermined* in meaning and constrained only by logical/mathematical axioms and rules. As a matter of fact, there actually was a traditional interpretation of *syncategoremata*, namely mental processes: demonstrations, calculations, *etc.* Neither Carnap nor Quine went that far, as both preferred to ignore the entire issue of interpreting the *syncategoremata* for deductive logic and mathematics. But then the problem remained apparently insoluble.

However, with respect to *inductive* logic, Carnap (1962) later went into considerable detail specifying an *epistemological* interpretation of logical probability in terms of subjective *belief* ("credence") – the same as the concept of subjective or personal probability of Bayesians. Thus, the major *syncategoremata* of *inductive logic* did indeed obtain an empirical interpretation for Carnap. The *syncategoremata* of *deductive logic* were not treated with similar care, as their interpretation was perhaps not as problematic as that of inductive inference[10] – although they are included in Carnap's credence concept, which is an extension of deductive inference.

It seems von Neumann had settled the issue of interpreting *syncategoremata* in his own mind much earlier – and much more radically! For him, the concepts and procedures of logic and mathematics were entirely reducible to physics, perhaps with psychology as a way station. Von Neumann's position was that of physicalistic reductionism, a version of materialism, such that mental processes of perception, memorizing and cognition were representable as – or "reducible to" – physical processes within the framework of some standard system of physics. This idea von Neumann took from Szilárd (1929), as shown in §2 below.

1.1 The Conversation between von Neumann and Carnap in Princeton, 1952

Carnap spent the years 1952–54 on sabbatical leave at the Institute for Advanced Study in Princeton, working mainly on extending his own theory of inductive or logical probability, particularly in collaboration with John Kemeny, just after having been mathematics assistant to Albert Einstein. Among the more interesting tasks undertaken by Carnap at the time was to apply his logical probability theory to information and entropy (well known from statistical mechanics). A controversy raged over the proper interpretation of these concepts: whether they were to be understood strictly as parameters of physical systems, or whether they should be taken as measures of human knowledge, *e.g.* to quantify the contents of scientific data or their theories. Carnap, together with one of his collaborators at the time, Yehoshua Bar-Hillel, insisted on cleanly separating the two cases, which were almost routinely being confused with each other. Hartley (1928) and Shannon (1948) carefully avoided confusion by insisting that their information theory was an *empirical* statistical theory about carrying capacity of communications channels, *not* about the epistemic content of scientific data – *i.e.*, about *real* information.

Von Neumann strongly objected to such a distinction, and a conversation with Carnap is described by Bar-Hillel (1964, p. 11f.):

The words 'cybernetics' (and its derivatives) and 'information' were surely two of the most used, and misused, members of the scientific vocabulary of the fifties. ... Since my first days at MIT, I was greatly irked by what seemed to me an almost deliberate misuse of this term and by the confusions created thereby. Certain curious expressions of Wiener, von Neumann, Weaver and others did nothing to disperse these confusions or the halo of mystery which sometimes surrounded this newly discovered "commodity" whose importance was put on a par with that of "energy", for instance. True enough, Shannon explicitly dissociated himself from those who interpreted his measures of information as measures of meaning or semantic content, but few were thereby discouraged from making just this interpretation, though they occasionally paid lip-service to Shannon's disclaimer.

My attempts to clarify the issue went in three directions: First, I worked out a whole new terminology for the elementary parts of what I proposed to call Theory of Signal Transmission (instead of the so utterly misleading Theory of Information, a term which seems to have been created, by what turned out to be a very unfortunate ellipsis, from Theory of Information Transmission, in correspondence with Hartley's pioneering paper of 1928). But this terminology never really caught on, and I did not even bother to publish my proposal. I still think that this is a pity.

Second, I substituted at the last minute for Carnap in the Tenth (and last) Macy Foundation Conference on Cybernetics that took place in Princeton in April 1953. In the talk I gave there, I combined a short presentation of our Theory of Semantic Information with an admittedly very tentative critique of the use of Information and related concepts such as Entropy in certain representations of modern physics, in particular those connected with Maxwell's demon. Many physicists, among them a man no less eminent than John von Neumann, were in the habit of expressing themselves, at least occasionally, as if entropy, as it occurs in statistical mechanics, was a measure of ignorance, of the lack of knowledge of one who observes a physical system. I could never accept this attitude, nor even really understand it. Talks with some of my physicist friends did not help, and I still remember with a shudder an almost traumatic experience, shared with Carnap. During one of my visits to him in Princeton, in 1952, von Neumann also came to see him, and we started discussing the talk I had heard von Neumann deliver shortly before at an AAAS meeting in St. Louis, in which he had proclaimed, among other things, a triple identity between logic, information theory and thermodynamics.[11] Carnap and I wondered with what degree of seriousness this "identity" was to be taken. We were quite ready to agree that there existed a certain formal analogy, up to a common partial calculus, and had indeed ourselves shown, in our *Outline* [Carnap & Bar-Hillel (1952)], how entropy-like expressions, such as the famous $-\Sigma\, p_i \log p_i$, occurred in (inductive) logic, but could not see how any stronger relationships could possibly be supposed to exist. After all, inductive logic was a formal and thermodynamics a real science. We tried to convince von Neumann that his way of presenting the analogy as an identity must lead to confusion. His calm reply was that he could see how logicians and methodologists might be worried by his statements but that *no* physicist would misunderstand them. I know that Carnap decided, after this talk, to postpone *sine diem* the publication of a long paper he had almost finished on the role of inductive logic in statistical mechanics. In this paper he had made some forceful and, to my mind, very illuminating criticisms of certain standard formulations in this theory. I myself omitted some of my remarks on physics in the printed version of my Princeton talk [Bar-Hillel (1952)].

In claiming as a matter of course a fundamental distinction between "formal" and "real" sciences, Bar-Hillel dogmatically maintains Carnap's well-known claim of a fundamental distinction between analytic and synthetic theories.

Carnap (1963, p. 36f.) also described events in Princeton, but more mutedly:

I had some talks separately with John von Neumann, Wolfgang Pauli, and some specialists in statistical mechanics on some questions of theoretical physics with which I was concerned. I certainly learned very much from these conversations; but for my problems in the logical and methodological analysis of physics, I gained less help than I had hoped for. At that time I was trying to construct an abstract mathematical concept of entropy, analogous to the customary physical concept of entropy. My main object was not the physical concept, but the use of the abstract concept for the purposes of inductive logic. Nevertheless, I also examined the nature of the physical concept of entropy in its classical statistical form, as developed by Boltzmann and Gibbs, and I arrived at certain objections against the customary definitions, not from a factual-experimental, but from a logical point of view. It seemed to me that the customary way in which the statistical concept of entropy is defined or interpreted makes it, perhaps against the intention of the physicists, a purely logical instead of physical concept; if so, it can no longer be, as it was intended to be, a counterpart to the classical macro-concept of entropy introduced by Clausius, which is obviously a physical and not a logical concept. The same objection holds in my opinion against the recent view that entropy may be regarded as identical with the negative amount of information. I had expected that, in the conversations with the physicists on these problems, we would reach, if not an agreement, then at least a clear mutual understanding. In this, however, we did not succeed, in spite of our serious efforts, chiefly, it seems, because of great differences in point of view and in language. I recognized the fundamental difference between our methodological positions when one of the physicists said: "Physics is not like geometry; in physics there are no definitions and no axioms."

To the latter flippancy, one is tempted to rejoin: if so, physics is not a science! For Aristotle had early characterized theoretical science as involving definitions and axioms *essentially*. Of course, what Carnap's physicist meant was to take a rather intuitive attitude toward these. But intuition is fallible and continually needs the help of *logical analysis*. If physics had no axioms, no one would talk about the Second Law of Thermodynamics, the Law of Gravitation or the Light Principle. If physics had no definitions, no one would worry about measurement units and scales.

Carnap's "long paper he had almost finished" (Bar-Hillel) ultimately appeared without major changes as Carnap (1977). Although this contains a wealth of interesting and controversial treatments of information and entropy, both from physical as well as logical orientations, neither it nor Shimony's introduction solve the dispute between von Neumann and Carnap, which is more fundamental and general than anything dealt with there. Carnap is right about claiming logic and physics are different, but he was wrong in implying that a "logical concept" like information *cannot* be physical at all. For further discussion, see §2.1 below, and a solution is attempted in §5.

2. HOW SZILÁRD INFLUENCED VON NEUMANN ON ENTROPY

In a highly influential paper, Szilárd (1929) treated the paradox of Maxwell's demon.[12] This demon clarifies primarily the Second Law of Thermodynamics within the framework of statistical mechanics and was introduced by Maxwell (1871) to explore hypothetical circumstances under which the Second Law of Thermodynamics might be falsified by demonic interventions on a micro-level. The Second Law states that, in a closed system, its entropy (a measure of how thoroughly its temperature is uniform, *i.e.* mixed) can never *decrease*: if effort is expended to "unmix" it in *one* region, that effort generates heat in *neighboring* regions, and the overall entropy rises. For example, although refrigerators cool their interiors, the overall heat and energy expended just outside by motors is even greater. In Boltzmann's attempt to reduce thermodynamics to particle mechanics, entropy was interpreted as a statistical measure of the degree of orderliness of the states (the kinetic energies, *i.e.* the temperatures) of the particles. Maxwell's point was that this statistical measure could not be interpreted as strictly identical to the macro-thermodynamic entropy.

Maxwell's hypothetical demon controls a valve (or gate) between two chambers containing particles originally uniform in temperature, and he opens the valve for fast-moving particles in one direction but not the other, so that the temperature in one chamber rises. The valve is assumed to be frictionless, so the demon expends no net work moving it. Szilárd (1929) was the first to show that, even if turning the valve incurred no net work, the demon must nevertheless expend "epistemological work", as it takes some physical effort to discriminate fast from slow particles. Szilárd equated such discrimination to measurement involving interaction with the system of particles, which incurs energy expenditure and consequent entropy gain at least compensating for the loss of entropy due to "unmixing" the fast from the slow particles. Szilárd was the first to measure information growth in the memory of intelligent beings in terms of *bits* – because entropy would be measured on a scale using logarithms of base 2 with respect to binarily discriminating the presence or absence of states of the system.

Although Szilárd (1929) remained unclear on various points (according to Leff & Rex 1990), he was the first to explicate more precisely than anyone had done before how cognitive states and processes – in particular memory and discrimination – could be treated within physics. It was crucial for understanding the *physical* workings of an intelligent agent to represent its observational and logical operations *within the physical theory itself*. Boltzmann (1877) had already conjectured on the widely-recognized relation between thermodynamic entropy and states of knowledge. But for many physicists such as John von Neumann and Wolfgang Pauli, Szilárd (1929) was a revelation they would never forget. Already von Neumann's seminal treatment of Quantum Mechanics of 1932 was decisively influenced by Szilárd (1929). For example, in treating macroscopic measurement, von Neumann (1932, p. 213) states (transl. E.K.):

Thus, at the end of the process, the molecule is back in volume V; but we now no longer know whether it is located on the left or right, although there is a compensating entropy decrease of $x \ln 2$ (in the reservoir). That is, we have exchanged our knowledge for the entropy decrease of $x \ln 2$.[202] That is: in volume V the entropy is the same as that in volume $V/2$ under the assumption one knows in which half of the container the molecule is located.

[v.N.'s note:] 202. L. Szilárd [in (1929)] has shown that one cannot obtain this "knowledge" without at least a compensating entropy increase of $x \ln 2$ – in general, $x \ln 2$ is the "thermodynamic value" of knowing which of two cases of an alternative obtains. – All attempts to carry out the process described above not knowing in which half of the container the molecule is located may be shown to be invalid, although they involve, under certain circumstances, quite complicated automata-mechanisms.

We here see that von Neumann quite early confidently regarded thermodynamic entropy decrease as *exchangeable* with – hence equivalent to – information increase, *i.e.* increase of knowledge. Thus it appears that *epistemological* states and processes may be understood as *physical* states and processes.

This position was later emphasized all the more strongly in von Neumann's work on automata toward the end of his life in the late 1940s and the early 1950s. For example, in a lecture series on the "Theory and Organization of Complicated Automata" held at the University of Illinois[13] in December, 1949, he referred to connections between thermodynamics and logic (or knowledge acquisition and manipulation, *i.e.* data storage and inference). Burks (1966, p. 59f.) reports on the third lecture, "Statistical Theories of Information":

In connection with entropy and information von Neumann referred to Boltzmann, Hartley, and Szilard. He explained at length the paradox of Maxwell's demon and how Szilard resolved it by working out the relation of entropy to information.

Von Neumann (1966, p. 62f.) summed up his views in the third lecture:

I have been trying to justify the suspicion that a theory of information is needed and that very little of what is needed exists yet. Such small traces of it which do exist ... indicate that, if found, it is likely to be similar to two of our existing theories: formal logics and thermodynamics. It is not surprising that this new theory of information should be like formal logics, but it is surprising that it is likely to have a lot in common with thermodynamics. ...

Thermodynamical concepts will probably enter into this new theory of information. There are strong indications that information is similar to entropy and that degenerative processes of entropy are paralleled by degenerative processes of information. It is likely that you cannot define the function of an automaton, or its efficiency, without characterizing the milieu in which it works by means of statistical traits like the ones used to characterize a milieu in thermodynamics.

... An automaton in which one part is too fast for another part, or where the memory is too small, or where the speed ratio of two memory stages is too large for the size of one, looks very much like a heat engine which doesn't run properly because excessively high temperature differences exist between its parts. ...

It is worth noticing that von Neumann here *hedged* the absolute claim of an "identity" between information and entropy (made later in the conversation with Carnap in Princeton reported on by Bar-Hillel in §1.1 above), here making it out to be a mere "similarity". This we may take to be a concession on his part: that there is perhaps a difference after all, but one which he could not express. This I will return to in §4.1 below. But his *main* point was that (epistemic) information theories must take thermodynamical theory seriously, and in this I'm convinced he's correct – whatever logicians may say.

Other authors on information theory, in contrast, carefully avoided entangling the communications engineering concept of information, developed to study the carrying capacity of telephone networks, with the semantic, *i.e.* epistemic, concept involving knowledge of an intelligent agent. For example, Cherry (1951, p. 383) states "It is important to emphasize, at the start, that we are not concerned with the meaning or the truth of messages; semantics lies outside the scope of mathematical information theory." In a classic article, Miller (1953) alleged many misuses of the term 'Information' in Information Theory. Shannon & Weaver (1949, §2.2) state:

The word *information*, in this theory, is used in a special sense that must not be confused with its ordinary usage. In particular, *information* must not be confused with meaning.

In fact, two messages, one of which is heavily loaded with meaning and the other of which is pure nonsense, can be exactly equivalent, from the present viewpoint, as regards information. It is this, undoubtedly, that Shannon means when he says that "the semantic aspects of communication are irrelevant to the engineering aspects." But this does not mean that the engineering aspects are necessarily irrelevant to the semantic aspects.

To be sure, this word information in communication theory relates not so much to what you *do* say, as to what you *could* say. That is, information is a measure of one's freedom of choice when one selects a message. If one is confronted with a very elementary situation where he has to choose one of two alternative messages, then it is arbitrarily said that the information, associated with this situation, is unity. Note that it is misleading (although often convenient) to say that one or the other message conveys unit information. The concept of information applies not to the individual messages (as the concept of meaning would), but rather to the situation as a whole, the unit of information indicating that in this situation one has an amount of freedom of choice, in selecting a message, which it is convenient to regard as a standard or unit amount.

It is possible to state exactly what the difference is: *meaning* presupposes validity of *specific* linguistic norms (codes) for communicating facts, in particular semantic relations between signs and objects of the natural world. These norms are assumed to be standard for all communication, but Shannon's information theory does not stipulate which codes to use *at all*. Carnap's theory of semantic information, on the other hand, makes *full use* of *specific* semantic relations, hence Carnap has a *genuine* information theory.

Von Neumann seemed perfectly aware of the conceptual unclarity of usages of 'information', with its so-to-speak hermaphroditic, opalescent nature, shifting between thermodynamics, engineering, and abstract epistemology, and so he restrained his publicly stated views – without further clarification. In private, however, he seems to have loved to muddy the waters and to behave as an impish demon on his own. When Shannon (1948) was completing his famous article founding modern information theory, he felt insecure about what to call Boltzmann's entropy function $-\Sigma\, p_i \log p_i$ when applying it in information theory. Tribus (1979 p. 2f.) [originally (1963)] relates

In the 1961 interview with Shannon, to which I referred, I obtained one anecdote which seems to me worth recording on this occasion. I had asked Dr. Shannon what his personal reaction had been when he realized he had identified a measure of uncertainty. Shannon said that he had been puzzled and wondered what to call his function. 'Information' seemed to him a good candidate as a name, but 'Information' was already badly overworked. Shannon said he sought the advice of John von Neumann, whose response was direct, "You should call it 'entropy' and for two reasons: first, the function is already in use in thermodynamics under that name; second, and more importantly, most people don't know what entropy really is, and if you use the word 'entropy' in an argument you will win every time!"

Denbigh (1981), opposing Brillouin's idea that entropy can measure subjective states of knowledge, also objected to von Neumann's influence on Shannon and replied as follows to Tribus's anecdote:

In my view von Neumann did science a disservice! There are, of course, good mathematical reasons why information theory and statistical mechanics both require functions having the same formal structure. They have a common origin in probability theory ... Yet this formal similarity does not imply that the functions necessarily signify or represent the same concepts. The term 'entropy' had already been given a well-established physical meaning in thermodynamics, and it remains to be seen under what conditions, if any, thermodynamic entropy and information are mutually interconvertible. They may be so in some contexts, but not in others. As Popper [(1974, p. 130)] has very clearly put it, it needs to be shown whether or not the P_i "are probabilities of the same attributes of the same system."

Influencing Carnap to drop the publication of his (1977) was perhaps just such a disservice, as it might have influenced research in the '50s in a way it no longer did in the '70s. For example, it could have influenced the work of Jaynes (1957) concerning Gibbs, which went back to 1951. Especially in case of doubt, lively debate is preferable to silence! Unfortunately, von Neumann was backed up by another powerful voice. Even if Carnap had been disposed to contradict von Neumann in 1952, Wolfgang Pauli may have finished off any remaining resistance in 1954 – see the next section.

2.1 Carnap's Logical Theory of Information

In Carnap & Bar-Hillel (1952) on logical information, a clean break away from the confusion to be found in much of the literature was attempted by simply pre-supposing *a priori* that the semantic (alternatively called "logical", "epistemic", or "abstract") concept of information is *different* from the physical (statistical-thermodynamical) concept. Since Carnap had by that time built up a large body of writings[14] on his *logical* theory of probability – clearly different from the empirical-statistical[15] theory –, this seemed inevitable. The logical notion of information is a very powerful tool of epistemology deserving of much wider attention among philosophers.[16] Equations obtained for logical information are identical, however, to ones on negative entropy (Brillouin's negentropy), except that in information-theoretical contexts, Carnap uses his logical probability function c, whereas in the physical equations statistical probability p is used. These of course are distinguished on the basis of the well-known view of Carnap (1945).

An outstanding feature of Carnap's logical information theory is the concept of the amount of information contributed to our knowledge by an observation or experiment. The exact definition depends on two main things: our conceptual framework H, and a probability (confirmation) function c used to weigh knowledge. One such definition is proposed by Carnap & Bar-Hillel (1952):

D11-2 $$\text{est}\,(\text{inf},\,H,\,e) =_{Df} \sum_{p} c\,(h_p,e) \times \inf\,(h_p/e)\,,\,^{17}$$

which measures, intuitively speaking, how much the partition of available hypotheses H will probably be reduced in size – *i.e.*, how much more definitely we know where the truth lies – when we learn e. This measure is naturally maximized if the evidence e is strong enough to "clinch the case" in favor of a particular hypothesis h_p, excluding all alternatives. Such functions and related ones were not new. Quite interesting, however, is the idea of Carnap & Bar-Hillel (1952, §12) to measure the 'efficiency of a conceptual framework', *e.g.* of L_1 (Language 1):

$$\text{ef}\,(\,L_1,\,\text{inf},\,e) =_{Df} \frac{\text{est}\,(\text{inf};\,H_1;\,e)}{\max_{i}\,[\,\text{est}\,(\text{inf};\,H_i;\,e)\,]}\,,\,^{18}$$

which measures, intuitively speaking, the relative explanatory power of framework L_1 compared to all "similar" alternative frameworks L_i, *i.e.*, how well L_1 does in "deciding" all open questions h_p in H_1 on data e, compared to the best alternative conceptual framework. This is relative to specific data e, so *overall* efficiency might also be taken with respect to a variety of data, weighted by their (absolute or relative) likelihood. However, if we set e to be the "total evidence" standardly "known" to science at present, that of course is what should be taken as the standard reference point for such efficiency estimates in any case.

This idea of using information measures to maximize the epistemic payoff of choosing a conceptual framework is virtually a "philosopher's stone" of ontology, as it gives science a crucial benchmark for systems of objects and properties. In cognitive science, the crucial importance of choosing a framework leading to easy explanations of data (the "frame problem") is widely accepted, especially in artificial intelligence.[19]

Using Carnap's concept of efficiency 'ef' appears to anticipate Jaynes's Maximum Entropy Principle for conceptual frameworks, because the most efficient framework (the one maximizing estimated information gain) will be the one with the least *a priori* information about the world – greatest ignorance, a Principle of Indifference –, *viz.* that with the maximum *a priori* entropy. This is where every conceptual structure (Carnap's "structure description", preferred for establishing *a priori* logical probabilities) has an equal *a priori* chance to be realized. Jaynes's (1957) program centering on his Maximum Entropy Principle to establish prior subjective probabilities for Bayesians seems to be incompatible with Carnap's program, however.[20]

Conceptual frameworks such as the simple partitions like H widely used in statistics for representing directly observable data are of course very elementary. For more sophisticated frameworks involving *theoretical* concepts, a more sophisticated apparatus is required. Philosophers such as Niiniluoto & Tuomela (1973) and Hauffe (1981) have applied the information concept to measure the "explanatory power" of theoretical conceptual frameworks to tell how much "economizing" of knowledge is made possible by introducing suitable theoretical concepts, *e.g.* relational structures, functors, fields, *etc.*

In his work on entropy, Carnap (1977, p. 72f.) approvingly smiles on Szilárd's solution of the paradox of Maxwell's demon, but frowns on Brillouin (1951, 1951a, 1953), who refined Szilárd's treatment:

These and other results of Brillouin's are certainly interesting and clarify the situation with respect to Maxwell's paradox in the direction first suggested by Szilard. However, when Brillouin proceeds to identify negentropy with amount of information, I cannot follow him any longer. He defines entropy by $k \ln P$, where P is the number of "possible structures" or "possible states" or "complexions"; thus he means presumably something like S_B^{II} (6-11) ... He does not seem to be aware that the definition of S which he uses (and which he ascribes to Boltzmann and Planck) makes S a logical rather than physical concept.

Carnap is correct in holding S_B^{II} to be a logical concept.[21] However, he does not show that it can *not also be a physical* concept. After all, it was precisely Szilárd's point that the epistemological ("logical") processes of Maxwell's demon could be "physicalized",[22] *i.e.* represented within and subject to physical theory! This point is virtually conceded by Shimony (1977, p. xviii) in the very attempt to support Carnap's "dualism":

> Finally, the incorrectness of conflating thermodynamic entropy with uncertainty in the information-theoretic sense can be illustrated by noting that in molecular biology the latter depends upon the number of isomers or well-defined variants of a type of molecule, a number which is independent of the thermodynamic entropy (Tisza and Quay [1963, VI]). Maintaining a clear conceptual distinction between thermodynamic entropy and information-theoretic uncertainty is, however, consistent with acknowledging that the price of acquiring information is always an increase of entropy, as Szilard discovered and Carnap lucidly explains in §10. A wonderful example is the increase of entropy due to the chemical reaction required for the "recognition" of an isomer (Monod [1971, p. 39]).

This is a very sharp point. But it only proves that information-theoretic uncertainty should not be conflated with *thermodynamic* entropy and does *not exclude* physical interpretation of uncertainty *entirely*! Shimony (1977, p. xiii) acknowledged that "There are many thermodynamic entropies, corresponding to different degrees of experimental discrimination and different choices of parameters ... ", so we may analogize further that entropy extends beyond thermodynamics to mixing-phenomena in general, even if they are not thermodynamic. Chemical isomer is after all a *physical* concept![23] Now we may also proceed to ask what Shimony (with Carnap) means by the hermaphroditic physical-epistemic "price of acquiring information" if not a genuine physical process? Is not the information-gain for which a noticeable energetic "price" of kT has been paid thereby *also* a *physical* process – as Szilárd worked so ingeniously to show? Just because a concept is used in logic or epistemology and assumed to obey axioms and rules of logic does not forbid its being physical. Operations of central-processing-unit registers are *simultaneously* physical (describable micro-physically in terms of electron-motions and macrophysically in terms of electronic engineering) *and* logical (in terms of symbol storage and manipulation). When we view these physical processes from an *empirical* point of view, we judge whether they in fact occur or not. Simultaneously, from a *logical* point of view we judge whether they are valid or not, *i.e.* whether they conform to rational ideas of correctness rather than to empirical data (this leads to my "solution" of the debate in §5 below).

The greatest clash of all is one imputed by Shimony between Carnap (1977, §8) and Jaynes (1967, p. 97). After reminding us that Carnap discovered that Gibbs's entropy concept was in fact logical and required re-interpreting to become a physical-statistical concept, Shimony (1977, p. xvii) continues:

> 6. A strong dissent to the foregoing reading of Gibbs is expressed by Jaynes, who considers him to be using probability in the epistemic sense of reasonable degree of belief

[1967, p. 97]; note that Jaynes sometimes characterizes this sense of probability as "subjective," but always in quotation marks, and indeed many of his remarks in this paper and elsewhere indicate a commitment to something like a logical concept of probability. Whether or not Jaynes is historically accurate regarding Gibbs, it is important to assess his claim that the appropriate probability concept for statistical mechanics is epistemic probability, and that the concept of entropy is accordingly a measure of "amount of uncertainty" [ibid., p. 97]. Carnap's exhibition of the discrepancy between $S_G^{\parallel}$ and S^{th} in cases 1 through 6 of §8 constitutes a serious objection to Jaynes's program. These cases are revealing, because in them the information upon which the computation of $\rho_i(U)$ is to be based is quite different from the usual kinds of information about the preparation of a system, $e.g.$, that it is in contact with certain reservoirs. Consequently, in these cases it is easy to distinguish epistemic probability from probability in the sense of propensity [the sense which Shimony argues that Gibbs actually used – Carnap notwithstanding], and it is difficult to borrow from the latter in evaluating from the former. An entirely different objection to Jaynes's program is that his maximum-entropy prescription for evaluating probabilities can be saved from inconsistency with the calculus of probability only by a highly implausible assumption about prior probabilities (Friedman & Shimony [1963, VI]). [At this point the citation from Shimony presented above starts. – E.K.]

As Carnap (1962) showed, the logical and "subjective" ("epistemic") concepts of probability are not distinct. Is Jaynes's "subjective" concept distinct from *physical concepts*? Shimony plausibly argues that Jaynes's epistemic entropy notion is incompatible with Carnap's statistical-physical explication of Gibbs's entropy, but again this does not prevent Jaynes's *epistemic* notion from *also being physical*. We will see in §5 below that, assuming psycho-physical reductionism, all epistemic (including logical) notions will be physical in some sense anyway!

After von Neumann dampened Carnap's conviction, Wolfgang Pauli directly urged Carnap not to publish his work on entropy. We have a cover letter from Pauli (1999, p. 541) motivating a lengthy critique of Carnap's first essay: [24]

Princeton, 22 March, 1954

Dear Mr. Carnap!

I have studied your manuscript a bit; however I must unfortunately report that I am *quite opposed* to the position you take. Rather, I would throughout take as physically most transparent what you call "Method II". In this connection I am not at all influenced by recent information theory; the works by Brillouin you cite at the end I have unfortunately not studied. The works by Szilard, on the other hand, have been known to me for a long time.

Since I am indeed concerned that the confusion in the area of foundations of statistical mechanics not grow further (and I fear very much that a publication of your work in its present form would have this effect), I have taken the trouble to formulate my own standpoint in the enclosed sketch (which you may gladly keep for yourself). It got to be longer than I intended (which also is because I can think better when I write than when I talk).

Now that I have taken some time with this matter I will unfortunately not be able to deal with it very much in the future, since I still have many other things to do in the short time I am still in Princeton. But I do hope to see you briefly on Thursday or Friday of this week (Tuesday and Wednesday I am in New York).

With cordial greetings,   Your W. Pauli

The long "sketch" enclosed with his letter discusses in particular the difficult measurement problem for entropy, and Pauli (1999, p. 545) comes to the negative conclusion (which must have stunned Carnap): [25]

> It is for me for this reason *impossible to understand your conclusion (p. 73) "that Gibbs did not reach* his aim of constructing a statistical concept corresponding to S_{th}." The cases investigated by you do not not correspond to [situations of] heat equilibrium or restricted quasi-equilibrium and therefore have nothing to do with thermodynamics.

Pauli can quite well insist that *thermodynamic* entropy refer to gradual temperature diffusion processes; however, the Boltzmann-Gibbs program of reducing thermodynamics to statistical mechanics *reduces* the classical entropy value to the statistic for energy-spread of the micro-ensemble. Pauli does not *explicitly* reject Carnap's separation of physical from epistemic concepts of entropy – rather, he seems to say that a "general" concept must use Carnap's *non-physical* Method II for its evaluation! In the following passage, Pauli (1999, p. 544) implies Boltzmann had an *epistemic* notion of entropy in mind from the beginning:

> But it was obvious that the more general concept can *not* be *"objective"*, since the applicability of the temperature concept (or the entropy concept) must be complementary* to the completeness of the (microphysical) *description* of the system in bounded times. ... An "objectively" completely described state must have entropy zero (or a trivial constant). Assuming this known, heat and work are not distinguishable, everything is reversible, *etc.* In this way one comes to what you have named *"Method II"*, *which appears to me to be the only satisfactory one*: the amount of entropy describes logically our subjective knowledge of the system, the thermodynamic entropy corresponds in particular to that information about systems (this appeared to the last century as "objective reality") obtainable by infinitely slow diffusion processes.
>
> * This has *nothing* directly to do with the Heisenberg Uncertainty Relation or with the magnitude of the energy quantum.

If every "completely described state must have entropy zero", then $\inf(h_p/e) = 0$, which is trivial when h_p is *included* in e. What Pauli claims here is that, assuming all states have the same *a priori* probability, they have the same "natural" orderliness. This assumption completely ignores the "logical information" implicit in the prior choice of a particular conceptual scheme – as if the conceptual schemes carefully developed over centuries of refinement by sophisticated theoreticians and engineers of experimental instruments were utterly arbitrary. Ordinary physical intuition rebels against this, declaring that *not* every "completely described state must have entropy zero". Assuming we are using a "sophisticated", not arbitrarily chosen conceptual scheme, it could assume maximum (non-zero) entropy for a *symmetrically* distributed state. Thus it is indeed reasonable to regard *some* states as "inherently" more orderly and more informative than others.

Pauli, like Jaynes (1965, p. 398), takes the "general" entropy concept to be *necessarily* epistemic ("subjective"), which seems to make it logical in Carnap's sense. But surely the entropy parameter used in the Second Law of Thermodynamics cannot be logically *a priori*! (I assume Pauli doesn't regard the Second Law as an axiom of logic!) "Infinitely slow cycle processes"[26] can and ought of course to be presupposed by a thermodynamicist to guarantee repeatable measurements of heat.[27] This requirement does not contradict Carnap's principles behind Method I, but rather supplement them in order to characterize concepts as thermodynamic. Method II, however, establishes probability distributions *normatively* over descriptions like D_i and does not claim to describe physical reality. There seems to be a serious confusion here. Finally, on the other hand, Carnap does not seem to realize that his "logical" information measures can be given *physical* interpretations once underlying physical "substrata" (von Neumann) are laid down for epistemological states and processes (see §4.2 below).

Taking an "objectivist" line, Grünbaum (1973, Ch. 19) criticizes Jaynes's (hence, presumably, Pauli's) view. Once any partition of space into cells preparatory to calculation of entropy is made, then "objectively completely described states" can indeed have non-vanishing entropy. The same thing holds for time and length: values assigned are based on (nature's feedback from) a "subjectively" determined conceptual framework and observation procedures using it. (In §4.3 below, I try to clear up some widespread confusions about the subjective-objective distinction, which apparently confounded Pauli and Jaynes.)

None of the authorities mentioned up to now convincingly showed *why* physical and logical entropy should be either identified or distinguished – although Carnap at least showed that the *methods* of attributing values for entropy are different for the two. Those *for* assimilation (*e.g.* von Neumann) confused the differing logical and empirical value-assignment methods, those *against* (*e.g.* Carnap) couldn't explain the reason for the obvious "analogy" between the concepts in the two apparently different theories.

3. VON NEUMANN'S VIEWS ON THE NATURE OF MATHEMATICS

Von Neumann's drive to "physicalize" logic, mathematics, and statistics would lead one to conjecture offhand that he should be considered an empiricist with regard to foundations of mathematics in the sense of Mill. Mill claimed mathematics is just an especially *general* branch of natural science. This view implies that the data we use to justify mathematical principles and rules are taken from empirical observation – thus leading us to incorporate mathematics (and logic, and presumably statistics, decision theory and other rationality sciences as well) into the natural sciences. This is essentially the position held by naturalists at the present time. Naturalists may then be sorely disappointed to learn that von Neumann definitely rejected (Mill-like) empiricism! Furthermore, he was in excellent company. Only a few years before von Neumann's famous lecture of 1947 which we shall quote below, Godfrey Harold Hardy (1940, Ch. 10) wrote:

A mathematician, like a painter or poet, is a maker of patterns. ...
   The mathematician's patterns, like the painter's or poet's, must be *beautiful*; the ideas, like the colours or the words, must fit together in a harmonious way. Beauty is the first test: there is no permanent place in the world for ugly mathematics.

Hardy then conceded that mathematics must have not only beauty but also "seriousness", meaning depth of *content*. At the same time, however, he was notorious for his eccentric spurning of *applied mathematics*, which he judged exceedingly *ugly*. Ironically, "seriousness" is really nothing other than applicability!
   Hardy's claim for the centrality of *beauty* is corroborated by von Neumann (1947, second-to-last paragraph) [= (1961, p. 9)]:[28]

... mathematical ideas originate in empirics, although the genealogy is sometimes long and obscure. But, once they are so conceived, the subject begins to live a peculiar life of its own and is better compared to a creative one, governed almost entirely by *aesthetic motivations*, than to anything else and, in particular, to empirical science. There is, however, a further point which, I believe, needs stressing. As a mathematical discipline travels far from its empirical source, or still more, if it is a second and third generation only indirectly inspired by ideas coming from "reality", it is beset with very grave dangers. It becomes more and more purely aestheticizing, more and more purely *l'art pour l'art*. This need not be bad, if the field is surrounded by correlated subjects, which still have closer empirical connections, or if the discipline is under the influence of men with exceptionally well-developed taste. But there is a grave danger that the subject will develop along the line of least resistance, that the stream, so far from its source, will separate into a multitude of insignificant branches, and that the discipline will become a disorganized mass of details and complexities. In other words, at a great distance from its empirical source, or after much "abstract" inbreeding, a mathematical subject is in danger of degeneration. At the inception, the style is usually classical; when it shows signs of becoming baroque, then the danger signal is up. ...

Thus von Neumann properly renders Hardy's point that great mathematics requires "seriousness" or "content": this *is* nothing other than connection with important applications. *Usefulness* was strongly upheld by von Neumann (1954) [= (1963, p. 448f.)]:

There are large areas of mathematics which have been practically very *useful*. This practicality, however, is sometimes a rather indirect kind of practicality.
   For instance, a mathematician usually means that a theory is directly useful if it can be used in theoretical physics. After which he still has to say that insight in theoretical physics itself is only useful if it is useful in experimental physics. After which you must say that a concept in experimental physics is, by ordinary criteria, useful if it is useful in engineering. Even after engineering you can make one more step. So all of these concepts of usefulness are rather limited, and we only mean by them that each science should have applications outside its own area, and that there is some general direction in this sequence of applications towards practical ones for immediate social use. However, if one doesn't quibble about the definition of usefulness, and means, for instance, that by the standards of the mathematician anything is useful which is not mathematics, then one must say that

large areas have been useful. Also, very large areas are really directly useful by the sum of all these criteria. ...

Now comes the climactic thrust: search for *elegance* is the essential driving force after all:

Now, it is very interesting that the majority of these things were developed with very little regard to usefulness, and very often without any suspicion that they might become useful later, for reasons of an entirely different character. It is a very characteristic situation. I might mention certain forms of algebra, in the field of matrices and operators, which were involved at times when there was no earthly reason to suspect that anywhere from twenty to a hundred years later they would play a role in (not yet existing) quantum mechanics. It is equally true for the discoveries in the area of differential geometry, for which there was absolutely no reason to expect that some day there would be a theory of general relativity, and that the theory of general relativity would make use of this type of geometry. Yet these things are quite vital. The examples could be multiplied.

I must say, however, there are also examples to the contrary. One very important example is that the calculus was certainly invented by Newton specifically for a specific purpose in theoretical physics.

But still a large part of mathematics which became useful developed with absolutely no desire to be useful, and in a situation where nobody could possibly know in what area it would become useful; and there were no general indications that it ever would be so. ...

Successes were largely due to forgetting completely about what one ultimately wanted, or whether one wanted anything ultimately; in refusing to investigate things which profit, and in relying solely on guidance by criteria of *intellectual elegance*, it was by following this rule that one actually got ahead in the long run, much better than any strictly utilitarian course would have permitted.

Furthermore, although mathematics is aesthetically motivated, it is reliable and objective, claims von Neumann (1954) [= (1963, p. 478)]:

... it is perfectly clear that mathematics furnishes something that is quite important, namely that it establishes certain standards of *objectivity*, certain standards of truth; and it is quite important that it appears to give a means to establish these standards rather independently of everything else, rather independently of emotional, rather independently of moral, questions. It is quite important to achieve this realization: That objective criteria of truth are possible, that such an aim is not self-contradictory, not in some sense in-human.

Von Neumann does not mention the controversial notion of intuition here, but we may conclude that beauty and elegance are to be judged by the mathematician's ("not in some sense inhuman") intuition.[29] Since intuition guides the mathematician in his choice of what he takes to be correct (or "reasonable") principles and rules of calculation and inference, we may say it amounts to insight into the rational. The field of methodology of science or *epistemology*, to be understood as having the same basic nature as logic (and of being largely co-extensive with statistics), would similarly be based on rational intuition.

All of this is relevant to the previous discussion on entropy, as the question was all along whether entropy and information were logical/epistemic or thermo-dynamic/physical concepts. If they are physical, propositions about them are to be tested empirically; if epistemic, propositions about them are based on aesthetics (rational intuition). Carnap insisted on a distinction between epistemic and physical theories of information, based on the perhaps dogmatic distinction between analytic and synthetic propositions of the Vienna Circle. Later on, however, Carnap (1968) came to view logic and epistemic theories as based on intuition – as discussed in Köhler (2001). This places Carnap in the same position as von Neumann in holding mathematics and physics to have quite different natures. Since von Neumann strongly agreed to such a distinction, why did he equate (epistemic) information with (physical) entropy? It was because he thought, as did Szilárd, that epistemic processes could be dealt with in physics. In philosophical jargon, this is psycho-physical reductionism, a widely-held view among many pioneers of computer science and artificial intelligence. Turing, McCulloch and Pitts, McCarthy, Minsky, Wiener, Rosenblueth, all believed that mental processes of perception and cognition ultimately had a physical basis – even when they are normatively determined like mathematical rules.

4. PSYCHO-PHYSICAL REDUCTIONISM

Von Neumann consistently avoided "philosophical" discussions of epistemologi-cal issues. He was all the more engaged in implementing logic and methodology in concrete engineering applications, as the standard history of computers by Goldstine (1972) shows. It is obvious that his efforts to explicate knowledge-acquisition using automata and nerve nets ultimately went back to Szilárd's "physicalizing" thermodynamic knowedge, as von Neumann (1966, p. 62f.), quoted in §2 above, himself indicated. In a lecture on the role of mathematics, von Neumann (1954) [= 1963, p. 486f.] gives a more general idea of his approach:

Another thing about which we can't tell today as much as we would like, but about which we know a good deal, is that it might have been quite reasonable to expect a vicious cycle when one tries to analyze the *substratum which produces science*, the function of human intelligence. The whole evidence of exploration in this area is that the system which occurs in intellectual performance, in other words in the human nervous system, can be investigated with physical and mathematical methods. Yet there is probably some kind of *contradiction* involved in imagining that, at any one moment, an individual should be completely informed about the state of his nervous apparatus at that particular moment. The chances are that the absolute limitations which exist here can also be expressed in mathematical terms, and only in mathematical terms.

Von Neumann then goes on to remind us that physical laws can all be formulated both *causally*, on the one hand, as well as *teleologically*, on the other; *i.e.* they are both formulable as differential equations tracking trajectories of objects or states through time (making earlier events causes of later ones), or they are formulable by equivalent integral equations describing overall (optimality) conditions on entire trajectories.[30] This fact makes it perhaps less surprising that machines, for example, can be built to think and perceive, since all natural processes can be described as *goal-directed* anyway – which is often held to apply to living beings alone. Von Neumann (1954) [= 1963, p. 484]:

The first approach is strictly causal, working from point to point in time. The second is strictly teleological, and defines only the total history by virtue of certain optimal properties. Yet the two are strictly equivalent; the actual history for movements that you derive from one is precisely that which you find from the other; and the question as to whether mechanics is causal or teleological (which in any other field would be viewed as an important substantive question calling for a yes or no answer) is manifest nonsense in mechanics, because it depends purely on how you choose to write the equations. I'm not trying to be facetious about the importance of keeping teleological principles in mind when dealing with biology; but I think one hasn't started to understand the problem of their role in biology until one realizes that in mechanics, if you are just a little bit clever mathematically, your problem disappears and becomes meaningless. And it is perfectly possible that if one understood another area the same thing might happen.

Here von Neumann warns biologists against overstressing goal-directed behavior, since this can always be reformulated *causally*. A prime example of von Neumann's drive to reduce apparently "emergent" biological functions to physics is his famous demonstration that automata can *reproduce themselves*, in von Neumann (1966). Similarly, psychologists do not monopolize *intelligence*, as simple circuit devices (*e.g.*, automata) and even thermodynamic systems have it!

Although some readers will interpret von Neumann here as *discounting* teleological formulations where he insists, as do Yourgrau & Mandelstam (1955), that the distinction between teleological and causal analyses is "meaningless", he can just as well be interpreted the other way around as *upgrading* "inanimate", causally driven nature to be in fact goal-driven, *i.e.* always having a purpose, sharing a feature of life.

Based on his general approach, one may say von Neumann was a psychophysical reductionist who thought human intelligence could in principle be presented and explained on a physical level – in particular, neurophysiologically, in terms of nerve nets. Between the physiology of nerves and the physics of electronic computer devices von Neumann recognized no fundamental difference in functional capability. Indeed, McCulloch & Pitts (1943) had proven that nerve nets are functionally equivalent to automata. It is presumably this approach which led von Neumann to reject Carnap's idea of trying to separate the epistemic notion of information from the physical notion of entropy: the latter was the *same thing as the former* in a psycho-physical reduction of logic to computer devices or to neurophysiology. More precisely, semantic information in the mind of a rational agent equals a certain negentropy of his nerve cells.

In taking von Neumann to be a reductionist, readers again should not be misled into believing he wanted to *equate* logic/mathematics with physics/statistical thermodynamics. He did not, although his statements are misleading. Carnap had the opposite problem: wanting to ban all confusion by cleanly *separating* logical from physical theories of information, he leaves readers at a loss to explain the puzzlingly *close analogy* between the notions. Thus, both thinkers leave their readers in quandaries.

4.1 Problems with von Neumann's "Physicalization" of Mathematics

No Information without Representation!
Battle cry of Wojciech Zurek [31]

If one considers more closely what "analyzing the substratum which produces science" would require, *i.e.* a "physicalization" of epistemic methods presupposed in standard theories of science, we run into difficulties which von Neumann blithely leaves out of his discussions on mathematics. (Analogous difficulties arise when extending mathematics to other epistemic theories such as statistics and decision theory.)

1. If logic (inference) and epistemic method in general (data handling, *i.e.* observation and statistical sampling) are *physicalized*, then purely empirical data appear *prima facie* sufficient to justify logic. But von Neumann himself stressed that the main warrants for mathematics are *aesthetic*. [This fact I call *Transcendence 2* in Köhler (2000).]

2. The "substratum which produces science" may be much bigger than von Neumann implies here, it may be quite superhuman – in fact, otherworldy –, if scientific knowledge is held to satisfy widely-demanded rationality criteria; hence cannot be *physicalized* in our universe. [This fact is what I call *Transcendence 1* in Köhler (2000).]

3. Von Neumann skirts the danger of repeating the same "humiliating" mistake he made when naïvely believing in Hilbert's Program in the '20s: trying to compress the hyper-complex problem of representing *and* applying mathematicized theories *within* themselves.

The first difficulty arises because of the modal difference between mathematics and physics, as mentioned. I deal with it in §5 below. The second difficulty arises because of the infinitely strong rationality assumptions implicit in the mathematics and statistics of practically all advanced scientific theories. The third problem is actually a special case of the second, as it arises when assuming, as the early Hilbert did, that mathematical operations could be adequately represented in finite physical objects surveyable by "concrete intuition". [32]

The second difficulty may be called the "representation problem": how "much" of standard science can be effectively represented by currently available computing tools; how much by tools theoretically attainable? The embarrassing fact is that not even elementary empirical theories using just ordinary number theory can be "shoe-horned" [33] into our entire cosmos, as no universal Turing machine will fit. This was shown by Gandy (1980), (1982). For the many scientific theories using some analysis, the situation is correspondingly worse, depending on how strong the analysis is that they use. Now we may conjecture that, since von Neumann had little compunction against using branches of higher mathematics based on less-certain set-theoretical axioms, and if he rigorously held to Zurek's battle cry (the Motto at the head of this section), then he would have had to "physicalize" stronger mathematical theories by assuming what I call "hyperworlds", which allow non-finitary processing. Such a move virtually forces us into speculative cosmology – or theology –, as it is no longer clear what a general physics of such hyperworlds means. In short, physicalizing epistemic theories making "unrestricted-rationality" assumptions is not necessarily *impossible*, but if the "physics" resorted to is not the one holding for "our universe", then we run into the danger of claiming something *tautological*, as it seems to be entirely open what "physics" is *in general*.

The "representation problem" frequently involves confusion between "subjective" and "objective", which is dealt with in §4.3 below.

The third difficulty was noticed by von Neumann himself in his perceptive remark (just cited above) about a possible "contradiction involved in imagining that ... an individual should be completely informed about the state of his nervous apparatus at that particular moment." Logically, this is a special case of the problem Hilbert posed to prove the consistency of mathematics within mathematics itself. Gödel (1931) showed that Hilbert's Program presupposed that, within metamathematics (itself *part of* mathematics!), mathematics could be *completely* represented – but that this very assumption made every consistency proof *impossible*. Von Neumann had in the '20s very enthusiastically committed himself to Hilbert's Program, and was never so "humiliated" in his life as when Gödel (1931) proved this couldn't be done.

On the other hand, Gödel's Proof itself, as well as later work on automata which was theoretically dominated by Gödel (1931), also showed how powerful automata are in executing arbitrary mental/epistemic functions of perception and cognition. In particular, Gödel's technique of "reflecting" arbitrary meta-mathematical relations and processes – theorems, proofs – as sequences of "Gödel numbers" on the object-level already showed that, if any finitary mental process is exactly described, it can be simulated by an algorithm called a "general recursive function". Gödel-numbering achieved just this simulation. Soon thereafter, Turing (1936), Post (1941), and McCulloch & Pitts (1943) independently proposed that their automata (all equivalent to Gödel numbering) provided adequate explications of the mind – understood as certain factors of intelligence, in particular perception and inference. Ironically, these very same theories showed that the mind's knowledge was *limited*, especially its ability at self-reflection. Gödel's Proof showed Hilbert's Program, aiming at *reflecting* mathematics *onto itself* and *proving its own consistency* by some sort of Münchhausen-

bootstrapping, was absolutely impossible. This von Neumann, already "humiliated" once by Gödel, had by then realized.

Most readers are tempted to regard the claim as trivial that automata can
simulate arbitrarily complex finitary behavior, assuming it is described exactly
enough. But in fact, describing behavior exactly in the first place constitutes
genuine scientific creativity. It is just such a *prima facie* superficial task which
von Neumann (1945) achieved in his famous explication of the "von Neumann
machine", regarded as the standard architecture for most post World-War-II
computers; similarly, Szilárd (1929) achieved a breakthrough in explicating
exactly how the memory and discrimination of Maxwell's demon is physically
described.

While most agree on automata's simulational ability, it may still be doubted
that they have full mental power, because they allegedly lack intentions: they
have no idea of the meaning of actions, *i.e.* they cannot truly "understand" words
of texts they read. However, Carnap (1955, §6) shows that the semantic notion of
"intension", an abstract form of intention, can also be defined for automata (there
called "robots"). For me, this resolves the debate on the ultimate mental
capability of automata, at least in principle.

4.2 Subjective Probability as "Hidden Parameter"

The idea that logic and mathematics are physical theories in the sense that all
their concepts can be represented as physical concepts may be a bit hard to
believe for many readers. It may help to consider another theory first, where a
related claim not only seems plausible but is even widely accepted: subjective
probability theory. In a paper relating subjective to logical probability, Carnap
(1962),[34] substantiates this claim; it is his most important contribution to the
interpretation of the concept of probability since his (1945). He shows that, if an
agent is thought to be perfectly rational, *i.e.* having perfect command of logic,
mathematics and estimation, then the betting quotients he assigns to hypotheses
(which represent subjective states of belief called "credence") will coincide with
Carnap's previously developed logical probabilities. What Carnap provides
thereby is an example of a relation between semantics and pragmatics, previously claimed by his collaborator at Chicago, Charles Morris (1938), such that every
semantical theory (restricted to relations between signs and objects) can be extended to a corresponding *pragmatic* theory (about relations among signs, objects, and language users). Pragmatic theories are psychological or sociopsycho-
logical and have the effect of giving empirical content to logical/mathematical
theory: a pragmatic theory of mathematics is about mental states and procedures
of actors. Obviously we can have a *descriptive* theory of *actual* sign-usage
(concept-usage) – whose truth must be supported by empirical data; or we can
have a *prescriptive* theory of *correct* sign-usage – whose validity is supported by

rational intuition, *i.e.* value-judgments by authoritative witnesses. The latter Carnap called "pure pragmatics".

Suddenly, the "abstract-logical" concept of probability obtains *empirical content* – even when used in a normative (prescriptive) theory – once it is applied to rational agents, whose degrees of belief ("credences") are identified with their betting quotients, assuming they are willing to place bets on what they believe. Carnap could have added that logical probabilities then become statistical measures representing *empirical* mixed strategies of game-players or decision-makers. Previously "objective" probability assignments made without regard to agents suddenly become "subjective" simply by explicitly mentioning agents where they were implicit before – without losing precision or validity. One may conclude that the "objective" and "subjective" theories actually only differ in terms of the perspective they take: the objective theory explicitly mentions no actors, the subjective theory does, but without any other difference. The same holds for all logic, mathematics, and statistics: they are all "abstract-logical-objective" (if formulated so that mental acts are only implicit) *as well as* "concrete-empirical-subjective" (if formulated to explicitly account for mental acts). As with von Neumann's point about the equivalence of causal vs. teleological formulations of physical laws, it's all relative to one's perspective.

It follows that Carnap's *logical probability* – and hence logical information – is very much also an *empirical* concept in the sense that it has *implicitly* to do with mental states and processes – because the equivalent (pragmatic-subjectivist) theory of credence has *explicitly* to do with mental states and processes.

Furthermore, logical probability is *physical* if the mental states and processes are reducible to neurophysiology or to electronic circuits, as von Neumann was convinced. If Carnap was not able to see as far as von Neumann about reducing mental states and processes, no wonder von Neumann objected!

Carnap's work on logical probability tended overall to be rather strongly aprioristic. Perhaps he needed a corrective against this tendency in a way his friend Otto Neurath loved to provide: our *"a priori"* logical guesses about the usefulness of conceptual schemes implicitly reveal our *past* experience with them. Neurath was famous for telling us that our knowledge is always in the middle of the stream of life, never beginning at zero. We choose some basic partitions (sample spaces) the way we do because they are *a posteriori* equiprobable, based *e.g.* on past statistical experience, but also on our accumulated knowledge about our senses or about the measurement devices used. That Carnap actually did agree with Neurath on this point is confirmed by the way he defined efficiency 'ef' of a conceptual framework to depend on *past experience*: the *a priori* act of choosing a conceptual scheme depends in turn on an *a posteriori* value of 'ef'. But Carnap is nonetheless right in distinguishing *a priori–prescriptive* (logical-rational) from *a posteriori-descriptive* (physical-empirical) methods to assign entropy values.

4.3 The Subjective–Objective Distinction: A Matter of Perspective

> Basically I will try to find out whether the subjectively expressed
> opinions are subjective or objective. If they are subjective, I will
> hold to my objective ones. If they are objective, I will consider
> and perhaps let the objective subjectively expressed opinions of
> the players influence my objective ones.
>
> Erich Ribbeck[35]

Over and over again, authors writing on entropy and information have grappled
with the confused and "opalescent" distinction between the two notions. The
main trouble lies in the conceptually hermaphroditic *empistemological* nature of
information – because information may alternatively be viewed *either* "abstractly" (with respect to its "objective content") *or* "concretely" (with respect to
the physics/chemistry/neurophysiology of the signal-recording and -processing).
This characteristic "dualism" of the *information* concept is quite distinct from the
well-known ambiguity of *entropy* due to the latter's dependency on how the
partition of the phase space and the probability distribution over it are set up.[36]
Information is a measure of knowledge about a system S which is *explicit* only in
an epistemological theory E_S displaying the conceptual framework and logical
relations between concepts and propositions, *e.g.* alternative "models" and their
mutual consistency, about the system S. Such an ("abstract") theory E_S is of
course practically *never* related to any concrete system S at all! (That has typically been the approach taken by mathematicians and logicians – who avoid
concrete details of calculation and proof –, and was obviously the attitude taken
by Carnap in his discussion with von Neumann.) Since physicists are interested
in explaining *concrete* systems like S, they of course direct their attention to
them and characteristically neglect a detailed analysis of E_S, which they leave to
logicians and philosophers of science – unless they become ambitious, like
Szilárd and von Neumann! S is the *object* being investigated and E_S is some
abstract, disembodied *subject* – the intelligence/mind/consciousness/episteme –
responsible for the investigation.

Szilárd and von Neumann were, perhaps without realizing it, emulating Hilbert's Program for mathematics, which attempted to incorporate the (abstract)
metamathematical problem of proving consistency *within* (concrete) mathematics. They wanted to incorporate "meta-physical" knowledge of S by an intelligence E_S *within some physical system* S_E.[37] Now comes a big surprise! To do this
requires in turn an even *higher* intelligence E': that of a theoretical physicist
attempting to model E_S in S_E. (We may say E is the epistemology of the *applied
physicist – e.g.* of Maxwell's demon –, whereas E' is more like that of the *theoretical physicist*. E' would have to be modeled in an even higher system, a
meta-system $S_{E'}$.) What makes physicists seasick is the continual, poorly-
negotiated shifting between object-level and meta-level, not realizing that
entirely different levels of theoretical strength are required, that *different conceptual frameworks* for S_E and $S_{E'}$ are involved, and hence that *two* usually different

agents $\langle E, E' \rangle$ and *three* levels of systems $\langle S, S_E, S_{E'} \rangle$ are involved. (In fact, S_E will almost always have to be a "hyper-physical" system, as it, and $S_{E'}$ all the more, is necessarily *too big* to fit into "our" universe. Hilbert did not successfully negotiate the transition either, and he got a lesson from Gödel.) Szilárd and von Neumann failed to properly distinguish all these systems, which is tolerable as a first approximation, since they were not worried about completeness: Maxwell's demon is not required to self-reflect, *i.e.* represent his own mental states, just to reflect the surrounding thermodynamic system. But of course, if the latter *were* required of Maxwell's demon, there would be trouble – as von Neumann was aware.

Now we are prepared to define an exact distinction between subjective and objective knowledge. Usage in physics, as well as mathematics and statistics, holds that a theory is called *subjective* when it explicitly displays logical background terms (*syncategoremata* like 'and', 'not', 'therefore', 'probable to degree …') as *mental* states or processes – even though these are rendered "objective" immediately when they are so displayed. This terminology needs careful attention. From the point of view of E, E's initial attention is to the *object S*, but when E shifts attention from S to S_E, E suddenly becomes self-conscious of E's own *subjective* states. From this point of view, E moves attention from the *object S* to the *subject S_E*.

Another analysis gives the opposite result, however. E in fact must change position to E', because stronger concepts are required in order to adequately re-present all of E than E has standing alone. Correspondingly, with respect to von Neumann's concrete "substrata", we shift from S_E to $S_{E'}$, whereby $S_{E'}$ now models E's *subjective* knowledge about S *objectively*, *i.e.* from an *objective* point of view *outside* of E with respect to how the subjective knowledge is obtained in the first place. It is at the *higher* level E' that "subjective knowledge" is "objectified", *i.e.* explicitly validated as reliable. It is undoubtedly wise to initially regard *all* straightforward knowledge claims by E about S using object-theoretical displays as fallible (hence "subjective" in the ordinary, pejorative sense) and, only after checking within the context of some S_E (including experimental apparatus) to certify that E indeed followed specifications, hence allowing us to consider E's knowledge reasonably certain (*i.e.* "objective" in the ordinary, euphemistic sense). In this sense, knowledge which is explicitly declared as subjective in the first place is *relatively objective*. Nevertheless, I will stick with the standard usage of physics, mathematics, and statistics in calling a theory subjective if it explicitly displays logical terms as states or processes of some mind.

The distinction just made is obtained precisely by going from an object-theory to a meta-theory in the sense of Tarski's semantics. The object-theoretical presentation is *objective* – because it naïvely ignores the subjective conceptual framework; the meta-theoretical presentation is *subjective* – because it soberly becomes "aware" of the subject, making *it* an object. In this connection it is important to realize that the usual connotation of "subjective" (*viz.*, possibly *fallible*) is captured by this distinction quite exactly, since (sober, "aware") claims of error are inherently meta-theoretical: only on a higher level can errors of the object-level be displayed and possibly corrected. It is primarily because of this typical connotation that "subjective concepts" and "subjective knowledge" are treated with disdain. Such disdain is witnessed when someone applies the adjective 'alleged' to any proposition, or when he uses "scare quotes" (*e.g.* when Jaynes systematically uses '"subjective" probabilities' – apparently intending logical probabilities – by using scare quotes). But in fact, the reliability, solidity, or "objectivity" of knowledge is not changed in the least by merely shifting from the object-level to the meta-level, because all we do there is make some things explicit that are otherwise implicitly assumed. The only difference is that, because the epistemological background has been made explicit, we suddenly become aware of possible error and of ways to correct it, therefore helping us to become "more objective". In any case, the mere use of explicitly subjective probabilities in Quantum Mechanics (QM) should not raise the slightest suspicion – quite the contrary! Of course, in a subjectivized QM, probability distributions are assigned to ideal rational agents who sample particle-assemblies, whereas in the standard objective QM they are assigned to particle assemblies directly (or are interpreted as waves). But the resulting parameter values for measurements are the same, so the two representations are empirically equivalent. The conceptual difference is that the subjectivist assignment is meta-theoretical, the other is object-theoretical. As Carnap (1962) showed, subjective probabilities are just as hard and precise as "objective" logical probabilities; to be sure, both of these are explicitly meta-theoretical in the first place, hence their relation is not so controversial as that between the two interpretations of QM, which involves a move to a meta-theory.

Attentive readers may wonder why I have not mentioned the most widespread explication of subjectivity: independence from mind. The problem with it is that nothing is excluded by this constraint, since presumably *everything* is thinkable, or, more "soberly" put, everything depends on concepts, and concepts are mental.[38]

In conclusion: the notorious distinction between "objective" and "subjective" behaves perfectly well and is easy to understand once we get used to the logical distinction between object theories and meta-theories. There is no mystery, and knowledge can always be displayed either way, hence is, in a sense, always *both* objective *and* subjective, depending on which perspective is chosen.

5. TWO MODALLY DISTINCT INFORMATION THEORIES

I now summarize my solution to the dispute between von Neumann and Carnap. Von Neumann was right in the sense that there are *not* two different (types of) information or entropy *concepts*. Carnap was right in the sense that a distinction can indeed be found between logical and physical information *theories* – as they are standardly intended. The problem was that we ordinarily distinguish theories in the first instance by recognizing differences in their concepts, and only in the second instance by recognizing differences in propositions – in virtue of their concepts. In both logical and physical theories, the entropy concept is the same, or more precisely, of the same type: *disorderliness* (called 'uncertainty' if used epistemologically). The difference between the logical/epistemological and the physical theory thus lies not in a difference of concepts, but rather in a difference in *modality* – that the one is normative and the other is empirical. That means the normative theory uses aesthetic intuition to determine (*a priori*) "correctness", whereas the empirical theory uses observation of (*a posteriori*) "reality".

Consider an analogous engineering situation concerning a bridge. We have two different theories, each describing a bridge. The *concept* of being a bridge is clear: it is physical object spanning gaps. But one theory describes a bridge (stone arch, suspension, whatever) exactly following (*a priori*) building codes, the other theory describes a bridge based on measurements taken (*a posteriori*) in the field.

Since the normative approach can lead to value-assignments different from those obtained empirically (as is clear from Carnap's Method II and Method I), one might argue that we then have two different concepts, as concepts are usually distinguished whenever they constitute different (types of) procedures to "grasp their objects" (Frege). But all such concepts of entropy (like concepts of bridge) *nevertheless* belong together because the "objects they grasp" all have essential features in common: they all estimate the degree of orderliness of data in some underlying memory space. *This* is what justifies saying that logical information *is* entropy: it *would be* the entropy of the traces on the media an intelligent being *would* use in recording and processing those data.

Von Neumann was right in the sense that the *concept* of information can always be considered physical entropy – somehow. His claim is of course somewhat egregious, as it ignored serious difficulties. For even if physical data and theory are represented in conceptual frameworks using relatively weak constructive mathematical methods (which they are usually *not* – except by radical "endophysicists"), they can hardly be recorded in models of existing physical objects.[39] Even constructively admissible models of data, according to any constructive standard, are already much too large to fit into a finite universe, or even into an infinite universe limited by relativistic constraints. Hence information measures of *their* content are not *ordinary physical concepts* in the sense that they refer to situations in "our" cosmos. In fact, a "physical world" able to execute processes prescribed by the strong classical mathematics used widely throughout modern physics is not the "real world", but a *super-world*.[40] An information theory requiring *such* computations and inferences is of course *not* an empirical theory (even assuming it is "about" empirical processes), because, just like mathematics, it is *not supported by empirical observations* but is supported by *value judgments* of an ultimately aesthetic nature about how "convincing", how "clear", how "interesting" the epistemological processes are

which make use of that information concept. Our aesthetically motivated standard of rationality for mathematics leads us to assume "hyper-physical" objects
and processes.

What neither von Neumann nor Carnap ever stated – which would have
settled the misunderstanding between them – was that the information theory
intended to measure the epistemic content of empirical science is *normative*
(whereby it will be an entropy measure of the state space of "hyper"-physical
devices *normally* implied when representing and processing theories of empirical
science), whereas the Shannon information theory of statistical thermodynamics
is *descriptive*. The distinction between "logical" and "physical" is misleading
when applied to concepts and should be replaced by the distinction between normative and empirical moods of propositions; the distinction between "objective"
and "subjective" is always relative to a particular object-level/meta-level
distinction. *Clearly different is only their motivation and hence mode of justification.* Carnap felt this and made it an often dogmatic cornerstone of his thought.
Von Neumann also knew there was a fundamental difference between natural
science and mathematics, which he justified more persuasively than Carnap ever
did – for Carnap never clearly made beauty the cornerstone of logical truth. But
von Neumann did not realize that this fundamental difference was supremely
relevant to the issue of distinguishing moods of "logical" and "physical"
information *theories*.

NOTES

*   Early versions of this paper were presented at a conference on "John von Neumann and the
     Foundations of Quantum Mechanics" at Lóránd Eötvös University in Budapest, 23–25 Feb.
     1999, and at the conference "Logica '99" at Liblice castle organized by the Czech Academy of
     Sciences in 22–24 June 1999. I am grateful for discussions on the topic of this paper with Abner
     Shimony, Michael Stöltzner and Miklós Rédei.

1.   Köhler (2000), (2001).
2.   Although I hold that there are *not* two inherently different (classes of) *concepts* involved, I will
     follow the practice of Carnap and others for the time being in pretending that physical and logi-
     cal probabilities are distinct concepts. It is more accurate to speak instead of different physical
     and logical probability *theories*. But even this is misleading, as the real difference lies *neither* in
     the concepts *nor* the propositions (hence theories) involved, but in the *modalities* of logical and
     physical theories: normative vs. factual. See footnote 9.
3.   In Carnap's (1934, §§51, 52), an empirical ("descriptive") theory is a set of non-analytic
     generalized sentences, *i.e.*, sentences which cannot be determined valid or invalid by the lan-
     guage rules which specify axioms and inference rules. They then require natural laws or empiri-
     cal protocol sentences for their support (Carnap assumed completeness for the combined system
     of logic and mathematics).
4.   (The classical distinction between analytic and synthetic propositions was mainly influenced by
     Descartes, Leibniz and Kant.) Although Carnap began his philosophical career as a Kantian, for
     whom science included an important category of synthetic–*a priori* propositions, Carnap (1934)
     held that this category could in effect be eliminated entirely, as all of Kant's synthetic–*a priori*
     propositions could reasonably be reclassified as either synthetic–*a posteriori* (*i.e.* empirical in a
     broad sense) or analytic–*a priori* (*i.e.* logical or mathematical) propositions. The history of this
     famous distinction is rather frought with complications, but one may say that the major problem
     with Kant's treatment lay in his ignorance of the actual extent of *logical* theory. This point was

emphasized by the "logicists" Bolzano, Frege and Russell, who showed that logic (assuming it included set theory or an extensional theory of concepts and relations – and hence functions) also included arithmetic and much or all of analysis, hence geometry as well.

5. *I.e.*, such that the determination of the truth value of a sentence depended on particular empirical terms occurring in a sentence; if not, then the validity of a sentence could be calculated alone from knowledge of the logical terms occurring.

6. *E.g.*, 'and', 'or', 'not', 'all', 'satisfies the property of' (alternatively, 'is a member of'), 'stands in the relation to', *etc.* Of course, the term 'syncategorematic' itself negatively prejudices the question of interpreting *syncategoremata*, as it implies that these have *no* meaning *except* when used with meaningful "categorematic" terms, *e.g.* species and genuses.

7. Culminating in Laplace's ratio of "the number of favorable to the total number of equally possible cases". Carnap advanced this idea, first by rigorously formulating conceptual frameworks within modern logic, second by employing measures of concepts and propositions called 'logical widths' *m* defined over such frameworks. Then the logical probability of a proposition *h* relative to some evidential support for it *e* became

$$\frac{m(e \cdot h)}{m(e)}.$$

8. Since both Tarski and Quine, like the early Carnap, rejected logical intuition as justification for logical rules and axioms, they were predestined not to find any reason for an ultimate distinction between "logical truth" (analyticity) and "empirical truth" (factuality).

9. (This view goes back to Kant, who claimed that logic requires no intuition.) If logic and mathematics *had* their own special content, *i.e.*, a *subject matter* clearly independent from empirical science, *e.g.* "abstract structures" or the like, then this fact could be used to distinguish *syncategoremata* from empirical terms in general, because then the *syncategoremata* would refer to the special content of logic and mathematics. My claim is that logic and mathematics *have* content, *viz.* inferences and calculations, but that they are *not special*, *i.e.* non-empirical, as they are mental processes. Logic and mathematics nevertheless differ from empirical theories in being *normative*, *i.e.* supported by rational intuition instead of ordinary observation.

10. To be sure, controversies of epistemological applications of deductive inference abound in constructive logic and mathematics.

11. In lectures at the University of Illinois in 1949, von Neumann already took this position. See the citation in §2 below from von Neumann (1966, p. 62f.).

12. An authoritative and multifaceted overview of Maxwell's demon, with a large collection of papers on the subject, may be found in Leff & Rex (1990).

13. Here in Urbana IL was located one of seven electronic computers, the ILLIAC, being built in parallel with and following von Neumann's IAS computer at the Institute for Advanced Study, developed 1946–52. Two others in this series were the RAND Corporation's JOHNNIAC, and the IBM model 701, which went on to establish IBM's dominance in the mainframe business. Burks and Goldstine were von Neumann's principle collaborators in Princeton.

14. The three most important publications are Carnap (1950), Carnap (1952), and Carnap & Stegmüller (1959). Important extensions were offered by Hintikka (1966), Hintikka & Niiniluoto (1976), Pietarinen (1972), and Niiniluto & Tuomela (1973); all summed up by Kuipers (1978). Further refinements are in Carnap & Jeffrey (1971) and Jeffrey (1980). Carnap's logical view was presaged by Laplace (1812), von Kries (1886), Meinong (1915), Keynes (1921), Johnson (1920–24), and Jeffreys (1939). Trained in physics and engineering, Watanabe (1969) offers a wealth of methodologically interesting ideas on inference, learning and perception in the Shannon–Brillouin tradition, but with many references to Carnap and Hempel.

15. 'Statistical' is usually considered synonymous with 'empirical' or 'physical', but this is misleading, for the logical theory of Carnap is just as "statistical" as the empirical theory. If statistics is defined as the theory of data-handling and statistical inference – *i.e.*, sampling, estimation and hypothesis-testing –, that makes it a *normative* theory, not empirical, and Carnap's inductive logic *is* such a theory. It also provides sampling methods, rules of estimation and, instead of acceptance and rejection rules for hypotheses, a more informative Bayesian-style probability-assignment rule.

16. This claim may surprise some readers, as Carnap's program of logical probability is widely regarded as dead or passé. But Carnap's logical approach is essential for obtaining the prior prob-

abilities used in any Bayesian approach, and most obviously for measuring the *explanatory power of theories*, i.e. the degree of economization of data-representation afforded by *theoretical concepts*. Rational (logical) assignments of prior probabilities are required for measuring the information content of (formalized) scientific theories, whose explanatory power may then be measured. See Niiniluoto & Tuomela (1973) and Hauffe (1981). Bayesian accounts of prior probabilities which reject Carnap's approach have fundamental problems when ignoring or downplaying structural features of conceptual frameworks toward which Carnap's program is oriented.

One factor explaining the lack of interest in Carnap's program is that it has produced few interesting mathematical results. The λ-continuum Carnap (1952) discovered with Kemeny, and some results concerning de Finetti's exchangeability principle presented in Jeffrey (1980), seem not quite interesting enough. However, Humburg (1986) and (1987) discovered that Carnap's λ-continuum is in fact restricted to a narrow range of values, e.g. $3 < \lambda < 3.8$ for binary sample spaces $k = 2$. If a theorem were to be proven to explain this highly interesting fact, Carnap's program would very likely gain wide recognition.

The importance of a logical information theory based on logical probability may be seen in historical retrospect. Both Mach and Boltzmann had claimed that the essential function of theories is to *economize* knowledge representation and acquisition. Ironically, Mach seriously compromised this endeavor by severely restricting the kinds of theoretical concepts he would admit, causing him, e.g., to reject atomic theory and relativity theory. Logical information theory in the sense of Carnap can directly measure how much economizing power Mach *loses* in rejecting "invisible entities" and not-directly-observable principles. For a related discussion on a measure of "efficiency" of conceptual schemes, see Carnap's definition of 'ef' below.

17. Definition D11-2 of Carnap & Bar-Hillel (1952). We suppose our knowledge to be represented by h_p (= hypothesis p), a partition of alternative, mutually exclusive hypotheses by $H = \{h_1, ..., h_p, ..., h_n\}$ (this is called the 'sample space' in statistics and represents – the base level of – our conceptual framework), an observation by e (evidence, or 'data' in statistics), logical probability by c (confirmation), the amount of information 'inf' based on c, and the estimation function 'est' based on c. 'inf' is defined in the way familiar to information theoreticians: $\inf (h) = - \text{Log } m (h)$, where Log is the logarithm to the base 2, and m is a logical measure, called 'logical width', of any proposition h represented in the system; m can be considered the *a priori* logical probability of the proposition. Similarly, conditional information is defined in the familiar way: $\inf (h / e) = \text{Log } m (e) - \text{Log } m (e \cdot h)$.

18. Here, H_1 is the set of all propositions using attributes Carnap calls 'Q-predicates' of L_1, which is called the outcome set in statistics, i.e. the set of all possible elementary situations things can be in. The alternative conceptual frameworks L_i are those capable of "covering" the "same area of" empirical situations as L_1, but many of them not so efficiently with respect to the data e. (Since this definition depends on particular e, we may want to consider other concepts such as "average efficiency" over a variety of e's; however, if e is taken to stand for the important case of current "total evidence", it represents the sum of our present empirical data and ought indeed be taken as standard.)

Carnap's sketch of 'ef' is quite informal in Carnap & Bar-Hillel (1952). Shimony pointed out in a personal communication that, since the c-functions used to calculate 'inf' will differ for different L_i, it requires a meta-meta-theory to compare different c-functions. The same holds for Humburg's work on calculating optimal λ-functions mentioned in footnote 16.

19. Langley, Simon et al. (1987) is a culmination of Simon's program in this field.

20. Work by Friedman & Shimony (1971), Dias & Shimony (1981), Seidenfeld (1979), Skyrms (1985) and Shimony (1985) show that Jaynes's principle – at least in the way previously formulated – is incompatible with standard Bayesian conditionalization, i.e. does not allow epistemic probability to be revised after witnessing new data. However, all participants in this debate agree that the specific conceptual frameworks presupposed are crucially important for gaining knowledge. The problem is whether a simple but powerful rule can be found which relates the logical structure of frameworks to epistemic probability without depending too much on specific knowledge of particular frameworks.

Shimony (1970, §V) emphasizes the importance of local empirical knowledge of statistical practice, casting into doubt aprioristic, universal approaches to inductive inference.

21. S_B^{II} is the entropy of Boltzmann (1896) as determined by "Method II" of Carnap (1977, §6), which Boltzmann used. Carnap opposed this method, which in fact predetermines entropy to be a function of the *a priori* probability of whether any of N molecules are in one of K cells of a partition of space. By this method, entropy represents the *uncertainty* of whether any *specific* distribution of particles D^{ind} obtains or not (D^{ind} states for every molecule in S which of the K cells it is in). This uncertainty may be expressed by the number m of the components of a *disjunction* D_i which states that the true distribution is one of m D^{ind}'s – whereby the more components the disjunction has, the greater the uncertainty about the true distribution. A degree of uncertainty is then defined as (surprise!) a logarithmic function, $-\ln m + $ const. I quote the definition in Carnap (1977, §6, p. 38ff.):

Method II, which seems to be most frequently used when an extension of S_B is made, proceeds as follows. For a given system of N molecules and a given system Ω^μ of K cells, let D_i be a disjunction of m distinct D^{ind} ($1 \leqq m \leqq Z = K^N$); the cell numbers need not be the same in the several components of D_i). Method II is characterized by defining $H^{II}(D_i)$ by a function of the form $-\ln m + $ const., and hence $S_B^{II}(D_i)$ by $k (\ln m - $ const.). The constant is chosen in various ways, sometimes as zero. Let us take it in such a form that S_B^{II} is in accordance with S_B in the case of a D^{st}. Therefore we define:

(6-11)(a) $\qquad S_B^{II}(D_i) \quad =_{Df} k \left\{ \ln m - N \ln \dfrac{N^K}{V^\mu} \right\}$,

(b) $\qquad\qquad\qquad = k (\ln m - N \ln + N \ln V^\mu).$

Any D_i^{st} is logically equivalent to the disjunction of the Z corresponding D^{ind}. Therefore

(6-12) $\qquad S_B^{II}(D_i^{st}) \quad = k \left\{ \ln Z (D_i^{st}) - N \ln \dfrac{N^K}{V^\mu} \right\}$.

$\qquad\qquad\qquad\qquad\qquad\qquad …$

However, Method II has also very serious disadvantages. We shall show that S_B^{II} is in most cases not in agreement with thermodynamic entropy S_{th}. Moreover, S_B^{II} is not a purely physical concept but a purely logical concept. This will become clear by the consideration of some examples.

$\qquad\qquad\qquad\qquad\qquad\qquad …$

Example 3. Suppose we know that the gas g was in a state of thermodynamic equilibrium during a time interval around t_1; we know the volume of the vessel and the volumes of the parts separated by walls, the number N of molecules and the numbers for the parts, and the molecular mass m; furthermore, a cell system Ω^μ with V^μ and K is specified. Now let us consider various descriptions; they are meant as typical descriptions of g for the interval around t_1.

$\qquad\qquad\qquad\qquad\qquad\qquad …$

(3d) Now we consider, in contrast, the situation with S_B^{II}. Let the statement " $S_B^{II} = r_1'$ " be given, with a specified real number r_1'. S_B^{II} is defined for any disjunction D_m of m D^{ind} with any m ($1 \leqq m \leqq Z$). Therefore the given statement means: "There is a number m and an (unspecified) D_m which holds for g at t_1 and for which $S_B^{II} = r_1'$." Now we see from (6-11) that, since k is known and V^μ, K, and N are supposed to be given, S_B^{II} depends merely on m; and conversely, from the given S_B^{II}-value r_1', we can determine the value of m, say

$\qquad m_1 \left[viz., m_1 = e^{r_1'/k} \left(\dfrac{NK}{V^\mu} \right)^N \right]$.

Suppose we find $m_1 = 1000$. Then the given statement means:
"There is a disjunction of one thousand D^{ind} which holds for g at t_1." The statement does not tell us *which* D^{ind} belong to the disjunction; it says nothing about the content of these D^{ind}, the cell numbers or anything else. It says merely that the (unspecified) disjunction consists of one thousand D^{ind}; this is a logical characteristic concerning logical strength or amount of information. The statement can also be formulated in this way: "The one (unspecified) D^{ind} which holds for g at t_1 belongs to an (unspecified) class of one thousand D^{ind}." This holds obviously for every D^{ind} and for every possible state of g. It is a tautology, a purely logical truth. Therefore S_B^{II} *is a purely logical concept. If it is to be introduced at all, the term* "entropy" should not be used for it.

In this and in other examples, it is thus clear that S_B^{II} is *not* based, as is the standard concept of thermodynamic entropy of Clausius, on measurements of temperature. But Carnap forgot that his own logical probability notion can be interpreted as an *empirical* (albeit normative!)

attribute of a decision maker. The number m refers to a representation in the brain of Maxwell's demon, which Szilárd physicalized. For more, see §§4.2, 4.3.

22. To avoid a widespread confusion, I call a concept 'physical' in this article iff it is reducible by acceptable procedures to standard physical concepts. (It should be realized that this is relative 1. to the current state of the art of observational technology, 2. to current logical/mathematical knowledge of proofs and definitions.) A concept is 'empirical' iff reducible to standard empirical concepts. However, in this article, a proposition is called 'empirical' only if its statement is intended *factually*, whereas exactly the same proposition is called 'logical' (or 'mathematical', or 'statistical', or ... , or 'rational' in general) only if its statement is intended *normatively*. In this sense a physical proposition (one which contains nothing but physical concepts and *syncategorematicae*) can be non-empirical if stated in a normative mood.

23. This is the well-known Gibbs paradox: two gases are exposed to each other at the same temperature and begin to mix, implying an increase of entropy. But being at the same temperature, their *thermodynamic* entropy remains constant! In a personal communication, Shimony pointed out a crucial difference between temperatures and gases or chemical isomers as state variables: the latter do not allow of transitions *between* them, whereas temperatures are "mutually accessible". This might justify distinguishing epistemic information from *thermodynamic* entropy, but not from *other* types of physical entropy – like the entropy of gas mixtures, or, most *apropos*, the entropy of memory chips and brain neurons. Furthermore, demanding of thermodynamic state transitions that they allow continuous transitions is not necessarily unique, since other non-thermodynamic systems may be found which *also* admit of continuous state transitions. Isomers are at least step-wise transmutable into each other (admittedly not under "normal conditions" without catalysts). And epistemic states allow neighborhood comparisons as well, so that they too can be step-wise transformed into each other. The only entirely genuine mark of a thermodynamic concept therefore seems to me to be that it is defined in terms of *temperature*.

24. My translation from the German. Pauli's ten-page "sketch" is about the *first essay* of Carnap (1977): "A Critical Examination of the Statistical Concept of Entropy in Classical Physics", in which Gibbs is treated.

25. The passage quoted by Pauli from Carnap's manuscript, p. 73, is at Carnap (1977, p. 54). In Pauli (1999), an unfortunate misprint seems to have occurred, as the central concept S_{th} (thermodynamic entropy) is incorrectly transcribed as 'S_m'.

26. Pauli's German is 'unendlich langsame Kreisprozesse'.

27. Albeit the once arch-positivist Pauli then has the problem that entropy will be a *theoretical* empirical concept, not *immediately* observable – a problem it shares with von Mises's (another arch-positivist's) statistical probability.

28. Especially noteworthy terms, to which I later recur, I set from here on in boldface. E.K.

29. I deal with intuition at length in Köhler (2001).

30. Physical laws of a "teleological" form using optimality criteria are called *variational principles*, e.g. the "principle of least action". In the age of rationalism they were thought to reveal the will of God, *e.g.* by Leibniz, Fermat, Maupertuis, Euler. Yourgrau & Mandelstam (1955) present a well-rounded historical and analytical treatment; they discount theological interpretations.

31. Discussion remark made at the conference "The Foundational Debate. Complexity and Constructivity in Mathematics and Physics" held by the Institute Vienna Circle, Vienna 1994.

32. For a discussion of Hilbert's Program, Gödel's Incompleteness Proof and Turing Machines, see Webb (1980). Nelson (1982) authoritatively discusses automata and more recent developments in philosophy of mind and cognitive science. Von Neumann had been influenced by Turing, but his work on computer architecture and especially cellular automata was even more strongly influenced by McCulloch & Pitts.

33. I mean by this the capability of a *Laplace demon*: 1. that each state description representing a situation covered by a theory should be represented by an address (or register) in a transfinite memory bank of an ideal computer or brain used to reflect on everything in the universe covered by that theory; 2. that the computer or brain be able to calculate "in real time" all effects from arbitrarily assumed boundary conditions: *e.g.* for any complete state description, all other states of the universe implied by these together with the laws should be "immediately known" to the computer. Both conditions involve enormous capacities and can only be described as *theological*. Frank (1932, Ch. II) provides a classic treatment of Laplace's demon.

34. Revised and expanded in Carnap (1971).
35. Response to player complaints about their coach after the failure of the German national soccer team to reach the advanced playoffs of the European Football Championship Games, June 2000, quoted in DER STANDARD, Vienna, 3 July 2000. One may say that Ribbeck has well internalized the Hegelian dialectic of transcendental idealism.
36. A nice summary of the ambiguity of Boltzmann-Gibbs entropy (the "Gibbs paradox") is provided by Jauch & Báron (1972, § 2), who, however, entirely fall prey to the *other* ambiguity, as they seem to deny that epistemological information can be "physicalized" at all and that the entire approach of Szilárd is to be rejected! Their argument is that a *purely physical mechanism* allegedly *without consciousness* (and thereby presumably without "intelligence", hence without "true" information?) can achieve the same effect as Szilárd's "intelligent being". But it was Szilárd's contention that intelligent ("conscious") agency is inherently capable of physical modeling in the first place. Jauch and Báron obviously do not realize that the concept of consciousness is *itself* highly ambiguous. Two major classes of consciousness are 1. reflective cognition on a high level involving propositionalization, and 2. mere sensory awareness in a "stream of consciousness" (as in William James). Jauch & Báron (1972) have succeeded in showing that all the intelligence needed by Szilárd's 'intelligent being' is on a *low level, i.e.* essentially on the level of a knee-jerk. And yet, even knee-jerks reveal consciousness on the level of a sensory stream! In a classic treatise, Culbertson (1963), perfectly grasping von Neumann's intentions (even those of Carnap (1955, §6) regarding 'intensions in robots'), shows in great detail how such low-level consciousness can be modeled in nerve nets. That high-level propositionalized consciousness can be physicalized is even easier to demonstrate, as is immediately apparent from the sophisticated logical abilities of present-day information-processing systems, *e.g.* "expert systems". (The heated debates about these are about how intelligent they are, not about whether they have consciousness.)
37. The famous theory of measurement by von Neumann (1932, Ch. VI), who incorporates it into Quantum Mechanics, should be considered along the same line, since the process of measurement is an intellectual task: observation! Since QM itself presupposes the very parameters dealt with by a QM–measurement theory, the question of completeness and the possibility of an inconsistency arises. But it was not mentioned, although von Neumann had just been "humiliated" by Gödel on the related problem concerning Hilbert's Program. However, Abner Shimony tells me that this chapter was written by Eugene Wigner – later Shimony's teacher at Princeton –, who was close to *both* von Neumann and Szilárd.
38. In the spirit of Bolzano and Frege – but not in their letter –, I regard concepts as *processes* (Frege's "Arten des Gegebenseins", alternatively called intentions, intensions, or constructions) for grasping objects. Grasping objects, however, is an inherently *mental* act, hence a theory such as that of Tichý (1988) or Zalta (1988) is a normative theory of mind. Bolzano and Frege were severely mistaken in claiming logic was antipsychological; what they should have said was that logic is *normative* and *thereby* distinct from *empirical* psychology.
39. Myrvold (1995) showed that important propositions of Quantum Mechanics are unavailable when attempting to place constructive restrictions on the mathematics of QM.
40. I call them "hyperworlds" in Köhler (2000). In analyzing Zeno's paradoxes, Benacerraf (1963) calls infinitely complex processes executable in finite time "super-tasks"; see also Grünbaum (1969) and Benardete (1964).

REFERENCES

Yehoshua Bar-Hillel (1952): "Semantic Information and its Measures", Transactions of the Tenth Conference on Cybernetics, New York; reprinted in Bar-Hillel (1964).
Yehoshua Bar-Hillel (1955): "An Examination of Information Theory", *Philosophy of Science*, **22**, 86–105; reprinted in Bar-Hillel (1964).
Yehoshua Bar-Hillel (1964): Language and Information. Selected Essays on Their Theory and Application, Addison-Wesley, Reading MA.

Yehoshua Bar-Hillel and Rudolf Carnap (1952): "Semantic Information", *Brit. Journ. f. t. Philos. of Sc.*, **4**, 147–157.
Paul Benacerraf (1963): "Tasks, Super-Tasks, and the Modern Eleatics", *Journ. o. Philos.* LIX, 765–784.
José Benardete (1964): *Infinity. An Essay in Metaphysics*, Oxford University Press, Oxford.
Ludwig Boltzmann (1872): "Weitere Studien über das Wärmegleichgewicht unter Gasmolekülen", *Kaiserliche Academie der Wissenschaften (Wien) Sitzungsberichte*, II. Abt., **66**, 275. [First appearance of ' $-\Sigma\, p_i \log p_i$ ']
Ludwig Boltzmann (1877): "Bemerkungen über einige Probleme der mechanischen Wärmetheorie", in Boltzmann (1909).
Ludwig Boltzmann (1896): *Vorlesungen über Gastheorie* I, Johannes Ambrosius Barth, Leipzig.
Ludwig Boltzmann (1909): *Wissenschaftliche Abhandlungen*, II, ed. by Fritz Hasenöhrl, J.A. Barth, Leipzig.
Léon Brillouin (1951): "Maxwell's Demon Cannot Operate: Information and Entropy I", *Journal of Applied Physics*, **22**, 334–337.
Léon Brillouin (1951a): "Physical Entropy and Information II", *Journ. o. Applied Phys.*, **22**, 338–343.
Léon Brillouin (1953): "The Negentropy Principle of Information", *Journal of Applied Phys.*, **24**, 1152–1163.
Léon Brillouin (1962): *Science and Information Theory*, Academic Press, New York.
Arthur W. Burks (1966): "Editor's Introduction", in von Neumann (1966).
Rudolf Carnap (1934): *Logische Syntax der Sprache*, Springer-Verlag, Vienna; translated and extended as *Logical Syntx of Language*, Routledge & Kegan Paul, London 1936.
Rudolf Carnap (1936): "Wahrheit und Bewährung", *Actes du Congrès international de philosophie scientifique*, Actualité scientifiques et industrielles **388**, Hermann & Cⁱᵉ, Paris; translated and combined with part of Carnap (1946) as "Truth and Confirmation" in Feigl & Sellars (1949).
Rudolf Carnap (1942): *Introduction to Semantics*, Harvard University Press, Cambridge MA.
Rudolf Carnap (1945): "Two Concepts of Probability", *Philosophy and Phenomenological Research*, **5**, 513–532; reprinted in Feigl & Sellars (1949).
Rudolf Carnap (1946): "Remarks on Induction and Truth", *Philos. and Phenomenolog. Research*, **6**, 590–602.
Rudolf Carnap (1950): *Logical Foundations of Probability*, University of Chicago Press, Chicago; ²1962.
Rudolf Carnap (1952): *The Continuum of Inductive Methods*, Univ. of Chicago Press, Chicago.
Rudolf Carnap (1953): "What Is Probability?", *Scientific American*, **189**,3 (Sept.); reprinted in the longer, original version in Madden (1960).
Rudolf Carnap (1955): "Meaning and Synonymy in Natural Languages", *Philosophical Studies*, **7**, 33–47.
Rudolf Carnap (1962): "The Aim of Inductive Logic", in Ernest Nagel, Patrick Suppes and Alfred Tarski (eds.): *Logic, Methodology and Philosophy of Science*, Stanford University Press, Stanford 1962.
Rudolf Carnap (1963): "Intellectual Autobiography", in Paul A. Schilpp (ed.): *The Philosophy of Rudolf Carnap*, Open Court Publishing Co., La Salle IL.
Rudolf Carnap (1968): "Inductive Logic and Inductive Intuition", in Lakatos (1968).
Rudolf Carnap (1971): "Inductive Logic and Rational Decisions", in Carnap & Jeffrey (1971).
Rudolf Carnap (1971a): "A Basic System of Inductive Logic I", in Carnap & Jeffrey (1971).
Rudolf Carnap (1977): *Two Essays on Entropy*, ed. and introduced by Abner Shimony, Univ. of California Press, Berkeley.
Rudolf Carnap (1980): "A Basic System of Inductive Logic II", in Jeffrey (1980).
Rudolf Carnap and Jehoshua Bar-Hillel (1952): "An Outline of a Theory of Semantic Information", Technical Report No. **247**, Research Laboratory of Electronics, MIT; reprinted in Bar-Hillel (1964). This item actually did not appear until 1953, and is sometimes assigned that year.
Rudolf Carnap and Wolfgang Stegmüller (1959): *Induktive Logik und Wahrscheinlichkeit*, Springer-Verlag, Vienna.
Rudolf Carnap and Richard Jeffrey (1971): *Studies in Inductive Logic and Probability* I, Univ. of California Press, Berkeley.

E. Colin Cherry (1951): "A History of the Theory of Information", *Proceedings of the Institute of Electrical Engineers*, **98** (III), 383–393; reprinted with minor changes as "The Communication of Information", *American Scientist*, **40** (1952), 640–664.

James T. Culbertson (1963): The Minds of Robots. Sense Data, Memory Images, and Behavior in Conscious Automata, University of Illinois Press, Urbana IL.

Martin Davis (ed.) (1965): *The Undecidable*, Raven Press, Hewlett NY.

Kenneth Denbigh (1981): "How Subjective Is Entropy?", *Chemical Britannica*, **17**, 168–185.

Kenneth Denbigh and J.S. Denbigh (1985): *Entropy in Relation to Incomplete Knowledge*, Cambridge University Press, Cambridge.

Werner DePauli-Schimanovich, Eckehart Köhler and Friedrich Stadler (eds.) (1995): *The Foundational Debate. Complexity and Constructivity in Mathematics and Physics* (Vienna Circle Institute Yearbook **3**), Kluwer Academic Publishers, Dordrecht, Holland.

Penha Maria Cardoso Dias and Abner Shimony (1981): "A Critique of Jaynes' Maximum Entropy Principle", *Advances in Applied Mathematics*, **2**, 172–211.

Herbert Feigl and Wilfried Sellars (eds.) (1949): *Readings in Philosophical Analysis*, Appleton-Century-Crofts, New York.

K. Friedman and Abner Shimony (1971): "Jaynes's Maximum Entropy Prescription and Probability Theory", *Journal of Statistical Physics*, **3**, 381–384.

Philipp Frank (1932): *Das Kausalgesetz und seine Grenzen*, Springer-Verlag, Vienna; reprinted by Suhrkamp stw734, Frankfurt am Main 1988.

Robin Gandy (1980): "Church's Thesis and Principles for Mechanisms", in Jon Barwise, H. Jerome Keisler & Kenneth Kunen (eds.): *The Kleene Symposium*, North-Holland Publ.Co., Amsterdam.

Robin Gandy (1982) "Limitations to Mathematical Knowledge", in Dirk van Dalen, D. Lasker & J. Smiley (eds.): *Logic Colloquium '80*, North-Holland, Amsterdam.

Josiah Willard Gibbs (1902): *Elementary Principles in Statistical Mechanics*, Yale Univ. Press, New Haven CN.

Kurt Gödel (1931): "Über formal unentscheidbare Sätze der Principia mathematica und verwandter Systeme I ", *Monatshefte für Mathematik und Physik*, **38**, 173–198; translated in Davis (1965).

Herman H. Goldstine (1972): *The Computer from Pascal to von Neumann*, Princeton Univ. Pr., Princeton.

Adolf Grünbaum (1969): "Can an Infinitude of Operations be Performed in a Finite Time?", *British Journal for the Philosophy of Science* **20**, 203–218.

Adolf Grünbaum (1973): *Philosophical Problems of Space and Time*, 2nd, ext. ed., Reidel Publ. Co., Dordrecht.

Godfrey Harold Hardy (1940): *A Mathematician's Apology*, Cambridge University Press, Cambridge; reprinted with foreword by C.P. Snow, ibid. 1969.

R.V.L. Hartley (1928): "Transmission of Information", *Bell System Technical Journal*, **7**, 535–563.

Heinz Hauffe (1981): *Der Informationsgehalt von Theorien. Ansätze zu einer quantitativen Theorie der wissenschaftlichen Erklärung*, Springer-Verlag, Vienna.

Steve J. Heims (1980): *John von Neumann and Norbert Wiener. From Mathematics to the Technologies of Life and Death*, MIT Press, Cambridge MA.

Jaakko Hintikka (1966): "A Two-Dimensional Continuum of Inductive Methods", in Hintikka & Suppes (1966).

Jaakko Hintikka and Juhani Pietarinen (1966): "Semantic Information and Inductive Logic", in Hintikka & Suppes (1966).

Jaakko Hintikka and Patrick Suppes (eds.) (1966): *Aspects of Inductive Logic*, North-Holland Publ. Co., Amsterdam.

Jaakko Hintikka and Ilkka Niiniluoto (1976): "An Axiomatic Foundation for the Logic of Inductive Generalization", in Przełęcki et al. (1976).

Jürgen Humburg (1986): "A Novel Axiom of Inductive Logic which Implies a Restriction of the Carnap Parameter λ", in Werner Leinfellner and Franz Wuketits (eds.): *The Tasks of Contemporary Philosophy. Proceedings of the 10th International Wittgenstein Symposium, Kirchberg am Wechsel 1985*, Verlag Hölder-Pichler-Tempsky, Vienna.

Jürgen Humburg (1987): "A Novel Axiom of Inductive Logic which Implies a Restriction of the Carnap Parameter λ", paper presented at the 8th International Congress of Logic, Methodology and Philosophy of Science, Moscow 1987; somewhat extended version of Humburg (1986).

J.M. Jauch and J.G. Báron (1972): "Entropy, Information and Szilard's Paradox", *Helvetica Physica Acta*, **45**, 220–232; reprinted in Leff & Rex (1990).

Edwin T. Jaynes (1957): "Information Theory and Statistical Mechanics, I, II", *Physical Review*, **106**, 620–630; **108**, 1771–190; reprinted in Jaynes (1983).

Edwin T. Jaynes (1965): "Gibbs vs. Boltzmann Entropies", *Am. J. o. Phys.*, **33**, 391–398; also in Jaynes (1983).

Edwin T. Jaynes (1967): "Foundations of Probability Theory and Statistical Mechanics", in Mario Bunge (ed.): *Delaware Seminar in the Foundations of Physics*, Springer, Heidelberg/Berlin/New York; reprinted in Jaynes (1983).

Edwin T. Jaynes (1978): "Where Do We Stand on Maximum Entropy?", in Levine/Tribus (1978); reprinted in Jaynes (1983).

Edwin T. Jaynes (1983): *Papers on Probability, Statistics and Statistical Physics*, ed. by Roger Rosenkrantz, Reidel, Dordrecht.

Richard Jeffrey (1980): *Studies in Inductive Logic and Probability* **II**, Univ. of California Press, Berkeley.

Harold Jeffreys (1939): *Theory of Probability*, Oxford Univ. Press, Oxford.

W.E. Johnson (1920/21/24): *Logic* **I/II/III**, Cambridge Univ. Press, Cambridge.

John G. Kemeny (1952): "Extension of the Methods of Inductive Logic", *Philosophical Studies*, **3**, 38–42.

John G. Kemeny (1953): "The Use of Simplicity in Induction", *Philosophical Review*, **62**, 391–408.

John G. Kemeny (1953): "A Logical Measure Function", *Journal of Symbolic Logic*, **18**, 289–308.

John G. Kemeny (1955): "Fair Bets and Inductive Probabilities", *Journ. o. Symb. Logic*, **20**, 263–273.

John G. Kemeny (1959): *A Philosoher Looks at Science*, Van Nostrand, Princeton.

John Maynard Keynes (1921): *A Treatise on Probability*, Macmillan & Co., London.

Eckehart Köhler (2000): "Gödel's Platonism", in Köhler et al.: *Kurt Gödel. Wahrheit und Beweisbarkeit*, Österreichischer Bundesverlag–Hölder-Pichler-Tempsky, Vienna, forthcoming.

Eckehart Köhler (2001): "Gödel on Intuition, and How Carnap Abandoned Empiricism", forthcoming.

Johannes von Kries (1886): *Die Principien der Wahrscheinlichkeitsrechnung*, Freiburg i.B.; 2nd ed. Tübingen 1927.

Theo A.F. Kuipers (1978): *Studies in Inductive Probability and Rational Expectation*, Reidel Publ. Co., Dordrecht.

Pat Langley, Herbert Simon, Gary Bradshaw and Jan Zytkow (1987): *Scientific Discovery. Computational Explorations of the Creative Process*, The MIT Press, Cambridge MA.

Imre Lakatos (ed.) (1968): *The Problem of Inductive Logic*. Proceedings of the International Colloquium in the Philosophy of Science, London, 1965, Vol. 2, North-Holland Publ. Co., Amsterdam.

Pierre Simon de Laplace (1812): *Théorie analytique des probabilités*, Paris.

Harvey S. Leff and Andrew F. Rex (eds.) (1990): *Maxwell's Demon. Entropy, Information, Computing*, Princteon University Press, Princeton, NJ.

Raphael D. Levine and Myron Tribus (eds.) (1979): *The Maximum Entropy Formalism*, MIT Press, Cambridge MA.

Edward H. Madden (ed.) (1960): *The Structure of Scientific Thought*, Houghton Mifflin, Boston.

James Clerk Maxwell (1871): *Theory of Heat*, Longmans, Green, London. [The demon is on p. 328.]

Warren S. McCulloch and Walter H. Pitts (1943): "A Logical Calculus of the Ideas Immanent in Nervous Activity", *Bulletin of Mathematical Biophysics*, **5**, 115–133.

Alexius Meinong (1915): *Über Möglichkeit und Wahrscheinlichkeit. Beiträge zur Gegenstandstheorie und Erkenntnistheorie*, Johann Ambrosius Barth, Leipzig.

George A. Miller (1953): "What Is Information Measurement?", *American Psychologist*, **8**, 3–11.

Richard von Mises (1919): "Grundlagen der Wahrscheinlichkeitsrechnung", *Math. Zeitschrift*, **5**, 52–99.

Richard von Mises (1928): *Wahrscheinlichkeit, Statistik und Wahrheit*, Springer-Verlag, Vienna.

Jacques Monod (1971): *Chance and Necessity*, transl. from French by A. Wainhause, Knopf, New York.

Charles Morris (1938): "Foundations of the Theory of Signs", in Otto Neurath, Rudolf Carnap and Charles Morris (eds.) *Foundations of the Unity of Science* I,2, Univ. of Chicago Press, Chicago 1955.

Wayne C. Myrvold (1995): "Computability in Quantum Mechanics", in DePauli-Schimanovich et al. (1995).

Johann von Neumann (1927): "Thermodynamik quantenmechanischer Gesamtheiten", *Göttinger Nachrichten*, 273–291.

Johann von Neumann (1932): *Mathematische Grundlagen der Quantenmechanik*, Springer-Verlag, Berlin; transl. by Robert Beyer as *Mathematical Foundations of Quantum Mechanics*, Princeton Univ. Press, Princeton NJ 1955.

John von Neumann (1945): "First Draft of a Report on the EDVAC", Moore School of Electrical Engineering, University of Pennsylvania, Philadelphia PA; extracts reprinted in Brian Randell (ed.) *The Origins of Digital Computers*, Springer-Verlag, Berlin/Heidelberg/New York 1973.

John von Neumann (1947): "The Mathematician", in Robert B. Heywood (ed.): *Works of the Mind*, University of Chicago Press; reprinted in von Neumann (1961, I) and in Newman (1956, 4).

John von Neumann (1954): "The Role of Mathematics in the Sciences and in Society", *Graduate Alumni* VI,27, 16–29, Princeton; reprinted in von Neumann (1963, VI).

John von Neumann (1958): *The Computer and the Brain*, Yale University Press, New Haven.

John von Neumann (1961–63): *Collected Works*, I–VI, Pergamon Press, Oxford.

John von Neumann (1966): *Theory of Self-Reproducing Automata*, edited, completed and introduced by Arthur W. Burks, University of Illinois Press, Urbana, IL.

Ilkka Niiniluoto and Raimo Tuomela (1973): *Theoretical Concepts and Hypothetical-Inductive Inference*, D. Reidel Publishing Co., Dordrecht.

Wolfgang Pauli (1999): *Wissenschaftlicher Briefwechsel Wolfgang Paulis mit Bohr, Einstein, Heisenberg und anderen*, vol. 4.2, ed. by Karl von Meyenn, Springer-Verlag, Berlin.

Juhani Pietarinen (1972): *Lawlikeness, Analogy, and Inductive Logic*, Acta Philosophica Fennica XXVI, North-Holland Publ. Co., Amsterdam.

Karl R. Popper (1974): "Intellectual Autobiography", in Paul A. Schilpp (ed.): *The Philosophy of Karl Popper*, Open Court Publishing Co., La Salle IL.

Emil Post (1941): "Absolutely Unsolvable Problems and Relatively Undecidable Propositions – Account of an Anticipation", in Davis (1965).

Marian Przełęcki, Klemens Szaniawski and Ryszard Wócicki (eds.) (1976): *Formal Methods in the Methodology of the Empirical Sciences*, Reidel Publ. Co., Dordrecht.

Willard Van Orman Quine (1951): "Two Dogmas of Empiricism", *Philosophical Review*, 60, 20–43; reprinted in Quine (1961).

Willard Van Orman Quine (1961): From a Logical Point of View. Logical-Philosophical Essays, Harvard Univ. Press. Cambridge MA.

Hans Reichenbach (1919): "Der Begriff der Wahrscheinlichkeit für die mathematische Darstellung der Wirklichkeit", *Die Naturwissenschaften* 7, 482–483.

Hans Reichenbach (1920): "Über die physikalischen Voraussetzungen der Wahrscheinlichkeitsrechnung", *Zeitschrift für Physik*, 2, 150–171; translated in Reichenbach (1978).

Hans Reichenbach (1920a): "Philosophische Kritik der Wahrscheinlichkeitsrechnung", *Die Naturwissenschaften*, 8, 46–55; translated in Reichenbach (1978).

Hans Reichenbach (1978): *Selected Writings 1909–1953*, II, Reidel Publishing Co., Dordrecht.

Teddy Seidenfeld (1979): "Why I Am Not an Objective Bayesian: Some Reflections Prompted by Rosenkrantz", *Theory and Decision* 11, 413–440.

Claude E. Shannon and Warren Weaver (1949): *The Mathematical Theory of Communication*, University of Illinois Press, Urbana IL.

Brian Skyrms (1985): "Maximum Entropy Inference as a Special Case of Conditionalization", *Synthese* 63, 55–74.

Leo Szilárd (1925): "Über die Erweiterung der phänomenologischen Thermodynamik auf fluktuierende Phänomene", *Zeitschrift für Physik*, 32, 753–788; translated as "On the Extension of Phenomenological Thermodynamics to Fluctuation Phenomena", in Szilard (1972).

Leo Szilárd (1929): "Über die Entropieverminderung in einem thermodynamischen System bei Eingriffen intelligenter Wesen", *Zeitschrift für Physik*, 53, 840–856; translated as "On the Decrease of Entropy in a Thermodynamic System by the Intervention of Intelligent Beings" by Anatol

Rapoport and Mechthilde Knoller in *Behavioral Science*, **9**, 301–310; the latter reprinted in Szilard (1972) and in Leff & Rex (1990).

Leo Szilárd (1972): *The Collected Works of Leo Szilard: Scientific Papers*, ed. by B.T. Field and G. Weiss, MIT Press, Cambridge MA.

Abner Shimony (1970): "Scientific Inference", in Robert Colodny (ed.): *The Nature and Function of Scientific Theories*, Univ. of Pittsburgh Press, Pittsburgh.

Abner Shimony (1977): "Introduction" to Carnap (1977).

Abner Shimony (1985): "The Status of the Principle of Maximum Entropy", *Synthese*, **63**, 35–53. [See also Friedman & Shimony (1971), and Dias & Shimony (1981).]

Karl Svozil (1995): "A Constructivist Manifesto for the Physical Sciences – Constructive Re-Interpretation of Physical Undecidability", in DePauli-Schimanovich et al. (1995).

Alfred Tarski (1936): "Der Wahrheitsbegriff in den formalisierten Sprachen", Studia Philosophica, 1, 261–405; transl. by Joseph H. Woodger as "The Concept of Truth in Formalized Languages" in A. Tarski: *Logic, Semantics, Mathematics*, Oxford Univ. Press, Oxford 1956.

Pavel Tichý (1988): *The Foundations of Frege's Logic*, de Gruyter, Berlin/New York.

Laszlo Tisza and P. Quay (1963): "The Statistical Thermodynamics of Equilibrium", *Annals of Physics*, **25**, 48–90.

Myron Tribus (1963): [Remarks on the history of information theory], *Boelter Anniversary Volume*, McGraw-Hill, New York.

Myron Tribus (1979): "Thirty Years of Information Theory", in Levine & Tribus (1979).

Alan M. Turing (1936): "On Computable Numbers with Applications to the Entscheidungsproblem", *Proceedings of the London Mathematical Society*, **42**, 230–265.

Satosi Watanabe (1969): Knowing and Guessing. A Quantitative Study of Inference and Information, Wiley, New York.

Hao Wang (1996): *A Logical Journey. From Gödel to Philosophy*, MIT Press, Cambridge MA.

Wolfgang Yourgrau and Stanley Mandelstam (1955): *Variational Principles in Dynamics and Quantum Theory*, Pitman & Sons, London.

Edward N. Zalta (1988): *Intensional Logic and the Metaphysics of Intentionality*, MIT Press, Cambridge MA.

Wojciech H. Zurek (1984): "Maxwell's Demon, Szilard's Engine and Quantum Measurements", in G.T. Moore and M.O. Scully (eds.): *Frontiers of Nonequilibrium Statistical Physics*, Plenum Press, New York 1984; reprinted in Leff & Rex (1990).

Wojciech H. Zurek (1989): "Thermodynamic Cost of Computation, Algorithmic Complexity and the Information Metric", *Nature*, **341**, 119–124.

Department of Business Administration
University of Vienna
Brünner Straße 72
A-1210 Vienna
Austria

STEPHEN J. SUMMERS

ON THE STONE – VON NEUMANN UNIQUENESS THEOREM AND ITS RAMIFICATIONS

1. INTRODUCTION

In the mid to late 1920s, the emerging theory of quantum mechanics had two main competing (and, initially, mutually antagonistic) formalisms – the wave mechanics of E. Schrödinger [61] and the matrix mechanics of W. Heisenberg, M. Born and P. Jordan [27][2][3].[1] Though a connection between the two was quickly pointed out by Schrödinger himself – see paper III in [61] – among others, the folk-theoretic "equivalence" between wave and matrix mechanics continued to generate more detailed study, even into our times. One outgrowth of this was associated with the canonical commutation relations (CCR):

$$PQ - QP = \frac{h}{2\pi i} \mathbf{I} , \tag{1}$$

which had begun to play such an important role in quantum theory [9][27][2][3] and were particularly central in the matrix mechanics approach.

Schrödinger found a representation of (1) in the context of his wave mechanics in paper III of [61]. Given in modern language, his Q is the multiplication operator

$$(Q\Psi)(x) = x\Psi(x) , \ x \in \mathrm{R} ,$$

on $L^2(\mathrm{R})$ and P is the differential operator

$$(P\Psi)(x) = -i\frac{h}{2\pi} \frac{d\Psi}{dx}(x) , \ x \in \mathrm{R} ,$$

on $L^2(\mathrm{R})$. Born and Jordan [2] had found another with P and Q formal matrices with infinitely many entries. Jordan [33] subsequently made a heuristic argument to the effect that these two representations of (1) are, in fact, equivalent in the sense described below. If that were indeed the case, it would be a very powerful confirmation that the physical content of matrix mechanics and wave mechanics coincided, since all physically relevant quantities can be expressed in terms of P and Q. And that, in turn, would enable physicists to employ with confidence whichever approach was most convenient.

However, much work remained to be done before this assertion could be mathematically well-formulated and then proven rigorously. First, quantum theory

M. Rédei and M. Stöltzner (eds.),
John von Neumann and the Foundations of Quantum Physics, 135–152.

needed to be formulated in Hilbert space, a crucial step begun by D. Hilbert himself [30],[2] made explicit by von Neumann in [44], and reached culmination in von Neumann's book [46].

Then, because there is no realization of P and Q satisfying (1) as bounded operators on Hilbert space [71][74][3], one needed to address the fact that (1) could not be understood as an operator equation on all of Hilbert space. This difficulty was side-stepped by reformulating the problem [70]: formally, if P and Q satisfy (1), then, with $U(a)$ and $V(a)$ defined by

$$U(a) = e^{\frac{-i2\pi aP}{h}} \text{ and } V(a) = e^{\frac{-i2\pi aQ}{h}} ,$$

it follows that, for any $a, b \in \mathbf{R}$,

$$U(a)V(b) = e^{\frac{i2\pi ab}{h}} V(b)U(a) . \tag{2}$$

This is the Weyl form of the CCR for one degree of freedom. $U(a)$ and $V(a)$ are, formally, unitary operators and therefore bounded; hence, (2) *may* be understood as an operator equation on all of the Hilbert space of states. In the Schrödinger representation we have

$$(U(a)\Psi)(x) = \Psi(x - a) \text{ and } (V(b)\Psi)(x) = e^{\frac{-i2\pi bx}{h}} \Psi(x) , \tag{3}$$

for any $\Psi \in L^2(\mathbf{R})$, and these are, indeed, unitary operators on $L^2(\mathbf{R})$.

In 1930, M.H. Stone [66] stated[4] and, in 1931, von Neumann [45] proved the following theorem. Note that a representation of the Weyl form of the CCR is said to be irreducible if the only subspaces of the Hilbert space $\mathcal{H}$ of states left invariant by the operators $\{U(a) \mid a \in \mathbf{R}\} \cup \{V(a) \mid a \in \mathbf{R}\}$ are $\{0\}$ and $\mathcal{H}$ itself.

Theorem 1 *If $\{\tilde{U}(a) \mid a \in \mathbf{R}\}$ and $\{\tilde{V}(a) \mid a \in \mathbf{R}\}$ are (weakly continuous[5]) families of unitary operators acting irreducibly on a (separable[6]) Hilbert space $\mathcal{H}$ such that*

$$\tilde{U}(a)\tilde{U}(b) = \tilde{U}(a + b) \quad , \quad \tilde{V}(a)\tilde{V}(b) = \tilde{V}(a + b) ,$$
$$\tilde{U}(a)\tilde{V}(b) = e^{\frac{i2\pi ab}{h}} \tilde{V}(b)\tilde{U}(a) ,$$

then there exists a Hilbert space isomorphism[7] $W : \mathcal{H} \to L^2(\mathbf{R})$ such that

$$W\tilde{U}(a)W^{-1} = U(a) \text{ and } W\tilde{V}(a)W^{-1} = V(a) ,$$

for all $a \in \mathbf{R}$, where $U(a)$ and $V(a)$ are the Weyl unitaries in the Schrödinger representation defined in (3).[8] If $\{\tilde{U}, \tilde{V}, \mathcal{H}\}$ is not irreducible but $\mathcal{H}$ is separable, then $\mathcal{H}$ decomposes into a direct sum of countably many closed subspaces, on each of which the restriction of $\{\tilde{U}, \tilde{V}\}$ is once again unitarily equivalent to the Schrödinger representation $\{U, V, L^2(\mathbf{R})\}$.

Hence, every irreducible Weyl representation of the CCR for one degree of freedom is unitarily equivalent to the Weyl form of the Schrödinger representation, and this is true, up to multiplicity, for reducible representations, as well. It therefore follows that the physical content of the irreducible representation $\{\tilde{U}, \tilde{V}, \mathcal{H}\}$ is identical to that of the Schrödinger representation $\{U, V, L^2(\mathbb{R})\}$. This theorem is usually referred to in the literature as the Stone-von Neumann uniqueness theorem.[9]

This discussion has been presented for one degree of freedom, but it may be reformulated for any finite number of degrees of freedom, and, there again, any irreducible representation of the Weyl form of the CCR for n degrees of freedom is unitarily equivalent to the corresponding Schrödinger representation (with analogous results in the reducible case) [45]. Hence, if one is considering a quantum system with only finitely many degrees of freedom, then it matters not which representation one chooses to work in, and the "equivalence" of matrix mechanics and wave mechanics is even more tightly knit. This seemed to satisfy the founders of quantum mechanics, though, much later, mathematical physicists found some clouds in this apparently brilliant sky when they refocussed their attention on the dynamical variables P and Q, as we shall see.

Returning to the unbounded operators P and Q, it should be noted that F. Rellich [53], followed by many authors (see [51][34] for references and recent results), provided sufficient conditions on the canonical conjugates P and Q in a representation of the CCR (1) which ensured that they are unitarily equivalent to the corresponding operators in the Schrödinger representation. The strategy ordinarily adopted was to find conditions on P and Q so that they may be exponentiated in such a way that (2) holds, and then to appeal to Theorem 1. As a useful and representative result of this type, we mention J. Dixmier's theorem, once again stated here only for one degree of freedom.

Theorem 2 ([11]) *Let P and Q be closed symmetric operators in a Hilbert space $\mathcal{H}$. Let $\mathcal{D}$ be a dense, linear subspace of $\mathcal{H}$ contained in the domains of both P and Q such that $P\mathcal{D} \subset \mathcal{D}$ and $Q\mathcal{D} \subset \mathcal{D}$. If (1) holds on $\mathcal{D}$ and the restriction of $P^2 + Q^2$ to $\mathcal{D}$ is essentially self-adjoint, then $\mathcal{H}$ decomposes into a direct sum of closed subspaces, on each of which the restrictions of P and Q are unitarily equivalent to the corresponding operators in the Schrödinger representation.*

On the other hand, there are many results (see [51] and [57] for references) to the effect that even if P and Q are essentially self-adjoint on a common invariant dense domain $\mathcal{D}$, on which they satisfy (1), they need *not* be unitarily equivalent to the Schrödinger representation. In fact, K. Schmüdgen [57] has produced an uncountable set of pairwise inequivalent representations of this type! Of course, by Theorem 1, when these operators are exponentiated, the resulting unitaries do *not* satisfy (2). Are all of these examples physically pathological? And even if so, could there be others which are not? The answer to this latter question is positive. H. Reeh [52] has provided such an example arising in the description of a charged particle in the exterior of an infinitely long cylinder with a magnetic flux running through it. This is therefore a physically meaningful representation of the CCR with finitely many degrees of freedom (two, after the idealization of letting the radius of the cylinder go to zero) which is not unitarily equivalent to the corresponding Schrödinger representation. Seventy years ago, this example would have been a bombshell; however, now that the developments described in the next section have accustomed us to the nonequivalence of physically relevant representations, Reeh's example[10] was hardly noticed. Nonetheless, even physicists should be a bit more careful when they proclaim the equivalence of the Heisenberg and Schrödinger representations in their quantum mechanics lectures.

We have been led to representations of the Weyl form of the CCR through the physically motivated interest in representations of conjugate P's and Q's. However, physically interesting applications have been found for representations $\{U, V, \mathcal{H}\}$ in nonseparable Hilbert spaces which have no connection with unbounded operators satisfying (1) at all – see the recent preprint [7] for references. In such representations the functions $a \mapsto U(a)$ and $a \mapsto V(a)$ are not weakly continuous; these representations are called nonregular. In [7] is given a generalization of Theorem 1 to the case of weakly measurable nonregular representations, which is sufficient to subsume the known physical models. We shall say no more about this interesting line of development here.

In this introduction, the mathematical level of the discussion has been deliberately held low. This will not be possible in the balance of the paper. We shall first consider the consequences of the fact that the analogue of Theorem 1 for infinitely many degrees of freedom is false; indeed, in that case, there is an enormously infinite number of unitarily inequivalent representations of the CCR in the Weyl form and, therefore, also of the original CCR. This fact was only slowly and painfully realized, because physicists chose to ignore the restriction in the hypothesis of the Stone-von Neumann uniqueness theorem. We shall indicate how this obduracy was overcome and what mathematical physicists have discovered in their exploration of this rich set of inequivalent representations in both its mathematical and physical aspects. We shall also discuss the correct generalization of Theorem 1 to infinitely many degrees of freedom. Finally, to trace another line of influence of the Stone-von Neumann uniqueness theorem, we shall briefly describe certain generalizations and their role in the harmonic analysis of locally compact groups, which has found particular application in such diverse fields as number theory, imaging science, communication theory and data/signal analysis. However, given the limitations of space imposed upon us, we have here no ambitions of completeness.

2. INFINITELY MANY DEGREES OF FREEDOM

Though the hypothesis of Theorem 1 clearly restricts its import to finitely many degrees of freedom and close examination of von Neumann's proof makes it evident that the argument loses its mathematical validity when extended to infinitely many degrees of freedom, physicists have always trusted their physical "intuition" more than mathematical proof. Indeed, that which a physicist calls a proof is often viewed by a mathematician as a plausibility argument, at best. Physicists are, however, often justified in not waiting for the mathematicians, whose concern for rigor they regard with impatience, to firmly bolster the physicists' ideas. If they were to do so, the natural sciences would not have advanced as rapidly as they have. Significantly, physicists have a source of conviction which mathematicians do not: mathematically unconstrained speculations can be checked, to a certain extent, in the laboratory. Nonetheless, important aspects of physicists' theories of nature – their attempts to formalize the physical intuitions gleaned from the complex feedback loop between theory and experiment – have often enough either remained vague or revealed themselves to be incorrect, if not nonsensical.

An example of this is the physicists' long-lived belief, based upon their experience with systems having finitely many degrees of freedom and the Stone-von Neumann uniqueness theorem, that the choice of representation of the CCR was merely a matter of convenience – one only needed to keep track of the number of degrees of freedom. It was realized quite early that quantum field theory necessitated infinitely many degrees of freedom in its canonical variables (see already [10]). When dealing with infinitely many degrees of freedom, they worked exclusively in the representation of the CCR associated with a Hilbert space containing a dense set of states describing only finitely many particles. This representation emerged heuristically in the first papers on quantum field theory by Heisenberg and W. Pauli [28] and was later formalized more completely by V. Fock [16] (see [8] for the first mathematically rigorous and Poincaré covariant presentation of this representation, now usually called the Fock[11] representation). Since the Fock representation, using annihilation and creation operators and a distinguished vacuum vector, is so well-known, and it is equally well-known that the Schrödinger representation can be re-written as a Fock representation with only finitely many annihilation and creation operators, we shall not interrupt the flow of our story with the details (but see [4] or [12], if necessary).

The Fock representation was therefore viewed as the natural generalization of the Schrödinger representation to infinitely many degrees of freedom and inherited its royal mantle of distinction. Hence, quantum field models were written in the Fock representation by theoretical physicists, insofar as a representation was actually specified, with the firm belief that it was the only representation they needed.

It is an interesting aside that von Neumann apparently did not appreciate systems of infinitely many degrees of freedom. He wrote in his treatment of radiation in [46]:

Now it is inconvenient formally and of doubtful validity to admit systems with infinitely many degrees of freedom, or wave functions with infinitely many arguments.[12]

In what is effectively the Fock representation, he therefore considered N degrees of freedom, computed energy spectrum, and then let $N \to \infty$. In order to compute this spectrum, von Neumann performed a canonical transformation[13] to obtain a second representation of the CCR in which the transformed Hamiltonian has a simpler form. For finite N this transformation is, by Theorem 1, unitarily implementable. However, in the limit $N \to \infty$ the transformation is not unitarily implementable and the representations are unitarily inequivalent. In other words, without realizing it, von Neumann himself worked with unitarily inequivalent representations of the CCR. His argument about the energy spectrum is therefore suspect.[14] From the very beginning of the subject, quantum field theory was plagued by divergences; when one source of infinity was heuristically taken care of, yet another was stumbled upon. This became such an apparently insurmountable problem, that some of the founders became quite pessimistic (particularly Bohr and Dirac) and decided that yet another conceptual revolution would be required to transcend quantum field theory and avoid its apparently inherent problems. However, some researchers had not yet given up on the possibility of getting sensible answers from quantum field theory and were trying to discern and then engage the various sources of these infinities from increasingly profound starting points.

Of direct relevance to our story, L. van Hove examined a simple model and argued that the origin of the divergences of perturbation theory (which is always carried out in Fock space) could be located in the fact that the state vectors of the interacting model were "orthogonal" to the state vectors in Fock space. In modern terms, what he argued was that the folium of states[15] of the interacting model was disjoint from the folium of states of the Fock representation. An immediate consequence of this observation would have been that the interacting representation for his model was unitarily inequivalent to the Fock representation. He did not quite get to this point.[16]

Also in the early 1950's, K.O. Friedrichs [17] undertook an influential attempt to reduce the hand-waving typical of quantum field theory up to that time. For our purposes here, the result of greatest interest was his construction of some representations of the CCR for infinitely many degrees of freedom which were *not* unitarily equivalent to the Fock representation. As he wrote:

Accordingly, there are different – non-equivalent – realizations of the basic field operators, and consequently different – non-equivalent – kinds of fields, a fact which seems worth noticing.[17]

In point of fact, he constructed representations in which the number operator does not exist (cf. the discussion further below).[18]

Though it would appear that not many theoretical physicists did take notice of Friedrichs' results, at least a handful of mathematical physicists and mathematicians were paying attention. In particular, in the following year L. Gårding and A.S. Wightman [22], taking their cue from Friedrichs and trying to classify representations of the CCR using properties of a number operator,[19] proved that there exists a large class of inequivalent representations of the CCR for infinitely many degrees of freedom.[20] Indeed, it slowly emerged that there exists an unimaginably infinite number of inequivalent representations – the space of unitary equivalence classes of such representations cannot even admit a separable Borel structure [40][19]. The task of classifying these representations would thus appear to be hopeless.

Another researcher who reacted to the examples of van Hove and Friedrichs was R. Haag. Aware of these preceding works, he presented an argument to the effect that the interaction representation, widely in use in quantum field theory on the basis of its prior success in quantum mechanical scattering theory, did not exist unless there was no interaction at all! This important assertion found a number of mathematically rigorous formulations and proofs, which can, perhaps, be summarized into two types, represented in [67] and [12]. We state Haag's theorem in a somewhat restricted form along the lines of [67].

Theorem 3 (Haag's Theorem) *Let $\phi(x)$ be a free hermitian scalar field[21] of mass $m > 0$, and let $\psi(x)$ be an irreducible local Poincaré-covariant field. If $\phi(x)$ and $\psi(x)$, resp. the canonical conjugates $\dot{\phi}(x)$ and $\dot{\psi}(x)$, are unitarily equivalent at some time t, then $\psi(x)$ is also a free field of mass m.*

Of course, the indicated hypothesis holds for the "free" field and the "interacting" field in the interacting representation. This was extremely inconvenient for the then-standard scattering theory for quantum fields. But it is clear that Haag's theorem also implies that the representations which are of physical interest, precisely *because* they involve interaction, are to be found among those *inequivalent* to the Fock representation. Therefore, by 1955, both the existence and the necessity of using representations inequivalent to the Fock representation had been firmly established – though not established in all theorists' minds: as late as 1961, a standard text on quantum field theory [62] could present the old scattering theory in the interaction representation with no mention of Haag's theorem[22].

Before we turn to a recounting of the progress made in constructing representations inequivalent to the Fock representation, we answer the natural question: which representations *are* equivalent to the Fock representation? It was evident to Friedrichs that a necessary condition for this equivalence is the existence of a number operator in the representation. A series of papers followed Friedrichs' lead and gave successively more general, rigorously proven content to the assertion "a representation of the CCR is unitarily equivalent to the Fock representation if and only if the number operator exists as a densely-defined self-adjoint positive operator in the representation." However, as was emphasized by J.M. Chaiken [5], this result is very sensitive to the definition of "number operator".[23]

The work of Gårding and Wightman did not provide an explicit construction of inequivalent representations. Wightman and S.S. Schweber [73] later constructed some classes of inequivalent representations of the CCR, as did I.E. Segal (see a later account [63] and the references given there). Many further classes of inequivalent representations have been constructed and brought under mathematical control since then. We mention the infinite product representations [35], coherent representations [36], quasi-free representations [55], quadratic representations [50] and higher-order representations [15]. These various classes of representations have found physical application and will surely prove to be of further use in the future.

But the most ambitious and difficult constructions of representations of the CCR have been carried out under the rubric "constructive quantum field theory."

This work was motivated by the desire to mathematically construct the sort of representations the quantum field theorists were tacitly referring to; in other words, to give some mathematical meaning to the quantum field models at the center of the theorists' discourse. This latter goal has been approached from two different directions – on the one hand, various axiom systems have been erected which hope to subsume basic principles common to large classes of quantum fields: then theorems are proven to establish physically interesting properties of all quantum fields satisfying the given axioms; and on the other hand, concrete models have been constructed to show that the axiom systems are not vacuous: of course, in this connection, valiant efforts have been made to construct the standard models of the quantum field theorists. The axiomatic approach will not be further discussed here.[24] Instead, we shall briefly indicate those results of constructive quantum field theory which are of direct relevance to the topic at hand.

We first discuss J. Glimm and A. Jaffe's construction of the $(\phi^4)_2$-model [20]. Let $\mathcal{H}_0$ be the Fock space for a scalar hermitian Bose field $\phi(x, t)$ of mass $m > 0$. Let $\pi(x, t) = \partial\phi(x, t)/\partial t$ and $\mathcal{D} \subset \mathcal{H}_0$ be the dense set of finite-particle vectors in $\mathcal{H}_0$. Then, for every f in a dense subspace $\mathcal{S}(\mathrm{R})$ of $L^2(\mathrm{R})$, the operator $\phi(f) \equiv \phi(f, 0) = \int \phi(x, 0)f(x)dx$ is essentially self-adjoint on $\mathcal{D}$ and $\phi(f)\mathcal{D} \subset \mathcal{D}$ (similarly for $\pi(f)$). Then one has on $\mathcal{D}$ the CCR[25]

$$\phi(f)\pi(g) - \pi(g)\phi(f) \;\; = \;\; i < f, g > \mathbf{I} , \tag{4}$$
$$\phi(f)\phi(g) - \phi(g)\phi(f) = \;\; 0 \;\; = \pi(f)\pi(g) - \pi(g)\pi(f) , \tag{5}$$

for all $f, g \in \mathcal{S}(\mathrm{R})$. When exponentiated, these operators provide a Weyl representation of the CCR. For each bounded open subset $\mathbf{O} \subset \mathrm{R}$, denote by $\mathcal{A}(\mathbf{O})$ the von Neumann algebra[26] generated by the Weyl unitaries

$$\{e^{i\phi(f)}, e^{i\pi(f)} \mid f \in \mathcal{S}(\mathrm{R}) , \; \mathrm{supp}(f) \subset \mathbf{O}\} .$$

Note that, though there are many C^*-algebras associated with the CCR in the Fock representation,[27] they all have the same weak closure.

The total energy

$$H_0 = \frac{1}{2}\int \; : (\pi(x, 0)^2 + \nabla\phi(x, 0)^2 + m^2\phi(x, 0)^2) : \, dx$$

of this field is a positive quadratic form on $\mathcal{D} \times \mathcal{D}$ and therefore determines uniquely a self-adjoint operator, which we also denote by H_0. With $g \in L^2(\mathrm{R})$ nonnegative of compact support, Glimm and Jaffe showed that, for each $\lambda > 0$, the cut-off interacting Hamilton operator

$$H(g) \equiv H_0 + \lambda \int \; : \phi(x, 0)^4 : g(x)dx$$

is essentially self-adjoint on $\mathcal{D}$,[28] and its self-adjoint closure, also denoted by $H(g)$, is bounded from below. By adding a suitable multiple of the identity we may take

0 to be the minimum of its spectrum. Then, 0 is a simple eigenvalue of $H(g)$ with normalized eigenvector $\Omega(g) \in \mathcal{H}_0$.

For any $t \in \mathbf{R}$, let $\mathbf{O}_t$ denote the subset of R consisting of all points with distance less than $|t|$ to $\mathbf{O}$. By choosing the cutoff function g to be equal to 1 on $\mathbf{O}_t$, then for any $A \in \mathcal{A}(\mathbf{O})$ the operator

$$\sigma_t(A) \equiv e^{itH(g)} A e^{-itH(g)}$$

is independent of g and is contained in $\mathcal{A}(\mathbf{O}_t)$. For any bounded open $\mathcal{O} \subset \mathbf{R}^2$ and $t \in \mathbf{R}$, let $\mathbf{O}(t) = \{x \in \mathbf{R} \mid (x,t) \in \mathcal{O}\}$ be the time t slice of $\mathcal{O}$. We define $\mathcal{A}(\mathcal{O})$ to be the von Neumann algebra generated by $\bigcup_s \sigma_s(\mathcal{A}(\mathbf{O}(s)))$.[29] Finally, we let $\mathcal{A}$ denote the closure in the operator norm of the union $\bigcup \mathcal{A}(\mathcal{O})$ over all open bounded $\mathcal{O} \subset \mathbf{R}^2$. Hence, σ_t is an automorphism on $\mathcal{A}$ and implements the time evolution associated with the interacting field. Similarly, "locally correct" generators for the Lorentz boosts and the spatial translations can be defined, resulting in an automorphic action on $\mathcal{A}$ of the entire Poincaré group in two spacetime dimensions.

For each $A \in \mathcal{A}$, we set $\omega_g(A) = < \Omega(g), A\Omega(g) >$ to define the locally correct vacuum state ω_g of the interacting field. Taking a limit as the cutoff function g approaches the constant function 1, Glimm and Jaffe showed that $\omega_g(A) \to \omega(A)$, for each $A \in \mathcal{A}$, defines a new (locally normal) state ω on $\mathcal{A}$ which is Poincaré invariant. By the GNS construction one then obtains a new Hilbert space $\mathcal{H}$, a representation ρ of $\mathcal{A}$ as a C^*-algebra acting on $\mathcal{H}$, and a vector $\Omega \in \mathcal{H}$ such that $\rho(\mathcal{A})\Omega$ is dense in $\mathcal{H}$ and

$$\omega(A) = < \Omega, \rho(A)\Omega > , \text{ for all } A \in \mathcal{A} .$$

In addition, one obtains a strongly continuous unitary representation of the Poincaré group in two spacetime dimensions under which the algebras $\rho(\mathcal{A}(\mathcal{O}))$ transform covariantly. The axioms of both the algebraic [25] and the field approach [67] have been verified for this model.

It is in this representation $(\rho, \mathcal{H})$ that the field equations for this model find a mathematically satisfactory interpretation [60]. And it is to the physically significant quantities in this representation that perturbation theory in λ is asymptotic – see the discussion in [21]. For this and other reasons, ω is interpreted as the exact vacuum state in the interacting theory, and its folium of states contains the physically admissible states of the interacting theory.

The generators of the strongly continuous Abelian unitary groups $\{\rho(e^{it\phi(f)}) \mid t \in \mathbf{R}\}$ and $\{\rho(e^{it\pi(f)}) \mid t \in \mathbf{R}\}$ satisfy the CCR (4). However, this representation of the CCR in $\mathcal{H}$ is not unitarily equivalent to the initial representation in Fock space. Indeed, by taking different values of the coupling constant λ in the above construction, one obtains an uncountably infinite family of mutually inequivalent representations of the CCR (4) (see [18])![30]

Similar constructions with similar results have been carried out for general polynomial interactions $P(\phi)$ and for the Yukawa model, both in two spacetime dimensions. For the sake of technical convenience, these constructions were redone in the Euclidean approach, and many additional models were constructed in that manner.[31] Although the program of constructing the standard models of quantum field theory in four spacetime dimensions is not completed, the lessons taught by the constructions of interacting field models in lower dimensions cannot be overlooked. In particular, quantum fields with different interactions are associated with mutually inequivalent representations of the CCR, which are in turn inequivalent to the Fock representation. The choice of representation would thus appear to be quite significant. Tersely summarized, one could say that the kinematics of the physical system fixes the CCR-algebra and the dynamics determines the representation.[32]

Though the original problem was stated in terms of unitary equivalence, there are, in fact, weaker notions of equivalence which are also of physical relevance. A notion of physical equivalence introduced by Haag and D. Kastler [26] will be discussed next.

Since one can carry out only finitely many experiments which themselves have only a finite accuracy, the experimental situation strictly limits our ability to test the many idealizations which are implicit in any physical theory and which are particularly strongly present in quantum mechanics and quantum field theory. These limitations on measurement and the statistical interpretation of the basic objects in the theory induce a natural topology on the set of states on the algebra of observables $\mathcal{A}$. Let $\{A_i\}_{i=1}^n \subset \mathcal{A}$ be a set of observables of a system which has been prepared in the state ω. Let $\{a_i\}_{i=1}^n$ be their measured average values to within the (respective) errors $\{\epsilon_i\}_{i=1}^n$. Hence one has the n inequalities

$$|\omega(A_i) - a_i| < \epsilon_i , \ i = 1, \ldots, n .$$

On the other hand, recall that the $\sigma(\mathcal{A}^*, \mathcal{A})$-topology on the set $\mathcal{A}^*$ of all continuous, linear complex-valued functions on $\mathcal{A}$ is generated by the seminorms $N_A(\omega) \equiv |\omega(A)|$, for each $A \in \mathcal{A}$. In other words, the $\sigma(\mathcal{A}^*, \mathcal{A})$-topology is the locally convex topology with basis of neighborhoods at the origin given by

$$\{\mathcal{N}_{\{A_i\}_{i=1}^n, \{\epsilon_i\}_{i=1}^n} \mid n \in \mathbf{N}, \{A_i\}_{i=1}^n \subset \mathcal{A}, \{\epsilon_i\}_{i=1}^n \subset (0, \infty)\} ,$$

where

$$\mathcal{N}_{\{A_i\}_{i=1}^n, \{\epsilon_i\}_{i=1}^n} = \{\omega \in \mathcal{A}^* \mid |\omega(A_i)| < \epsilon_i, i = 1, \ldots, n\} .$$

Thus we see that any experiment (or set of experiments, necessarily finite) determines the state of the system only up to a neighborhood in the $\sigma(\mathcal{A}^*, \mathcal{A})$-topology.

For purely mathematical reasons, J.M.G. Fell introduced the following notion of equivalence of representations. If $(\rho, \mathcal{H})$ is a representation of $\mathcal{A}$, then its kernel is given by $\mathrm{Ker}(\rho) = \{A \in \mathcal{A} \mid \rho(A) = 0\}$.

Definition([14]) *Two representations $(\rho_1, \mathcal{H}_1)$ and $(\rho_2, \mathcal{H}_2)$ of $\mathcal{A}$ are said to be weakly equivalent if $\mathrm{Ker}(\rho_1) = \mathrm{Ker}(\rho_2)$.*

Unitary equivalence implies weak equivalence, but the converse is false. Fell showed that two representations $(\rho_1, \mathcal{H}_1)$ and $(\rho_2, \mathcal{H}_2)$ of $\mathcal{A}$ are weakly equivalent if and only if given every state ω_1 on $\mathcal{A}$ determined by a density matrix on $\mathcal{H}_1$ and given any $\sigma(\mathcal{A}^*, \mathcal{A})$-neighborhood $\mathcal{N}$ of ω_1, there exists a state $\omega_2 \in \mathcal{N}$ determined by a density matrix on $\mathcal{H}_2$. In other words, the $\sigma(\mathcal{A}^*, \mathcal{A})$-closure of the folium of states associated with $(\rho_1, \mathcal{H}_1)$ coincides with the $\sigma(\mathcal{A}^*, \mathcal{A})$-closure of the folium associated with $(\rho_2, \mathcal{H}_2)$.

Therefore, if the kernels of two representations of $\mathcal{A}$ coincide, then it is physically impossible to determine which representation one is in (and conversely)! But if $(\rho, \mathcal{H})$ is a representation of $\mathcal{A}$, then $\mathrm{Ker}(\rho)$ is a norm-closed two-sided ideal of $\mathcal{A}$. Thus it follows that whenever $\mathcal{A}$ is simple, every representation of $\mathcal{A}$ must be faithful, and hence all representations of a simple algebra $\mathcal{A}$ are physically equivalent. What is more, in quantum field theory the quasilocal algebras $\mathcal{A}$ are typically simple!

So, have we returned to the physicists' original point of view – the choice of representation is just a matter of convenience, even in systems with infinitely many degrees of freedom? Not exactly! Let us posit, once again, that we have chosen observables $\{A_i\}_{i=1}^n \subset \mathcal{A}$ and made measurements with results $\{a_i\}_{i=1}^n$ to within errors $\{\epsilon_i\}_{i=1}^n$, thereby determining a $\sigma(\mathcal{A}^*, \mathcal{A})$-neighborhood $\mathcal{N}$ of the actual state ω, normal in the true physical representation $(\rho, \mathcal{H})$. If $\mathcal{A}$ is simple, then Fell's theorem entails that we can find a state $\rho_\mathcal{N}$, normal in any other fixed representation of $\mathcal{A}$, which is contained in $\mathcal{N}$ and therefore yields predictions conforming with the results of this experiment. But the moment we improve the experiment, *i.e.* reduce the errors, or we change the experiment to include another set of observables (but still preparing the system in the same original state), then the neighborhood $\mathcal{N}$ changes (though ω does not change), and we must find *another* approximate state in the given "wrong" representation to reproduce the results of the new experiment. In other words, in order to have (correct) predictive power beyond the particular experiments to which one fitted the approximate state, one *needs* the correct state in the correct representation. This, surely, is not merely a matter of convenience![33]

J. Manuceau's [42] and J. Slawny's [65] observation that the minimal C^*-algebra $\mathcal{A}_0$ associated with the CCR is simple and hence all representations of $\mathcal{A}_0$ are isomorphic can be seen as the *correct* generalization of Theorem 1 to infinitely many degrees of freedom. The existence of this algebraic isomorphism implies unitary equivalence in the finite case but *not* in the infinite case. This independence of representation enables the rigorous study of the canonical transformations commonly employed in theoretical physics (by von Neumann himself – see further above) as automorphisms on the C^*-algebra $\mathcal{A}_0$. In a given representation of $\mathcal{A}_0$, the given automorphism may or may not be unitarily implementable. If it is, then the original Hamiltonian operator will have the same spectrum as the transformed (diagonalized) one; if not, then there need be no relation between the spectra of these operators.

To close this section, we remark that the significance of the Stone-von Neumann uniqueness theorem is further emphasized by the fact that for the other important types of algebraic relations – such as the canonical anticommutation relations and, more recently, supersymmetric commutation relations, p-adic commutation relations, and the deformed commutation relations of quantum groups – one of the first questions addressed is the validity of the counterpart of Theorem 1 in the given setting. For further reading, we mention the papers [58][37][23][32].

3. GENERALIZATIONS TO THE HARMONIC ANALYSIS OF LOCALLY COMPACT GROUPS

In 1949, G.W. Mackey [39] provided a generalization of the Stone-von Neumann uniqueness theorem to the setting of locally compact groups, which itself found many applications in mathematics and elsewhere and which may justifiably be seen as yet another impact of von Neumann's work. For simplicity, we shall restrict our attention to Abelian groups, though Mackey formulated and proved an analogous result for arbitrary locally compact groups. With G a locally compact Abelian group and μ a Haar measure on G, one can naturally define the Hilbert space $L^2(G, d\mu)$. If G^* is the topological character group of G, i.e. each $\tau \in G^*$ is a continuous homomorphism from G into the multiplicative group of complex numbers of modulus 1, and G^* is endowed with the natural induced topology (so that it, too, becomes a locally compact Abelian group), then the analogue of the Weyl form of the Schrödinger representation is described by the following unitary operators on $L^2(G, d\mu)$:

$$(U_S(g)\Psi)(x) = \Psi(g^{-1}x) \text{ and } (V_S(\tau)\Psi)(x) = \tau(x)\Psi(x) , \tag{6}$$

for any $\Psi \in L^2(G, d\mu)$. (Compare with (3).)

Theorem 4 ([39]) *Let G be an arbitrary separable[34] locally compact Abelian group, and let G^* be its topological character group. Let U be a weakly continuous representation of G in the (separable) Hilbert space $\mathcal{H}$. If V is a weakly continuous representation of G^* on $\mathcal{H}$ such that $U(g)V(\tau) = \tau(g)V(\tau)U(g)$, for all $g \in G$ and $\tau \in G^*$, then $\mathcal{H}$ decomposes into a direct sum of at most countably many closed subspaces $\mathcal{H}_n$, each invariant under $\{U(g) \mid g \in G\} \cup \{V(\tau) \mid \tau \in G^*\}$. Moreover, letting U_n, resp. V_n, denote the restriction of U, resp. V, to $\mathcal{H}_n$, there exists a Hilbert space isomorphism $W_n : \mathcal{H}_n \to L^2(G, d\mu)$ with*

$$W_n U_n(g) W_n^{-1} = U_S(g) \text{ and } W_n V_n(\tau) W_n^{-1} = V_S(\tau) ,$$

for all $g \in G$ and $\tau \in G^*$.

Subsequently, arguments which were more elementary than Mackey's original proof were found, as well as some additional reformulations - see, e.g. [64][54]. Theorem 4 is often called the Stone-von Neumann-Mackey theorem.[35] It has been placed by Mackey into the context of his theory of induced representations and there was seen to be a consequence of his imprimitivity theorem. The interested reader is referred to [41] for an introduction to this circle of ideas.

Theorem 1 is obtained as a special case of Theorem 4 by choosing G to be the additive group of reals (for more than one degree of freedom, G is chosen to be the additive group of vectors $\mathbf{R}^n$). Note that, in that case, G^* is isomorphic to G itself.

From Theorem 4 follows one of the most useful theorems in Abelian harmonic analysis, which is in turn a generalization of the crucial Plancherel theorem in Fourier analysis (a special instance of Abelian harmonic analysis). (See [64] for a proof.)

Theorem 5 *Let G be a locally compact Abelian group. Given any element $f \in L^1(G, d\mu) \cap L^2(G, d\mu)$, its Fourier transform $\hat{f}$, defined by*

$$\hat{f}(\tau) = \int \tau(g) f(g) d\mu(g) ,$$

is in $L^2(G^*, d\mu^*)$, and the mapping $f \mapsto \hat{f}$ extends uniquely to a Hilbert space isomorphism from $L^2(G, d\mu)$ onto $L^2(G^*, d\mu^*)$ (with suitable normalization of the Haar measure μ^*).

From this then follows the generalized Riemann-Lebesgue lemma: the Fourier transform of an integrable function on a locally compact Abelian group G vanishes at infinity on G^*. It is surely evident by now how central a result the Stone-von Neumann-Mackey theorem is in Abelian harmonic analysis.

To take yet another perspective on this topic, consider the n-dimensional Heisenberg group, which is the universal covering group of the non-Abelian group of unitary operators on $L^2(\mathbf{R}^n)$ generated by the translations

$$T_p f(x) = f(x + p) , \quad p \in \mathbf{R}^n ,$$

and the multiplications

$$M_q f(x) = e^{iq \cdot x} f(x) , \quad q \in \mathbf{R}^n .$$

It is evident from the discussion in the introduction that the Stone-von Neumann uniqueness theorem may be used to classify the irreducible representations of the Heisenberg group. This again permits the proof of a corresponding Plancherel theorem, *etc.*. We refer the reader to [69] for a development of this theory, as well as indications of the many sorts of applications which have arisen. Here we only mention one buzzword: wavelets.

Finally, just to hint at further realms, we mention that Theorem 1 has also found applications to number theory (see, for example, [6]), function theory (see [59]) and invariant subspace theory(see [29]).

Acknowledgements: We have found [12] and [72] particularly useful during the preparation of this paper and further recommend them to the reader's attention.

NOTES

1. Significant portions of [2] were obtained independently by P.A.M. Dirac [9].
2. In the winter term of 1926-27, Hilbert gave lectures on the new quantum mechanics which were prepared in collaboration with his assistants, L. Nordheim and J. von Neumann. The notes of the lectures were written out by Nordheim, with von Neumann's assistance, and were published in [30].
3. Of course, the founders of quantum theory did not have these later results, but they had realized that all of their examples were unbounded.
4. with an indication of the elements of a proof.
5. As shown by von Neumann [47], this continuity assumption may be replaced by mere weak Lebesgue-measurability as long as the Hilbert space is separable.
6. Note that if the weakly continuous Weyl representation $\{\tilde{U}, \tilde{V}, \mathcal{H}\}$ is irreducible, then $\mathcal{H}$ must be separable.
7. For Hilbert spaces, an isomorphism is a one-to-one linear norm-preserving transformation from one Hilbert space onto the other.
8. If $\mathcal{H}$ is not separable and $\tilde{U}, \tilde{V}$ are not weakly continuous, then the conclusion of this portion of the theorem is false [13].
9. Many authors refer to it simply as the von Neumann uniqueness theorem.
10. There may well be other such physically motivated examples in the literature; we apologize in advance for not being aware of them.
11. or Fock-Cook, among mathematicians.
12. Nun ist es formal unbequem und bedenklich, Systeme mit unendlich vielen Freiheitsgraden bzw. Wellenfunktionen mit unendlich vielen Argumenten zuzulassen. – See page 141, resp. page 265, of [46].
13. in mathematical terms, a symplectic transformation.
14. We return to this point below. As the editors have kindly pointed out to the author, in an unpublished manuscript [49] based on his 1937 seminars in the Institute for Advanced Study, von Neumann acknowledges that very essential difficulties arise when treating systems having an infinite number of degrees of freedom and then, more significantly, makes an attempt to handle these difficulties.
15. *i.e.* the set of states determined by the density matrices on the Hilbert space of the given representation.
16. It is of interest to note that van Hove perceived a connection between his model and von Neumann's infinite product spaces [48]. With the benefit of hindsight, we see that he was anticipating the theory of infinite product representations of the CCR [35].
17. See p. 3 of [17].
18. It would seem that Friedrichs was not aware of either van Hove's example nor von Neumann's paper [48] when he did this work; they were mentioned only in the Comments and Corrections at the very end of the book [17], added after the work had been completed.
19. They also made use of von Neumann's paper on infinite products.
20. As straightforward an operation as multiplying all the P's by 2 and all the Q's by $\frac{1}{2}$ produces a representation of the CCR which is unitarily inequivalent to the initial representation.
21. and therefore irreducible, local and Poincaré-covariant.
22. and yet still cite [24] for other purposes!
23. See [4] for further references and [56] for a recent paper on those representations which have a "generalized number operator".
24. We refer the interested reader to [67][25] and also, for a historical overview, [72].
25. As is customary in quantum field theory, we adopt physical units in which $c = h/2\pi = 1$.
26. so-called because they were introduced in [43].

27. As opposed to the case of the canonical anticommutation relations, the C^*-algebra obtained here depends upon the choice of the dense subspace $\mathcal{S}(\mathbb{R})$ of test functions – see [4] for a detailed discussion of this point. However, once the dense subspace of test functions has been fixed, the closure $\mathcal{A}_0$ in the operator norm of the algebra generated by the set $\{e^{i\phi(f)}, e^{i\pi(f)} \mid f \in \mathcal{S}(\mathbb{R})\}$ has the property that to each, not necessarily continuous representation of the CCR, (4)-(5), there corresponds a representation of $\mathcal{A}_0$ [65]. Moreover, $\mathcal{A}_0$ is simple [42][65], so it is representation-independent – see the discussion further below.

28. Without the cutoff g, the interacting Hamilton operator is *not* densely defined in Fock space.

29. One can then show that the algebra $\mathcal{A}(\mathcal{O})$ coincides with the von Neumann algebra generated by bounded functions of the self-adjoint field operators $\int \phi(x,t) f(x,t) dx dt$, with test functions $f(x,t)$ having support in $\mathcal{O}$.

30. Of relevance to quantum field theory, but particularly to quantum statistical mechanics, which also must face systems with infinitely many degrees of freedom [4], is the observation that also equilibrium states at different temperatures are associated with mutually inequivalent representations [68]. The Fock representation is associated with temperature zero.

31. For an introduction to this work, as well as further references, see [21].

32. In fact, in a certain sense, the representation also determines the dynamics – see [1].

33. It is evident from this discussion that it is impossible to prove experimentally that a putative exact state is the correct one (and, thus, that the correct representation has been chosen). But at least it is logically possible to establish experimentally that it is *not* the correct one (if, in fact, it is not).

34. Loomis [38] later showed that the assumption of separability of G could be dropped.

35. It may be of interest to note that Theorem 4 was evoked in J. Slawny's proof [65] of the existence and properties of the minimal C^*-algebra associated with the CCR, which was mentioned in the previous section.

REFERENCES

[1] H. Araki, "Hamiltonian formalism and the canonical commutation relations in quantum field theory", *J. Math. Phys.*, **1**, 492–504 (1960).

[2] M. Born and P. Jordan, "Zur Quantenmechanik", *Zeitschr. f. Phys.*, **34**, 858–888 (1925).

[3] M. Born, W. Heisenberg and P. Jordan, "Zur Quantenmechanik II", *Zeitschr. f. Phys.*, **35**, 557–615 (1925).

[4] O. Bratteli and D.W. Robinson, *Operator Algebras and Quantum Statistical Mechanics*, Volume 1 (Springer Verlag, Berlin) 1979, Volume 2 (Springer Verlag, Berlin) 1981.

[5] J.M. Chaiken, "Finite-particle representations and states of the canonical commutation relations", *Ann. Phys.*, **42**, 23–80 (1967).

[6] P. Cartier, "Quantum mechanical commutation relations and theta functions", in: *Algebraic Groups and Discontinuous Subgroups*, edited by A. Borel and G.D. Mostow (American Mathematical Society, Providence, RI) 1966, pp. 361–383.

[7] S. Cavallaro, G. Morchio and F. Strocchi, "A Generalization of the Stone-von Neumann theorem to non-regular representations of the CCR-algebra", preprint (mp-arc 98-498).

[8] J.M. Cook, "The Mathematics of second quantization", *Trans. Amer. Math. Soc.*, **74**, 222–245 (1953).

[9] P.A.M. Dirac, "The Fundamental equations of quantum mechanics", *Proc. Roy. Soc, London*, **A109**, 642–653 (1925).

[10] P.A.M. Dirac, "The Quantum theory of the emission and absorption of radiation", *Proc. Roy. Soc. London*, **114**, 243–265 (1927).

[11] J. Dixmier, "Sur la relation $i(PQ - QP) = 1$", *Comp. Math.*, **13**, 263–270 (1958).

[12] G.G. Emch, *Algebraic Methods in Statistical Mechanics and Quantum Field Theory* (New York, John Wiley & Sons) 1972.

[13] G.G. Emch, "Von Neumann's uniqueness theorem revisited", in: *Mathematical Analysis and Applications*, Part A, edited by L. Nachbin (Academic Press, New York) 1981, pp. 361–368.

[14] J.M.G. Fell, "The Dual spaces of C^*-algebras", *Trans. Amer. Math. Soc.*, **94**, 365–403 (1960).

[15] M. Florig and S.J. Summers, "Further representations of the canonical commutation relations", *Proc. London Math. Soc.*, **80**, 451–490 (2000).

[16] V. Fock, "Konfigurationsraum und zweite Quantelung", *Zeitschr. f. Phys.*, **75**, 622–647 (1932).

[17] K.O. Friedrichs, *Mathematical Aspects of the Quantum Theory of Fields* (Interscience Publishers, New York) 1953; see also *Commun. Pure Appl. Math.*, **4**, 161–224 (1951), *ibid.*, **5**, 1–56 (1952), *ibid.*, **6**, 1–72 (1953).

[18] J. Fröhlich, "Verification of axioms for Euclidean and relativistic fields and Haag's theorem in a class of $P(\phi)_2$ models", *Ann. Inst. Henri Poincaré*, **21**, 271–317 (1974).

[19] J. Glimm, "Type I C^*-algebras", *Ann. Math.*, **73**, 572–612 (1961).

[20] J. Glimm and A. Jaffe, "A $\lambda\phi^4$ quantum field theory without cutoffs, I", *Phys. Rev.*, **176**, 1945–1951 (1968), II, *Ann. Math.*, **91**, 362–401 (1970), III, *Acta Math.*, **125**, 203–267 (1970), IV, *J. Math. Phys.*, **13**, 1568–1584 (1972).

[21] J. Glimm and A. Jaffe, *Quantum Physics*, (Springer Verlag, Berlin) 1981; Second, expanded edition appeared in 1987.

[22] L. Gårding and A.S. Wightman, "Representations of the commutation relations", *Proc. Nat. Acad. Sci.*, **40**, 622–626 (1954).

[23] H. Grosse and L. Pittner, "A Supersymmetric generalization of von Neumann's theorem", *J. Math. Phys.*, **29**, 110–118 (1988).

[24] R. Haag, "On quantum field theories", *Kong. Dan. Vidensk. Sels., Mat. Fys. Medd.*, **29**, no. 12 (1955).

[25] R. Haag, *Local Quantum Physics*, (Springer Verlag, Berlin) 1992; Second, expanded edition appeared in 1996.

[26] R. Haag and D. Kastler, "An algebraic approach to quantum field theory", *J. Math. Phys.*, **5**, 848–861 (1964).

[27] W. Heisenberg, "Über quantentheoretische Umdeutung kinematischer und mechanischer Beziehungen", *Zeitschr. f. Phys.*, **33**, 879–893 (1925).

[28] W. Heisenberg and W. Pauli, "Zur Quantendynamik der Wellenfelder", *Zeitschr. f. Phys.*, **56**, 1-61 (1929), II, *ibid.*, **59**, 168–190 (1929).

[29] H. Helson, *Lectures on Invariant Subspaces* (Academic Press, New York) 1964.

[30] D. Hilbert, "J. von Neumann and L. Nordheim, Über die Grundlagen der Quantenmechanik", *Math. Ann.*, **98**, 1–30 (1927).

[31] L. van Hove, "Les difficultés de divergence pour un modèle particulier de champ quantifié", *Physica*, **18**, 145–159 (1952).

[32] A. Iorio and G. Vitiello, "Quantum groups and von Neumann theorem", *Modern Phys. Lett. B*, **8**, 269–276 (1994).

[33] P. Jordan, "Über kanonische Transformationen in der Quantenmechanik", *Zeitschr. f. Phys.*, **37**, 383–386 (1926).

[34] P.E.T. Jørgensen and R.T. Moore, *Operator Commutation Relations* (D. Reidel, Dordrecht) 1984.

[35] J.R. Klauder, J. McKenna and E.J. Woods, "Direct-product representations of the canonical commutation relations", *J. Math. Phys.*, **7**, 822–828 (1966).

[36] J.R. Klauder and B.-S. Skagerstam, editors, *Coherent States: Applications to Physics and Mathematical Physics* (World Scientific, Singapore) 1985.

[37] A.N. Kochubei, "P-adic commutation relations", *J. Phys. A*, **29**, 6375–6378 (1996).

[38] L.H. Loomis, "Note on a theorem of Mackey", *Duke Math. J.*, **19**, 641–645 (1952).

[39] G.W. Mackey, "A Theorem of Stone and von Neumann", *Duke Math. J.*, **16**, 313–326 (1949).

[40] G.W. Mackey, "Borel structures in groups and their duals", *Trans. Amer. Math. Soc.*, **85**, 134–165 (1957).

[41] G.W. Mackey, *Induced Representations of Groups and Quantum Mechanics* (W.A. Benjamin, Inc., New York) 1968.

[42] J. Manuceau, "C^*-algébre de relations de commutation", *Ann. Inst. Henri Poincaré*, **8**, 139–161 (1968).

[43] F.J. Murray and J. von Neumann, "On rings of operators", *Ann. Math.*, **37**, 116–229 (1936); or in: J. von Neumann, *Collected Works*, Volume 3, edited by A.H. Taub (Pergamon Press, New York and Oxford) 1961, pp. 6–119.

[44] J. von Neumann, "Mathematische Begründung der Quantenmechanik", *Nachr. Akad. Wiss. Göttingen*, **1**, 1–57 (1927); or in: J. von Neumann, *Collected Works*, Volume 1, edited by A.H. Taub (Pergamon Press, New York and Oxford) 1961, pp. 151–207.

[45] J. von Neumann, "Die Eindeutigkeit der Schrödingerschen Operatoren", *Math. Ann.*, **104**, 570–578 (1931); or in: J. von Neumann, *Collected Works*, Volume 2, edited by A.H. Taub (Pergamon Press, New York and Oxford) 1961, pp. 221–229.

[46] J. von Neumann, *Mathematische Grundlagen der Quantenmechanik* (Springer Verlag, Berlin) 1932; English translation: *Mathematical Foundations of Quantum Mechanics* (Princeton University Press, Princeton) 1955.

[47] J. von Neumann, "Über einen Satz von Herrn M.H. Stone", *Ann. Math.*, **33**, 567–573 (1932); or in: J. von Neumann, *Collected Works*, Volume 2, edited by A.H. Taub (Pergamon Press, New York and Oxford) 1961, pp. 287–293.

[48] J. von Neumann, "On infinite direct products", *Compos. Math.*, **6**, 1–77 (1938); or in: J. von Neumann, *Collected Works*, Volume 3, edited by A.H. Taub (Pergamon Press, New York and Oxford) 1961, pp. 323–399.

[49] J. von Neumann, "Quantum mechanics of infinite systems", unpublished manuscript dated 1937 and held in the J. von Neumann Archive of the Manuscript Division of the Library of Congress, Washington, D.C.; first published in this volume.

[50] M. Proksch, G. Reents and S.J. Summers, "Quadratic representations of the canonical commutation relations", *Publ. Res. Inst. Math. Sci., Kyoto Univ.*, **31**, 755–804 (1995).

[51] C.R. Putnam, *Commutation Properties of Hilbert Space Operators* (Springer Verlag, Berlin) 1967.

[52] H. Reeh, "A Remark concerning canonical commutation relations", *J. Math. Phys.*, **29**, 1535–1536 (1988).

[53] F. Rellich, "Der Eindeutigkeitssatz für die Lösungen der quantenmechanischen Vertauschungsrelationen", *Nachr. Akad. Wiss. Göttingen, Math. Phys., Kl. II*, **11A**, 107–115 (1946).

[54] M.A. Rieffel, "On the uniqueness of the Heisenberg commutation relations", *Duke Math. J.*, **39**, 745–752 (1972).

[55] D.W. Robinson, "The Ground state of the Bose gas", *Commun. Math. Phys.*, **1**, 159–171 (1965).

[56] R. Schaflitzel, "Some particle representations of the canonical commutation relations", *Rep. Math. Phys.*, **25**, 329–344 (1988).

[57] K. Schmüdgen, "On the Heisenberg commutation relations", II, *Publ. Res. Inst. Math. Sci., Kyoto Univ.*, **19**, 601–671 (1983).

[58] K. Schmüdgen, "Operator representations of $\mathbb{R}_q^2$", *Publ. Res. Inst. Math. Sci., Kyoto Univ.*, **28**, 1029–1061 (1992).

[59] K. Schmüdgen, "An Operator-theoretic approach to a cocycle problem in the complex plane", *Bull. London Math. Soc.*, **27**, 341–346 (1995).

[60] R. Schrader, "Local operator products and field equations in $P(\phi)_2$ theories", *Fortschr. d. Phys.*, **22**, 611–631 (1974).

[61] E. Schrödinger, "Quantisierung als Eigenwertproblem, I", *Ann. Phys.*, **79**, 361–376 (1926), II, *ibid.*, **79**, 489–527, III, *ibid.*, **79**, 734–756, IV, *ibid.*, **80**, 109–139; reprinted in: E. Schrödinger, *Abhandlungen zur Wellenmechanik* (J.A. Barth, Leipzig) 1928; English translation: *Collected Papers on Wave Mechanics* (Blackie and Sons, London and Glasgow) 1928.

[62] S.S. Schweber, *An Introduction to Relativistic Quantum Field Theory* (Row, Peterson and Co., Evanston, IL) 1961.

[63] I.E. Segal, *Mathematical Problems of Relativistic Physics* (American Mathematical Society, Providence, R.I.) 1963.

[64] I.E. Segal and R.A. Kunze, *Integrals and Operators* (McGraw-Hill, New York) 1968; Second, enlarged edition published by Springer Verlag in 1978.

[65] J. Slawny, "On factor representations and the C^*-algebra of canonical commutation relations", *Commun. Math. Phys.*, **24**, 151–170 (1972).

[66] M.H. Stone, "Linear transformations in Hilbert space, III: Operational methods and group theory", *Proc. Nat. Acad. Sci. U.S.A.*, **16**, 172–175 (1930).

[67] R.F. Streater and A.S. Wightman, *PCT, Spin and Statistics, and All That* (Benjamin/Cummings Publ. Co., Reading, Mass.), 1964; Second, expanded edition published in 1978.

[68] M. Takesaki, "Disjointness of the KMS states of different temperatures", *Commun. Math. Phys.*, **17**, 33–41 (1970).

[69] M.E. Taylor, *Noncommutative Harmonic Analysis* (American Mathematical Society, Providence, R.I.) 1986.

[70] H. Weyl, *Zeitschr. f. Phys.*, **46**, 1–46 (1927); see also H. Weyl, *Gruppentheorie und Quantenmechanik* (Hirzel, Leipzig) 1928; English translation: *The Theory of Groups and Quantum Mechanics* (Dover, New York) 1949.

[71] H. Wielandt, "Über die Unbeschränktheit der Schrödingerschen Operatoren der Quantenmechanik", *Math. Ann.*, **121**, 21 (1949).

[72] A.S. Wightman, "Hilbert's sixth problem: Mathematical treatment of the axioms of physics", in: *Mathematical Developments Arising From Hilbert Problems*, edited by F. Browder, (American Mathematical Society, Providence, R.I.) 1976, pp. 147–240.

[73] A.S. Wightman and S.S. Schweber, "Configuration space methods in relativistic quantum field theory, I", *Phys. Rev.*, **98**, 812–837 (1955).

[74] A. Wintner, "The Unboundedness of quantum-mechanical matrices", *Phys., Rev.*, **71**, 738–739 (1947).

Department of Mathematics
University of Florida
Gainesville, Florida 32611
U.S.A.

MIKLÓS RÉDEI

VON NEUMANN'S CONCEPT OF QUANTUM LOGIC AND QUANTUM PROBABILITY*

1. INTRODUCTORY REMARKS

The idea of quantum logic first appears explicitly in the short Section 5 of Chapter III. in von Neumann's 1932 book on the mathematical foundations of quantum mechanics [31]; however, the real birthplace of quantum logic is commonly identified with the 1936 seminal paper co-authored by G. Birkhoff and J. von Neumann [5]. The aim of this review is to recall the main idea of the Birkhoff-von Neumann concept[1] of quantum logic as this was put forward in the 1936 paper. The review is motivated partly by two facts related to quantum logic: one, peculiar, is that the 1936 von Neumann concept is an almost totally neglected[2] topic in the enormous quantum logic literature [17]; the other, not very well-known, is that von Neumann was never completely satisfied with how he had worked out quantum logic.

The lack of papers analyzing von Neumann's concept is especially surprising in view of the fact that von Neumann's concept of quantum logic, as we shall see, is subtly but markedly different from what became later – and still is – the standard view. This raises the question of why von Neumann's interpretation has not received more attention from quantum logicians. There are probably more than one reason for this negligence, and it is not the aim of this review to find out all the likely ones. We are content with making the following remarks, which also specify the topic of the present review more precisely.

It is difficult – if not impossible – to understand von Neumann's concept of quantum logic exclusively on the basis of the 1936 Birkhoff-von Neumann paper: While proposing quantum logic in 1935-1936, von Neumann was simultaneously working on the theory of "rings of operators" (called von Neumann algebras in today's terminology), and in the year of the publication of the Birkhoff-von Neumann paper on quantum logic von Neumann also published a joint paper with J. Murray, a work that established the classification theory of von Neumann algebras [16]. We shall see that the results of this classification theory are intimately related to von Neumann's concept of quantum logic. To understand more fully some apparently counterintuitive features of von Neumann's idea of quantum logic, one has to take into account other, earlier results and ideas of von Neumann as well, however. In the second [29] of the three "foundational papers" [28]-[30] von Neumann worked out a derivation of the quantum mechanical probability calculus under the frequency interpretation of probability. That derivation – reproduced with appar-

153

M. Rédei and M. Stöltzner (eds.),
John von Neumann and the Foundations of Quantum Physics, 153–172.
© 2001 *Kluwer Academic Publishers. Printed in the Netherlands.*

ently small but revealing modifications in Chapter IV of his 1932 book [31] – was very problematic: it contained inconsistencies, of which von Neumann was more or less aware. The inconsistency led him to taking a critical attitude towards the Hilbert space formalism of quantum mechanics (see [24] for an analysis of von Neumann's critique) and to the hope that the mathematical formalism accomodating the quantum logic he proposed will also serve as a more satisfactory framework for quantum mechanics than the Hilbert space formalism.

Thus the 1936 von Neumann concept of quantum logic is related to deep mathematical discoveries in the mid thirties, to the history of quantum mechanics in the twenties, and to difficulties in connection with the frequency interpretation of quantum probability. So the issue is a convoluted one, and the present paper is thus rather an attempt to reconstruct the von Neumann concept than to just review it. The complexity of the problem is also reflected by the fact that, as we shall argue by citing evidence, von Neumann himself was never quite satisfied with how he had worked out quantum logic. In this review we also try to explain why. The essential point we make is that von Neumann wanted to interpret the algebraic structure representing quantum logic as the algebra of random events in the sense of a non-commutative probability theory. In a well-defined sense to be explained in this paper, this cannot be achieved if probabilities are to be viewed as relative frequencies. We claim that this was the main reason why von Neumann abandoned the frequency interpretation of quantum probability after 1936 in favor of a "logical interpretation" of probability, which von Neumann did not regard as very well developed and understood, however.

The organization of the paper is the following. In section 2 we recall the standard concept of quantum logic. Section 3 contains the reconstruction of the von Neumann concept. This section is divided into several subsections. Subsection 3.1 clarifies the relation of the modularity property of quantum logic to the existence of a well behaving (a priori) probability measure on quantum logic. Subsection 3.2 points out the invariance property of what von Neumann viewed as a priori probability on quantum logic, understood as the modular lattice of projections of a type II_1 factor. By recalling von Neumann's 1927-32 interpretation of quantum probability, subsection 3.3 shows why existence of an a priori probability in quantum mechanics was crucial for von Neumann's 1927-32 interpretation of quantum mechanics. Subsection 3.4 points out the inconsistencies in von Neumann's 1927-32 interpretation of quantum probability as modeled by Hilbert space. Subsection 3.6 analyses the importance of the "strong additivity" property of a probability measure from the point of view of the frequency interpretation of probability, and this subsection also shows how and why von Neumann abandoned the relative frequency view of quantum probability after 1936. Finally, Section 4 argues that von Neumann did not consider quantum logic a well-understood and sufficiently developed theory.

2. QUANTUM LOGIC - THE STANDARD VIEW

The main idea of quantum logic in the standard interpretation is very simple: Let Q be a selfadjoint operator (defined on the Hilbert space $\mathcal{H}$) representing an observable physical quantity, and let P^Q be its spectral measure. According to the standard interpretation of quantum mechanics the probability that Q takes its value in the Borel set d is equal to one if the system is prepared in a state vector ψ that lies in the spectral subspace $P^Q(d)$, a closed, linear subspace in $\mathcal{H}$. One can express this fact by saying that ψ *makes true the proposition* "Q has its value in d with probability one". Such a proposition will be denoted by $= Prop(Q, d)$. Identifying, as it is common in formal logic, a proposition with the set of interpretations making the proposition true, one can conclude that the set $\mathcal{P}(\mathcal{H})$ of all closed linear subspaces of $\mathcal{H}$ represent the set of quantum propositions of the form "Q has its value in the set d (with probability one)". Given two such propositions $Prop(Q, d)$, $Prop(Q', d')$ and their representatives $P^Q(d)$ and $P^{Q'}(d')$ the states $\psi \in P^Q(d) \cap P^{Q'}(d')$ make true *both* $Prop(Q, d)$ *and* $Prop(Q', d')$; so one is led to the idea of forming the proposition

$Prop(Q, d) \wedge Prop(Q', d') =$
"Q has value in d and Q' has value in d' with probability one"

and interpreting the intersection

$$P^Q(d) \cap P^{Q'}(d')$$

as the representative of $Prop(Q, d) \wedge Prop(Q', d')$.

Encouraged by this, one is tempted to say boldly that the (closed linear) subspaces of $\mathcal{H}$ not only represent single quantum propositions but they also represent the logical relations between them in the sense that the set of all (closed, linear) subspaces considered together with the set theoretical operations $\cap$ (meet), $\cup$ (union) and $\setminus$ (set theoretic complement) also represent the quantum propositional system:

$$P^Q(d) \cup P^{Q'}(d')$$

representing

$$Prop(Q, d) \vee Prop(Q', d')$$

where

$Prop(Q, d) \vee Prop(Q', d') =$
"Q has value in d or Q' has value in d' with probability one"

and $\mathcal{H} \setminus P^Q(d)$ representing the negation of $Prop(Q, d)$.

The idea, in this form, is flawed: whereas the intersection of two closed linear subspaces is again a closed linear subspace, the union of two linear subspaces and the set theoretical complement of a linear subspace are not linear subspaces. However, the closure $P^Q(d) \vee P^{Q'}(d')$ of the sum $\{x + y : x \in P^Q(d), y \in P^{Q'}(d')\}$

and the orthogonal complement $P(Q,d)^\perp$ are again closed linear subspaces; so one is led to this question: Does the set of all closed linear subspaces of $\mathcal{H}$ and its structure defined by the operations $\cap = \wedge$, $\vee$ and $\perp$ represent the logic of the quantum propositions? One can show (see [23] for details) that under a suitable *formal* definition of interpretation (in the sense of logic) $P(\mathcal{H})$ can indeed be viewed as the *analogue* of the Tarski-Lindenbaum algebra of a classical propositional logic. But only the analogue, since $P(\mathcal{H})$ is not a Boolean algebra with respect to the operations $\wedge, \vee$; $P(\mathcal{H})$ turns out to be only a lattice, an atomic, atomistic, orthomodular, complete lattice having the covering property. The complete lattice property of $P(\mathcal{H})$ means that the intersection (join) of any set of projections is a projection, and the intersection (join) is the largest (smallest) (with respect to the partial relation $\leq$ that is identical with the set theoretical containment) projection which is smaller (larger) than any projection in the intersection (join). The atomicity means: There exist smallest (with respect to $\leq$) non-zero projections in $P(\mathcal{H})$, namely the one-dimensional projections projecting onto the subspace spanned by a single element ξ in $\mathcal{H}$, and for every non-zero projection there exists an atom which is smaller. Complete atomicity means that every projection can be obtained as the join of all the atoms that are smaller. The covering property means that for every A if P is an atom such that $A \wedge P = 0$, then $A \vee P$ is the smallest projection that is larger than $A \vee P$. The orthomodularity property means that the following equation holds:

Orthomodularity:

$$\text{If } A \leq B \text{ and } A^\perp \leq C, \text{ then } A \vee (B \wedge C) = (A \vee B) \wedge (A \vee C) \qquad (1)$$

Orthomodularity is a weakening of the following *distributivity* law (which is *not* valid in $P(\mathcal{H})$):

Distributivity:

$$A \vee (B \wedge C) = (A \vee B) \wedge (A \vee C) \quad \text{for all } A, B, C \qquad (2)$$

But the orthomodularity property is not the finest weakening of distributivity: The *modularity* property

Modularity:

$$\text{If } A \leq B, \text{ then } A \vee (B \wedge C) = (A \vee B) \wedge (A \vee C) \qquad (3)$$

is stronger than orthomodularity. It is not difficult to prove (see eg. [23]) that $P(\mathcal{H})$ is modular if and only if $\mathcal{H}$ is *finite* dimensional as a linear space.

To sum up, quantum logic in the standard interpretation is the orthomodular, non-modular Hilbert lattice of projections $P(\mathcal{H})$ on an infinite dimensional Hilbert space $\mathcal{H}$, and this structure is interpreted as the Tarski-Lindenbaum structure of a quantum propositional logic. More generally, "abstract quantum logic" is an abstract orthomodular lattice (see eg. [1], [21]).

The idea that certain properties of a physical system can be characterized by distilling a logic from the theory describing the system in question was developed consciously in the 1936 paper by Birkhoff and von Neumann [5]. In this paper the authors first deduced a logic from classical mechanics. Their conclusion was that the logic of a classical mechanical system is a Boolean algebra, i.e. an orthocomplemented, *distributive* lattice of certain subsets of the phase space of a classical mechanical system. Concerning a quantum system they stated

Hence we conclude that the *propositional calculus of quantum mechanics has the same structure as an abstract projective geometry.* (emphasis in the original) [5], [37, p. 115] [3]

An abstract projective geometry is an orthocomplemented *modular* lattice. We wish to emphasize that the lattice of projections of an infinite dimensional Hilbert space is *not* modular, it is only orthomodular. Von Neumann and Birkhoff were fully aware of this,[4] and also of the fact that the Hilbert space needed to describe a quantum mechanical system is typically infinite dimensional. Yet, they insisted on the modularity of quantum mechanics; hence, for them quantum logic was not the non-modular Hilbert lattice. What did then von Neumann consider as proper quantum logic, and why, in particular, did he reject the obvious candidate for quantum logic, the Hilbert lattice?

3. THE BIRKHOFF-VON NEUMANN CONCEPT OF QUANTUM LOGIC

3.1 Modularity and a priori probability

To see why von Neumann insisted on the modularity of quantum logic, one has to understand that he wanted quantum logic to be not only the propositional calculus of a quantum mechanical system but also wanted it to serve as the event structure in the sense of probability theory. In other words, what von Neumann aimed at was establishing the quantum analogue of the classical situation, where a Boolean algebra can be interpreted both as the Tarski-Lindenbaum algebra of a classical propositional logic and as the algebraic structure representing the random events of a classical probability theory, with probability being an additive normalized measure on the Boolean algebra. A characteristic property of a classical probability measure is the following "strong additivity" [5] property:

$$\mu(A) + \mu(B) = \mu(A \vee B) + \mu(A \wedge B) \qquad (4)$$

In Section 3.6 we shall return to the question of why the strong additivity (4) is a crucial property of a probability measure. Suffice it to say at this point that in 1936 von Neumann required the existence of an "a priori" probability measure on quantum logic, a probability measure which, besides having the property (4), also is faithful in the sense that every non-zero event has a finite, non-zero probability.[6] Hence, according to von Neumann, quantum logic is supposed to be a lattice $\mathcal{L}$ on which there exists a finite "a priori quantum probability" i.e. a map d having finite, non-negative values and having the following two properties:

(i) $d(A) < d(B)$ if $A < B$

(ii) $d(A) + d(B) = d(A \wedge B) + d(A \vee B)$

A non-negative map d on a lattice having the two properties (i)-(ii) is called a *dimension function*. It is easy to prove, and von Neumann had known already very well, that if a lattice $\mathcal{L}$ admits a dimension function that takes on only finite values, then $\mathcal{L}$ is modular. Since a Hilbert lattice $\mathcal{P}(\mathcal{H})$ is not modular in general, there exists no finite dimension function on $\mathcal{P}(\mathcal{H})$, i.e. there exists no a priori probability on the quantum logic determined by the Hilbert space formalism. This is a rather surprising fact, and von Neumann viewed it as a pathological property of the Hilbert space formalism. It was largely because of this pathology that von Neumann expected the Hilbert space formalism to be superseded by a mathematical theory that he hoped would be more suitable for quantum mechanics.[7]

However, there does exist exactly one (up to constant multiple) function d on the lattice of projections of a Hilbert space that is faithful and satisfies both (i) and (ii): this is the usual dimension function d, the number $d(A)$ being the linear dimension of the linear subspace A. (Equivalently: $d(A) = Tr(A)$, where Tr is the trace functional.) But this d is not finite if the Hilbert space is not finite dimensional. So one realizes that the conditions (i)-(ii) can be satisfied with a finite d, if d is the usual dimension function and $\mathcal{L}$ is the projection lattice of a *finite* dimensional Hilbert space. The assumption of a finite dimension function (a priori probability) is thus consistent with the assumption that the lattice is non-distributive; consequently, requiring the existence of a well-behaving finite a priori probability does not exclude the existence of non-classical probability structures. But the modular lattices of finite dimensional linear spaces with the discrete dimension function are certainly not sufficient as a framework for quantum theory, since one needs infinite dimensional Hilbert spaces to accomodate quantum mechanics (for instance the Heisenberg commutation relation cannot be represented on a finite dimensional Hilbert space). Birkhoff and von Neumann had known this, and they also pointed out that it would be desirable to find models of quantum logic with a non-discrete dimension function. The fate of von Neumann's idea of quantum logic as a modular, non-distributive lattice thus hinges upon whether there exist modular lattices with a finite dimension function that are not isomorphic to the modular lattice of projections of a finite dimensional linear space. However, the question of whether such modular lattices exist remains unanswered in the 1936 Birkhoff-von Neumann paper [5]: one finds only a reference to the paper by Murray and von Neumann [16], where "a continuous dimensional model" of quantum logic is claimed to be worked out.

3.2 A priori probability and invariance

The paper [16] shows that there exist non-distributive, modular lattices different from the Hilbert lattice of a finite dimensional Hilbert space. Proving the existence of such a structure is part of what is known as the "dimension theory of von Neumann algebras", which has since 1936 become a classical chapter in the theory of operator algebras. The relevant – and surprising – result of this dimension theory is that there exists a modular lattice of non-finite (linear) dimensional projections on an infinite dimensional Hilbert space, and that on this lattice there exists a (unique up to normalization) dimension function d that takes on every value in the interval $[0, 1]$. The von Neumann algebra generated by these projections is called the "type II_1 factor von Neumann algebra" $\mathcal{N}$. (For the details of the dimension theory see eg. [26] or [15], for a brief review we refer to [19], [24], [10], [6].) Furthermore, it can be shown that the unique dimension function on the lattice of projections of a type II_1 factor comes from a (unique up to constant) trace τ defined on the factor itself – just like in the finite dimensional case, where too the dimension function is a restriction of the trace functional Tr to the lattice of projections. The difference between Tr and τ is that while Tr is determined (up to constant multiple) by the requirement of unitary invariance with respect to *all* unitaries: $Tr(VAV^*) = Tr(A)$ (for all unitary operator V on $\mathcal{H}_n$) the trace τ is determined (up to constant multiple) by unitary invariance with respect to every unitary *belonging to the algebra*: $\tau(VAV^*) = \tau(A)$ for every unitary $V \in \mathcal{N}$.

According to von Neumann it is the modular lattice of a type II_1 factor von Neumann algebra that is to be viewed as the proper quantum logic, and the dimension function d (or equivalently, the τ trace) should be interpreted as the a priori probability measure on this quantum logic. In the Introduction to the first paper on the rings of operators, where Murray and von Neumann give the quantum mechanical motivation of the project of investigating the different types of von Neumann algebras, the authors discuss the significance of the type II_1 case and write

Considering the immediate applicability of $T(A)$ [$= \tau(A)$] to quantum mechanics it is the 'a priori' expectation values of the observable A, which is correctly normalized here, but cannot be in I_∞... [16] [36, p. 11]

Note that the a priori probability on a type II_1 factor is not a uniform probability measure on the set of events (i.e. on the lattice): If $\{A_i\}$ is an infinite set of pairwise disjoint events (i.e. $A_i \leq A_j^\perp$ $(i \neq j)$), then obviously there cannot exist a non-zero, additive, normalized, uniform measure p on $\{A_i\}$ (i.e. a p such that $p(A_i) = c > 0$ for all A_i), no matter whether the events form a distributive lattice or not. "A priori", therefore, acquires a new meaning: it reflects the symmetry of the system, for the following reason. The dimension function (=probability) on the lattice of a type II_1 factor comes from a trace. Now, as we have pointed out, a trace is just the *unique* (positive, linear, normalized) functional that is invariant with

respect to all unitary transformations. Since the physical symmetries of the system
are generally expressed as representations of the symmetry group on the algebra
of observables by unitaries, the existence of a unique trace means physically that
the probability is determined uniquely as the only (positive, linear) assignment
of values in $[0, 1]$ to the events that is invariant with respect to any conceivable
symmetry.

3.3 Why von Neumann wanted to have a priori probability
in quantum mechanics

But why did von Neumann insist on the requirement of existence of an a priori
probability measure on quantum logic? He wanted to have such a measure so
badly that he was ready to abandon even his brainchild, the abstract Hilbert space
formalism of quantum mechanics, just to be in the position to have one such measure at hand. Why?

To answer this question, one has to recall the essence of von Neumann's 1927
and 1932 derivation of the probabilistic part of quantum mechanics under the relative frequency interpretation of probability.

By 1927 it was clear that the probability statements in quantum mechanics have
the following general form

$$Tr(P^Q(d_1)P^R(d_2)) \tag{5}$$

The formula (5) yields the (relative, i.e. unnormalized) probability that the value
of the physical quantity represented by the selfadjoint operator Q lies in the set d_1
provided that the value of the physical quantity represented by operator R (commuting with Q) lies in the set d_2. (Recall that P^Q and P^R are the spectral measures
of the operators Q and R respectively.) Von Neumann's aim in his 1927 paper [29]
was to derive formula (5), while interpreting probability as relative frequency. In
doing this von Neumann showed first that the probability of a quantum event represented by a projection P is always of the form $Tr(UP)$ with some "statistical
operator" U, a positive, linear operator on $\mathcal{H}$ that characterizes the statistical ensemble in which to compute the relative frequency of occurrence of P. The crucial
step in the derivation of formula (5) was then to obtain via *statistical inference*
the projection $P^R(d_2)$ as a specific statistical operator: von Neumann asked what
statistical operator should we *infer* on the basis of knowing only that the value of R
lies in the interval d_2. He pointed out that this problem is underdetermined ([31, p.
338]) as long as one does not make some additional assumptions: the sought-after
probability is not uniquely determined by the value it assigns to a single event (it
assigns value 1 to the event that R takes its value in d_2). However, he argued, if we
assume that the ensemble $\mathcal{E}'$ on whose elements R takes its value in d_2 is obtained
from the a priori ensemble (=ensemble characerized by the identity operator I as
statistical operator) by selection, then the statistical operator of $\mathcal{E}$ is $P^R(d_2)$. In
other words, von Neumann viewed $P^R(d_2)$ as the conditional statistical operator

of the identity operator, where the condition is exactly the event that R takes its value in d_2. In this interpretation formula (5) gives the unnormalized conditional probability of $P^Q(d_1)$ with respect to the condition that R takes its value in d_2 and where the prior probability of $P^Q(d_1)$ is $Tr(IP^Q(d_1))$.

3.4 Conceptual problems in von Neumann's 1927-1932 interpretation of probability in Hilbert space quantum mechanics

Von Neumann's interpretation of quantum probability as relative frequency and his interpretation and derivation of the statistical formula (5) have conceptual difficulties. One difficulty is related to the very notion of ensemble and the selection of an ensemble via quantum measurements. We shall return to this problem in the next section. Here we just wish to point out the problem that the statistical operator I yielding the a priori probabilities is not normalized, i.e. its trace is infinite, as is the trace of any infinite dimensional projection; in other words, there exists no non-zero, additive, finite probability on the set $P(\mathcal{H})$ of all projections which would give the a priori, uniform probability on the set of one dimensional projections. A consequence of the a priori probabilities not being finite is that the a priori "probabilities" *cannot* in fact be considered as relative frequencies (which are always ≤ 1) in any ensemble.

One reaction to the conceptual difficulty related to the existence of infinite probabilities can be to take the position that the numbers $Tr(A)$ are not probabilities in the sense of relative frequency. Taking this position means in effect that one abandons the relative frequency interpretation of probability in favor of an interpretation which could in principle be compatible with infinite probabilities. Von Neumann does indeed abandon the frequency interpretation from 1936 on – we return to this issue in the last section. But for von Neumann this option was out of the question in the years 1927-1932: in his paper [29] von Neumann speaks of the frequency interpretation of probability as *the* (i.e. unique) theory of probability. In the twenties and early thirties his view of probability was clearly shaped under the influence of von Mises' relative frequency interpretation. Although there is no mention of von Mises' name in the three fundamental papers [28, 29, 30], von Neumann probably had known von Mises' frequency interpretation [27] already in 1926-1927, and he refers explicitly to von Mises' 1928 book on probability [14] in footnote 156 of [31]. In this footnote von Neumann identifies what he takes as the quantum mechanical ensembles with what von Mises calls 'Kollektive'.

Another option is to abandon the idea that the statistical operator I gives the a priori uniform probability. But von Neumann himself makes clear how indispensable the assumption of the a priori probabilities/weights is if one wants to be able to derive the statistical formula (5) via statistical inference: he stresses that the assumption that the prior probability is given by the trace is absolutely crucial. That we have the outcome of the measurement of R in the ensemble $\mathcal{E}'$ to lie in d_2

... can be attributed ... to the fact that originally a large ensemble $\mathcal{E}$ was given in which the measurements were carried out, and then those elements for which the desired results occurred were collected into a new ensemble. This is then $\mathcal{E}'$. Of course everything depends on how $\mathcal{E}$ is chosen. This initial ensemble gives, so to speak, the a priori probabilities of the individual states of the system S. The whole state of affairs is well-known from the general theory of probability: to be able to conclude from the results of the measurements to the states, i.e. from effect to cause, i.e. to be able to calculate a posteriori probabilities, we must know the a priori probabilities. [31, p. 338]

So von Neumann could not do without the a priori probabilities as long as he wanted to found the statistical *Ansatz* using statistical inference. Facing the clash between the necessary but infinite a priori probability and the frequency interpretation of probability, and not wanting or being able to abandon either the a priori probability or the frequency interpretation, von Neumann was left with one option only, which is a radical one: To consider the appearance of infinite, not normalizable 'a priori probabilities' as a pathology (from the point of view of probability theory) of Hilbert space quantum mechanics, and to try to work out a well-behaved non-commutative probability theory, one in which there exists (normalized) a priori probability.

3.5 Back to the type II_1 case

This is a program which goes beyond Hilbert space quantum mechanics and which von Neumann successfully established while working on the theory of "rings of operators". The classification theory of operator rings, which was obtained in 1936 as a result of joint work with Murray, resulted in the discovery of the 'type II_1' configuration, an algebraic structure in which finite a priori probabilities emerge in a canonical manner. It was only because he had known this already at the publication of his quantum logic paper in 1936 that von Neumann could propose the idea of quantum logic as a modular lattice different from the Hilbert lattice of a finite dimensional Hilbert space.

An example of a type II_1 factor is the infinite tensor product of the algebra of two-by-two matrices M_2 with itself, where τ is the product of the normalized trace functional tr on M_2. The following table shows where the non-commutative probability theory defined by a type II_1 factor fits into the system of typical probability and measure theoretic structures.

type of classical probability space	type of non-commutative probability
$\{0, 1, 2, \ldots n\}$ $p(i) = \frac{1}{n}$ uniform permutation invariant finite a priori probability	$\mathcal{B}(\mathcal{H}_n), \mathcal{P}(\mathcal{H}_n), \dim\mathcal{H} = n$ $p(1 - dim.proj) = \frac{1}{n}Tr(P) = \frac{1}{n}$ uniform unitary invariant finite a priori probability
$\{0, 1, 2, \ldots \infty\}$ $p(i) = 1$ uniform permutation invariant NOT probability (∞)	$\mathcal{B}(\mathcal{H}), \mathcal{P}(\mathcal{H}), \dim\mathcal{H} = \infty$ $p(1 - dim.proj) = Tr(P) = 1$ uniform unitary invariant NOT probability (∞)
$[0, 1]$ μ Lebesgue measure translational invariant normalized a priori probability	$\mathcal{M}, \mathcal{P}(\mathcal{M}); \mathcal{M}$ type $\mathbf{II}_1$ factor τ trace unitary invariant normalized a priori probability
$\mathbb{R}$ μ Lebesgue measure translational invariant NOT probability (∞)	$\mathcal{M}, \mathcal{P}(\mathcal{M}); \mathcal{M}$ type $\mathbf{II}_\infty$ factor no unitary invariant trace no apriori probability

3.6 Strong additivity and relative frequency interpretation

Von Neumann's insistence on the frequency interpretation of probability in the years 1927-1932 makes understandable why he considered the strong additivity (4) a key feature of probability. Assume that the probability $\mu(X)$ ($X = A, B, A \wedge B, A \vee B$) is to be interpreted as relative frequency in the following sense:

1. There exists a fix ensemble $\mathcal{E}$ consisting of N events such that

2. for each event X one can decide unambigiously and

3. without changing the ensemble whether X is the case or not;

4. $\mu(X) = \frac{\#(X)}{N}$ where $\#(X)$ is the number of events in $\mathcal{E}$ for which X is the case.[8]

Under the assumptions 1.-4. it trivially follows that (4) holds since one can write

$$\frac{\#(A \cup B)}{N} + \frac{\#(A \cap B)}{N} =$$

$$\frac{\#((A \setminus A \cap B) \cup (B \setminus A \cap B) \cup A \cap B))}{N} + \frac{\#(A \cap B)}{N} =$$

$$\frac{\#(A \setminus A \cap B) + \#(B \setminus A \cap B) + \#(A \cap B) + \#(A \cap B)}{N} =$$

$$\frac{\#(A) + \#(B)}{N}$$

which is the strong additivity. Thus, if a map d on a lattice does not have the strong additivity (4) then the probabilities $d(X)$ cannot be interpreted as probability in the sense of relative frequency formulated above via 1.-4.; consequently, the lattice cannot be viewed as representing a collection of random events in the sense of a relative frequency interpreted probability theory specified by 1.-4. (with the understanding that $A \wedge B$ denotes the joint occurrence of events A and B). It can be shown that a normal state ϕ on a von Neumann lattice satisfies the strong additivity (4) if (and only if) it is a trace [18]. Thus, the only quantum probabilites that can be interpreted as relative frequencies via 1.-4. are the ones given by the trace.

Behind the mathematical fact that only traces satisfy the strong additivity lies the conceptual difficulty that assumptions 2. and 3. of the frequency interpretation of probability cannot be upheld in interpreting the elements of a von Neumann lattice as random quantum events and the lattice operation $A \wedge B$ as the joint occurence of A and B: 3. fails if "deciding" means "measuring", since measuring disturbs the measured system, hence also the ensemble; therefore, there is no single, fixed, well-defined ensemble in which to compute as relative frequencies the probabilities of *all* projections representing quantum attributes. Von Neumann was fully aware of this difficulty: One of his arguments against hidden variables is essentially the argument that if hidden parameters did exist, then it should be possible to resolve any ensemble into subensembles that are dispersion-free, but this is not possible if "resolving" means selecting subensembles by measurement, since if one selection ensures dispersion-freeness with respect to observable Q_1, the subsequent selection by measurement of this subensemble into a further subensemble in which another observable Q_2 has sharp value destroys the result of the first step (see [31, p. 304]) and also the paper [13] by F. Laudisa and R. Giuntini in the present volume). Yet, in his 1932 book von Neumann thought to be able to maintain an ensemble interpretation of quantum probability by getting around the problem that quantum measurements disturb the ensemble:

Even if two or more quantities R, S in a single system are not simultaneously measurable, their probability distributions in a given ensemble $[S_1, \ldots S_N]$ can be obtained with arbitrary accuracy if N is sufficiently large.

Indeed, with an ensemble of N elements it suffices to carry out the statistical inspections, relative to the distribution of values of the quantity R, not on all N elements $[S_1, \ldots S_N]$, but on any subset of M ($\leq N$) elements, say $[S_1, \ldots S_M]$ – provided that M, N are both large, and that M is very small compared to N. Then only the M/N-th part of the ensemble is affected by the changes which result from the measurement. The effect is an arbitrary small one if M/N is chosen small enough – which is possible for sufficiently large N, even in the case of large M... In order to measure two (or several) quantities R, S simultaneuosly, we need two sub-ensembles, say $[S_1, \ldots S_M], [S_{M+1}, \ldots S_{2M}]$ $(2M \leq N)$, of such a type that the first is employed obtaining the statistics of R, and the second in obtaining those of S. The two measurements therefore do not disturb each other, although they are performed in the same ensemble $[S_1, \ldots S_N]$ and they can change this ensemble only by an arbitrarily small amount, if $2M/N$ is sufficiently small – which is possible for sufficiently large N even in the case of large M . . . [31, p. 300]

Implicit in this reasoning is the assumption that the subensembles are representative of the large ensemble in the sense that the relative frequency of every attribute is the same both in the original and in the subensemble. This non-trivial assumption, known in von Mises' theory as the requirement of "randomness" concerning the ensembles that can serve as ensembles to compute probabilities as frequences, is crucial in von Mises' theory, and von Mises takes pains in giving it a precise formulation (see "Forderung II" in [27], [9, p. 61]). In particular, von Mises emphasizes that not every sort of selection procedure is admissible in selecting the subensemble in which the relative frequencies are supposed to be the same as in the original one. Von Neumann does not elaborate on the details and significance for his interpretation of quantum probability of the randomness requirement; apparently he did not see any problem with taking advantage of this non-trivial (and controversial) feature of von Mises interpreation.

However, even granting that an ensemble interpretation remains meaningful if one relaxes 3. in the specification of the frequency interpretation of probability in the way von Neumann does, the problem remains for von Neumann that 2. does not make sense at all in quantum mechanics if one takes the position that (i) $A \wedge B$ represents the joint occurrence of A and B, and (ii) the joint occurrence cannot be checked by measurement at all on whatever ensemble if A and B are not simultaneously measurable. Von Neumann's position regarding the simultaneous decidability of A and B if A and B are not commuting is somewhat ambiguous in 1932: In Section III. 4. of [31] he considers A and B to be "simultaneously measurable for the states" given by vectors lying in the closed linear subspace $A \wedge B$ even if A and B are not commuting ([31, p. 230-232], also see [13], especially Section 2.); however, discussing the issue of quantum logic in Section III. 5 in [31] he allows forming $A \wedge B$ and $A \vee B$ *only if* A and B are commuting (see p. 251 and the summary of Section III. 5 in [31] on p. 253). Hence, the notion of quantum logic von Neumann suggests in Section III. 5 of his 1932 book [31] is closest to the so-called "partial Boolean algebra" interpretation of the Hilbert lattice as quantum logic.

The Birkhoff-von Neumann paper's position on the meaning(fullness) of $A \wedge B$ for non-commuting A and B is more subtle:

The set-theoretical product of any two mathematical representatives of experimental propositions concerning a quantum mechanical system, is itself the representative of an experimental proposition. [5] [37, p. 109]

But, in general, the set-theoretical pruduct $A \wedge B$ is not taken to be the representative of the experimental proposition "A and B":

It is worth remarking that in classical mechanics, one can easily define the meet or join of any two experimental propositions as an *experimental proposition* – simply by having independent observers read off the measurements which either proposition involves, and combining the results logically. This is true in quantum mechanics only exceptionally – only when all the measurements involved commute (are compatible); in general, one can only express the join or meet of two given experimental propositions as a class of logically equivalent experimental propositions – i.e. as a *physical quality*. [5] [37, p. 112]

So, the idea in the Birkhoff-von Neumann paper is this: The representative of the experimental proposition

$$P(Q, d) = Q \text{ has its value in } d \text{ (with probability one)}$$

is the spectral projection $P^Q(d)$, and, given two such (non-commuting) projections $P^{Q_1}(d_1)$ and $P^{Q_2}(d_2)$, their meet $A = P^{Q_1}(d_1) \wedge P^{Q_2}(d_2)$ is the *equivalence class* of all the experimental propositions $P(Q, d)$ such that $P^Q(d) = A$; however, Birkhoff and von Neumann did not regard the single proposition "$P(Q_1, d_1)$ *and* $P(Q_2, d_2)$" experimentally meaningful .

Thus, on the one hand, both in the 1932 and in the 1936 concept of quantum logic the $A \wedge B$ is viewed in general *not* as an experimentally meaningful proposition stating that "both A and B is the case"; on the other hand, the 1936 requirement of the existence of an a priori probability measure on quantum logic satisfying the strong additivity and interpretable in terms of relative frequencies[9] means that $A \wedge B$ *is* in fact viewed as the *joint occurrance* of A and B as events.

The incompatibility of these two viewpoints may have been the deeper[10] reason why von Neumann abandoned the frequency interpretation of probability when talking about quantum logic after 1936. In an unfinished manuscript written about 1937 and entitled "Quantum logic (strict- and probability logics)" he writes:

This view, the so-called "frequency theory of probability" has been very brilliantly upheld and expounded by R. von Mises. This view, however, is not acceptable to us, at least not in the present "logical" context. [32] [37, p. 196]

Instead, von Neumann embraces in this unfinished note a "logical theory of probability", which he associates with J. N. Keynes, but which he does not spell out in detail. Von Neumann gives a more explicit formulation of this idea in the talk he delivered at the World Congress of Mathematicians in Amsterdam in 1954[11]:

Essentially if a state of a system is given by one vector, the transition probability in another state is the inner product of the two which is the square of the cosine of the angle between them. In other words, probability corresponds precisely to introducing the angles geometrically. Furthermore, there is only one way to introduce it. The more so because in the quantum mechanical machinery the negation of a statement, so the negation of a statement which is represented by a linear set of vectors, corresponds to the orthogonal complement of this linear space. And therefore, as soon as you have introduced into the projective geometry the ordinary machinery of logics, you must have introduced the concept of orthogonality. This actually is rigorously true and any axiomatic elaboration of the subject bears it out.

So in order to have logics you need in this set a projective geometry with a concept of orthogonality in it.

In order to have probability all you need is a concept of all angles, I mean angles other than 90°. Now it is perfectly quite true that in geometry, as soon as you can define the right angle, you can define all angles. Another way to put it is that if you take the case of an orthogonal space, those mappings of this space on itself, which leave orthogonality intact, leave all angles intact, in other words, in those systems which can be used as models of the logical background for quantum theory, it is true that as soon as all the ordinary concepts of logic are fixed under some isomorphic transformation, all of probability theory is already fixed.

What I now say is not more profound than saying that the concept of a priori probability in quantum mechanics is uniquely given from the start. You can derive it by counting states and all the ambiguities which are attached to it in classical theories have disappeared. This means, however, that one has a formal mechanism, in which logics and probability theory arise simultaneously and are derived simultaneously. [34, p. 21-22; p. 244f. in this volume.]

The projection lattice of a type II_1 factor as quantum logic shows clearly what von Neumann means: The unitaries, with respect to which the trace on the type II_1 factor (probability) is invariant, define isomorphisms of the lattice i.e. of the logic of the system, and so the probability appears as being fixed once the logic of the system is fixed (up to isomorphism).

4. CLOSING REMARKS

As we have seen von Neumann's proposal for quantum logic was part of a bold suggestion to replace Hilbert space quantum mechanics by a different mathematical framework based on the theory of von Neumann algebras, in particular on the theory of type II_1 algebras. This is a *very* bold suggestion indeed, and maybe this is why it was put forward by von Neumann in a very cautious manner, in footnotes and introductions to his papers[12]. By 1936 Hilbert space quantum mechanics had been established and accepted as one of the most successful physical theories ever, thus the idea that the Hilbert space quantum mechanics may not be adequate in all respects after all must have seemed extremely dubious to the physicists of the time. Especially because von Neumann's suggestion was not based on new empirical findings; rather, it was based partly on purely mathematical results and partly on two, philosophically flavored and motivated requirements: that probabilities must be interpreted as relative frequencies in a manner specified by Mises' theory and that the quantum probability statements must be interpretable as conditional probability statements with the prior probability given by a trace. This position regarding quantum probabilities was not maintainable if one considered the standard Hilbert lattice of an infinite dimensional Hilbert space as representing the random event structure; only the theory of type II_1 von Neumann algebras offered a non-trivial, non-finite (linear) dimensional example of a structure that satisfied von Neumann's requirements.

But von Neumann was not completely satisfied even with the theory of type II_1 factors and their lattices as quantum logic. A document of his frustration with quantum logic is his letter of July 2, 1945 to Silsbee, where he writes

It is with great regret that I am writing these lines to you, but I simply cannot help myself. In spite of very serious attempts to write the article on the "Logics of quantum mechanics" I find it completely impossible to do it at this time. As you may know, I wrote a paper on this subject with Garrett Birkhoff in 1936 ("Annals of Mathematics", vol. 37, pp. 823-843), and I have thought a good deal on the subject since. My work on continuous geometries, on which I gave the Amer. Math. Soc. Colloquium lectures in 1937, comes to a considerable extent from this source. Also a good deal concerning the relationship between strict and probability logics (upon which I touched briefly in the Henry Joseph Lecture) and the extension of this "Propositional calculus" work to "logics with quantifiers" (which I never so far discussed in public). All these things should be presented as a connected whole (I mean the propositional and the "quantifier" strict logics, the probability logics, plus a short indication of the ideas of "continuous" projective geometry), and I have been mainly interrupted in this (as well as in writing a book on continuous geometries, which I still owe the Amer.Math.Soc.Colloqium Series) by the war. To do it properly would require a good deal of work, since the subjects that have to be correlated are very heterogenous collection – although I think that I can show how they belong together.

When I offered to give the Henry Joseph Lecture on this subject, I thought (and I hope that I was not too far wrong in this) that I could give a reasonable general survey of at least part of the subject in a talk, which might have some interest to the audience. I did not realize the importance nor the difficulties of reducing this to writing.

I have now learned – after a considerable number of serious but very unsuccessful efforts – that they are exceedingly great. I must, of course, accept a good part of the responsibility for my method of writing – I write rather freely and fast if a subject is "mature" in my mind, but develop the worst traits of pedantism and inefficiency if I attempt to give a preliminary account of a subject which I do not have yet in what I can believe in its final form.

I have tried to live up to my promise and to force myself to write this article, and spent much more time on it than on many comparable ones which I wrote with no difficulty at all – and it just didn't work. [33]

Why didn't it work? Since von Neumann does not elaborate further on the issue in the letter – nor did he ever publish any paper after 1936 on the topic of quantum logic – all one can do is try to interpret von Neumann's published works to understand why he considered his efforts unsatisfactory. Our interpretation in this paper has been this. What von Neumann aimed at in his quest for quantum logic in the years 1935-36 was establishing the quantum analogue of the classical situation, where a Boolean algebra can be interpreted as being both the Tarski-Lindenbaum algebra of a classical propositional logic and the algebraic structure representing the random events of a classical probability theory, with probability being an additive normalized measure on the Boolean algebra satisfying the strong additivity, and where the probabilities can also be interpreted as relative frequencies. The problem is that there exist no "properly non-commutative" versions of this situation: The only (irreducible) examples of non-commutative probability spaces probabilities of which can be interpreted via relative frequencies are the modular lattices of the finite (factor) von Neumann algebras with the canonical trace; however, the non-commutativity of these examples is somewhat misleading because the non-commutativity is suppressed by the fact that the trace is exactly the functional that is insensitive for the non-commutativity of the underlying algebra. So it seems that while one can have both a non-classical (quantum) logic and a mathematically impeccable non-commutative measure theory, the conceptual relation of these two structures cannot be the same as in the classical, commutative case – as long as one views the measure as probability in the sense of relative frequency. This must have been the main reason why after 1936 von Neumann abandoned the relative frequency view of probability in favor of what can be called a "logical interpretation". In this interpretation, advocated by von Neumann explicitly in his address to the 1954 Amsterdam Conference, (quantum) logic determines the (quantum) probability, and vice versa, i.e. von Neumann sees logic and probability emerging simultaneously.

Von Neumann did not think, however, that this rather abstract idea had been worked out by him as fully as it should. Rather, he saw in the unified theory of logic, probability and quantum mechanics a problem area that he thought should be further developed. He finishes his address to the Amsterdam Conference with these words:

I think that it is quite important and will probably shade a great deal of new light on logics and probably alter the whole formal structure of logics considerably, if one succeeds in deriving this system from first principles, in other words from a suitable set of axioms. All the existing axiomatisations of this system are unsatisfactory in this sense, that they bring in quite arbitrarily algebraical laws which are not clearly related to anything that one believes to be true or that one has observed in quantum theory to be true. So, while one has very satisfactorily formalistic foundations of projective geometry of some infinite generalizations of it, including orthogonality, including angles, none of them are derived from intuitively plausible first principles in the manner in which axiomatisations in other areas are.

Now I think that at this point lies a very important complex of open problems, about which one does not know well of how to formulate them now, but which are likely to give logics and the whole dependent system of probability a new slam. [34, p. 22-23; p. 245 in this volume.]

NOTES

*   Work supported by AKP and OTKA (contract numbers T 025841, T 015606 and T 032771) and by the Dibner Institute MIT where I was staying as a Resident Fellow during the academic year 97/98.

1.   In what follows we refer to the Birkhoff-von Neumann concept as the "von Neumann concept". We do this partly for the sake of brevity, partly because it was especially von Neumann who should be credited with proposing the particular concept that got published in 1936.

2.   Notable exceptions are Bub's two papers [6], [7] and Birkhoff's brief remarks [4, p. 285] and in [3]. Karl Popper also discussed the 1936 version of quantum logic [20]; however, Popper's reaction does not seem to be based on a clear and proper understanding of the necessary technicalities involved, see [25] and [8] for a critique of Popper's views.

3.   When quoting from von Neumann's works that can be found in his Collected Works, we first refer to the original source (without giving page numbers) and then (in square brackets) to the volume of von Neumann's Collected Works that contains the source with the page number in that volume on which the quoted text can be found.

4.   To be more precise: Both knew that the Hilbert lattice is not modular in the infinite dimensional case (the 1936 paper contains explicit examples of infinite dimensional subspaces violating the modularity law); however, the orthomodularity property is not stated explicitly in [5] as a relation generally valid in Hilbert lattices, and it is not clear whether Birkhoff or von Neumann had been aware of the orthomodularity property.

5.   The inequality $\mu(A)+\mu(B) \geq \mu(A \vee B)$ is called "subadditivity". Clearly, if a measure μ satisfies (4), then it also is subadditive. It can be shown that the converse also is true in the framework of von Neumann algebras, see [18].

6.   In [5] the a priori probability is called the "a priori thermodynamical weight of states" [5], [5, p. 115]. For an explanation of this terminology see [23] Chapter 7.

7.   Other features of the Hilbert space formalism which he viewed as unsatisfactory include the pathological behavior of the set of *all* unbounded operators on a Hilbert space and the unphysical nature of the common product (composition) of operators.

8.   Strictly speaking one should write $\mu(X) = \lim_{N \to \infty} \frac{\#(X)}{N}$; however, the limit is not important from the point of view of the present considerations, so we omit it.

9.   Probability is clearly taken also in this paper in the sense of the ensemble interpretation, see [37, p. 109].

10.   The explicit reason von Neumann gives to explain why he thinks von Mises' theory to be inadequate for logical purposes involves pointing out that von Mises' requirement on the existence of the limit of relative frequences in finite ensembles does not have a strict mathematical sense. See [32] [37, p. 196].

11.   For a detailed analysis of von Neumann's Amsterdam talk see [22].

12.   See [5, 33. footnote] and [16, Introduction].

REFERENCES

[1]   E.G. Beltrametti and G. Cassinelli: *The Logic of Quantum Mechanics*. Addison Wesley, Massachusetts, 1981.

[2]   Enrico Beltrametti and B.C. van Fraassen (eds.): *Current Issues in Quantum Logic*. Plenum Press New York, 1981.

[3]   G. Birkhoff: "Lattices in Applied Mathematics", in *Lattice Theory* (Proceedings of the Second Symposium in Pure Mathematics of the American Mathematical Society, April, 1959, ed. by R.P. Dilworth), Providence, 1961.

[4]   G. Birkhoff: *Lattice Theory*, 3rd ed., American Mathematical Society, Providence, 1967.

[5]   G. Birkhoff and J. von Neumann: "The logic of quantum mechanics", *Annals of Mathematics* **37** (1936), 823-843; in [37].
[6]   J. Bub: "What does quantum logic explain?"; in [2].
[7]   J. Bub: "Hidden variables and quantum mechanics – a sceptical review", *Erkenntnis* **16** (1981), 275-293.
[8]   M.L.D. Chiara and R. Giuntini: *Popper and the logic of quantum mechanics*. (Unpublished manuscript).
[9]   Ph. Frank, S. Goldstein, M. Kac, W. Prager, G. Szegő and G. Birkhoff (eds.): *Selected Papers of Richard von Mises* Vol. II. American Mathematical Society, Rhode Island, 1964.
[10] S.S. Holland Jr.: "The current interest in orthomoduar lattices" in *Trends in Lattice Theory*, J.C. Abbott (ed.), Van Nostrand, New York, 1970; in [11].
[11] C.A. Hooker (ed.): *The Logico-Algebraic Approach to Quantum Mechanics, vol. I. Historical Evolution*. D. Reidel Publishing Co., Dordrecht Holland, 1975.
[12] C.A. Hooker (ed.): *The Logico-Algebraic Approach to Quantum Mechanics vol. II. Contemporary Consolidation*. D. Reidel Publishing Co., Dordrecht Holland, 1975.
[13] F. Laudisa and R. Giuntini: "The impossible causality: the no hidden variables theorem of John von Neumann"; in the present volume.
[14] Richard von Mises: *Probability, Statistics and Truth* (second English edition of *Wahrscheinlichkeit, Statistik und Wahrheit*, Springer, 1928. (Dover Publications, New York, 1981)
[15] G. Kalmbach, *Orthomodular Lattices*. Academic Press, London, 1983.
[16] F. J. Murray and J. von Neumann: "On rings of operators", *Annals of Mathematics* **37** (1936), 6-119; in [36].
[17] M. Pavicic: "Bibliography on quantum logic", *International Journal of Theoretical Physics* **31** (1992).
[18] D. Petz and J. Zemanek: "Characterizations of the trace", *Linear Algebra and its Applications* **111** (1988), 43-52.
[19] D. Petz and M. Rédei: "John von Neumann and the theory of operator algebras" in *The Neumann Compendium. World Scientific Series of 20th Century Mathematics Vol. I.*, F. Brody and T. Vámos (eds.). World Scientific, Singapore, 1995, pp. 163-181.
[20] K.R. Popper: "Bikhoff and von Neumann's interpretation of quantum mechanic", *Nature* **219** (1968), 682-685.
[21] P. Pták and S. Pulmannová: *Orthomodular Structures as Quantum Logic*. Kluwer Academic Publishers, Dordrecht, Boston, London, 1991.
[22] M. Rédei: "Unsolved problems in mathematics" J. von Neumann's address to the International Congress of Mathematicians, Amsterdam, September 2-9, 1954, *The Mathematical Intelligencer* **21** (1999), 7-12.
[23] M. Rédei: *Quantum Logic in Algebraic Approach*. Kluwer Academic Publishers, Dordrecht, Holland, 1998.
[24] M. Rédei: "Why John von Neumann did not like the Hilbert space formalism of quantum mechanics (and what he liked instead)", *Studies in the History and Philosophy of Modern Physics* **27** (1996), 493-510.
[25] E. Scheibe: "Popper and quantum logic", *The British Journal for the Philosophy of Science* **25** (1974), 319-342.
[26] M. Takesaki: *Theory of Operator Algebras, I.* Springer Verlag, New York, 1979.
[27] R. von Mises: "Grundlagen der Wahrscheinlichkeitsrechnung", *Mathematische Zeitschrift* **5** (1919), 52-99; in [9] 57-105.
[28] J. von Neumann: "Mathematische Begründung der Quantenmechanik", *Göttinger Nachrichten* (1927), 1-57; in [35] 151-207.
[29] J. von Neumann: "Wahrscheinlichkeitstheoretischer Aufbau der Quantenmechanik", *Göttinger Nachrichten* (1927), 245-272; in [35] 208-235.
[30] J. von Neumann: "Thermodynamik quantenmechanischer Gesamtheiten", *Göttinger Nachrichten* (1927), 245-272; in [35] 236-254.
[31] J. von Neumann: *Mathematische Grundlagen der Quantenmechanik*. Dover Publications, New York, 1943 (first American Edition). First edition: Springer Verlag, Heidelberg, 1932.

[32] J. von Neumann: *Quantum logics (strict- and probability logics)*. Unfinished manuscript, John von Neumann Archive, Libarary of Congress, Washington, D.C. Reviewed by A. H. Taub in [37] p. 195-197.

[33] J. von Neumann: *Letter to Dr. Silsbee, July 2, 1945*, John von Neumann Archive, Library of Congress, Washington, D.C. (Published in the current volume.)

[34] J. von Neumann: *Unsolved problems in mathematics*, Unpublished address to the the World Congress of Mathematics. Typescript John von Neumann Archive, Library of Congress, Washington, D.C. (First published in the current volume.)

[35] J. von Neumann: *Collected Works Vol. I. Logic, Theory of Sets and Quantum Mechanics* , A.H. Taub (ed.). Pergamon Press, 1962.

[36] J. von Neumann: *Collected Works Vol. III. Rings of Operators* , A.H. Taub (ed.). Pergamon Press, 1961.

[37] J. von Neumann: *Collected Works Vol. IV. Continuous Geometry and Other Topics*, A.H. Taub (ed.). Pergamon Press, 1961.

[38] J. von Neumann: "Continuous Geometries with Transition Probability", *Memoirs of the American Mathematical Society* **34**, No. 252 (1981), 1-210.

Department of History and Philosophy of Science
Loránd Eötvös University
Faculty of Sciences
Pf. 32
H-1518 Budapest 112
Hungary

ROBERTO GIUNTINI
FEDERICO LAUDISA

THE IMPOSSIBLE CAUSALITY:
THE NO HIDDEN VARIABLES THEOREM
OF JOHN VON NEUMANN

1. INTRODUCTION

The debate over the question whether quantum mechanics should be considered as a *complete* account of microphenomena has a long and deeply involved history, a turning point in which has been certainly the Einstein-Bohr debate, with the ensuing charge of incompleteness raised by the Einstein-Podolsky-Rosen (EPR) argument. In quantum mechanics, physical systems can be prepared in pure states that nevertheless have in general positive dispersion for most physical quantities; hence in the EPR argument, the attention is focused on the question whether the account of the microphysical phenomena provided by quantum mechanics is to be regarded as an exhaustive description of the physical reality to which those phenomena are supposed to refer, a question to which Einstein himself answered in the negative. However, there is a mathematical side of the completeness issue in quantum mechanics, namely the question whether the kind of states with positive dispersion can be represented as a different, dispersion-free kind of states in a way *consistent with the mathematical constraints of the quantum mechanical formalism*. From this point of view, the other source of the completeness issue in quantum mechanics is the no hidden variables theorem formulated by John von Neumann in his celebrated book on the mathematical foundations of quantum mechanics, the preface of which already anticipates the program and the conclusion concerning the possibility of 'neutralizing' the statistical character of quantum mechanics:

There will be a detailed discussion of the problem as to whether it is possible to trace the statistical character of quantum mechanics to an ambiguity (i.e. incompleteness) in our description of nature. Indeed, such an interpretation would be a natural concomitant of the general principle that each probability statement arises from the incompleteness of our knowledge. This explanation "by hidden parameters" [...] has been proposed more than once. However, it will appear that this can scarcely succeed in a satisfactory way, or more precisely, such an explanation is incompatible with certain qualitative fundamental postulates of quantum mechanics ([1], pp. ix-x).

Despite its alleged significance, however, von Neumann's theorem has had a rather unfortunate fate. After its formulation it was welcomed as the ultimate word on

M. Rédei and M. Stöltzner (eds.),
John von Neumann and the Foundations of Quantum Physics, 173–188.
© 2001 *Kluwer Academic Publishers. Printed in the Netherlands.*

causality and determinism 'in nature' and – as always happens with the ultimate words – it was not studied with due care. After the appearance of the first consistent hidden variables theories, and above all after their defence by John S. Bell in the early sixties, the domination was overthrown but the theorem and its conceptual framework continued to be more cited (and attacked!) than studied, and in our time a thorough analysis of them is still lacking. The present paper is an attempt to (begin to) fill the gap. In the section 2 we analyse the strategy employed by von Neumann to prove his non-existence statement, whereas in the section 3 we try to assess the foundational significance of the theorem, and we argue that in a historical perspective it well deserves the status of a scientific milestone attributed to it by its original praisers. Finally, in the section 4, we draw a comparison between von Neumann's and Einstein's notions of completeness.

2. A CONCEPTUAL ANALYSIS OF VON NEUMANN'S IMPOSSIBILITY THEOREM

In quantum mechanics the state of a system with n degrees of freedom is represented by a wave function $\phi(q_1, \ldots, q_n)$, whose square modulus gives the probability that the system is in the $(q_1, \ldots, q_n)$-point of its configuration space. If $\mathcal{R}$ is a physical quantity relevant to the system in the state ψ, the most general assertion that quantum mechanics is able to make is that the probability that $\mathcal{R}$ has a value in the interval Δ is given by

$$(E_R(\Delta)\psi, \psi), \tag{1}$$

where $E_R(\Delta)$ belongs to the spectral decomposition of the operator R representing $\mathcal{R}$.[1] This can be generalized to the case of l quantities $\mathcal{R}_1, \ldots, \mathcal{R}_l$, where the probability that $\mathcal{R}_1, \ldots, \mathcal{R}_l$ have a value in the intervals $\Delta_1, \cdots, \Delta_l$ is given by

$$\|E_{R_1}(\Delta_1) \cdots E_{R_l}(\Delta_l)\psi\|^2. \tag{2}$$

In formulating (2) (which von Neumann calls **P**), the order of the $\mathcal{R}_i$ (and hence of the $E_{R_i}(\Delta_i)$) was meant to be irrelevant: this is why von Neumann restricts the validity of (2) to *commuting* operators $R_1, \ldots, R_l$ (an assumption that turns $E_{R_1}(\Delta_1) \cdots E_{R_l}(\Delta_l)$ into a projection operator). The assertion (1) implies that the expectation value of a quantity $\mathcal{R}$ for a physical system in the state ψ is

$$Exp(\mathcal{R}, \psi) = (R\psi, \psi). \tag{3}$$

Von Neumann then aptly points out the two 'striking features' (p. 206) of the condition (2):

1. (**P**) [i.e. (2)] is statistical, and not causal, i.e., it does not tell us what values $\mathcal{R}_1, \ldots, \mathcal{R}_l$ have in the state ϕ, but only with what probability they take on all possible values.
2. The problem of (**P**) cannot be answered for arbitrary quantities $\mathcal{R}_1, \ldots, \mathcal{R}_l$, but only for those whose operators $R_1, \ldots, R_l$ commute with one another.

The strategy of those who regard 1 and 2 as simply provisional is then suggested by von Neumann himself. Concerning 1, he argues:

If we want to explain the non-causal character of the connection between ϕ [the wave function] and the values of the physical quantities following the pattern of classical mechanics, then this interpretation is clearly the proper one: In reality, ϕ does not determine the state exactly. In order to know this state absolutely, additional numerical data are necessary. That is, the system has other characteristics or coordinates in addition to ϕ. If we were to know all of these, then we could give the values of all physical quantities exactly and with certainty (p. 209).

And again, concerning 2, he emphasizes that

the most obvious step would be to assume that this is an incompleteness in (**P**) [i.e. (2)], and that there must exist a more general formula, containing this as a special case. Because even if quantum mechanics furnishes only statistical information regarding nature, the least we can expect from it is that it describe not only the statistics of individual quantities, but also the relations among several such quantities (p. 211).

In order to investigate such possibilities, von Neumann deems it necessary to outline the pattern of a typical measurement both of a single quantity and of several ones (provided they are simultaneously measurable). In the latter case, von Neumann focuses on the particular quantity $\mathcal{R}+\mathcal{S}$ and emphasizes a condition that will be later shown to be crucial to the whole impossibility theorem, namely that the expectation value of the sum of the operators R and S, representing respectively $\mathcal{R}$ and $\mathcal{S}$ equals the sum of the expectation values of R and S. Von Neumann considers the most general case, in which no restrictions are placed on the spectrum of the respective operators R, S, and argues:

A simultaneous measurement of $\mathcal{R}, \mathcal{S}$ is also a measurement of $\mathcal{R}+\mathcal{S}$ because the addition of the results of the measurements gives the values of $\mathcal{R}+\mathcal{S}$. Consequently, the expectation value of $\mathcal{R}+\mathcal{S}$ in each state ψ is the sum of the expectation values of $\mathcal{R}$ and of $\mathcal{S}$. It should be noted that this holds independently of whether (and which) correlations exist between them – because the law

Expectation value of the sum = Sum of the expectation values

holds in general, as is well known. Therefore, if T is the operator of $\mathcal{R}+\mathcal{S}$, then this expectation value is on the one hand $(T\psi, \psi)$, and on the other

$$(R\psi, \psi) + (S\psi, \psi) = ((R+S)\psi, \psi)$$

i.e., for all ψ

$$(T\psi, \psi) = ((R+S)\psi, \psi).$$

Therefore $T = R+S$. Consequently, $\mathcal{R}+\mathcal{S}$ has the operator $R+S$ (p. 226, our emphasis).

Interestingly, in a footnote just to this passage, von Neumann argues that the law holds for *any* $\mathcal{R}, \mathcal{S}$, although *it is not possible to prove it*, and that it is one of the main postulates of the theory.[2] As we will see shortly, von Neumann will attempt to ground such 'linearity' postulate for physical quantities on a *purely mathematical* linearity property satisfied by all Hermitean operators. But let us follow the line of von Neumann's reasoning.

The title of the fourth chapter of von Neumann's book, *Deductive Development of the Theory*, states explicitly the strategy to follow, which is to derive the statistical formula of quantum mechanics (3) from "few general qualitative assumptions" (p. 295). Given a physical quantity $\mathcal{R}$ in a state φ, its *dispersion* is defined as $\|R\varphi\|^2 - (R\varphi, \varphi)^2$, which in general is positive: "therefore there exists a statistical distribution of $\mathcal{R}$, even though φ is one individual state" (p. 295). When we also happen to ignore the state in which the system is, namely when φ is a mixture of the states $\varphi_1, \ldots, \varphi_n$ (with weights $w_1, \ldots, w_n$), then

$$Exp(\mathcal{R}, \varphi) = \sum_n w_n (R\varphi_n, \varphi_n). \tag{4}$$

Since in general $(R\varphi_n, \varphi_n) = \mathrm{Tr}(P_{\varphi_n} R)$, where P_{φ_n} is the projector on the unidimensional subspace spanned by φ_n, one obtains $Exp(\mathcal{R}, \varphi) = \sum_n w_n \mathrm{Tr}(P_{\varphi_n} R)$. The positive, trace one operator $U := \sum_n w_n P_{\varphi_n}$ can be introduced, so that we have finally

$$Exp(\mathcal{R}, \varphi) = \mathrm{Tr}(UR). \tag{5}$$

As stressed by von Neumann the formula (5) "characterizes the mixture of states just described completely, with respect to its statistical properties" (p. 296).

The procedure of derivation of (5) from general qualitative assumptions starts by addressing the problem of simultaneous measurability of several quantities with the aid of the notion of *statistical ensemble*.[3]

Let us consider then a physical system σ and some physical quantities $\{\mathcal{P}, \mathcal{Q}, \mathcal{R}, \ldots\}$ measurable on σ. In general, given a quantity $\mathcal{R}$, the distribution of $\mathcal{R}$ is calculated and to this aim certain ensembles $\Sigma = \{\sigma_1, \ldots, \sigma_N\}$ of N replicas (*Exemplaren*) of the system under scrutiny can be used. Each $\sigma_i \in \Sigma$ is prepared according to the same fixed preparing procedure. For a possible value r_i of $\mathcal{R}$, $N(r_i)$ represents the fraction of elements of Σ for which the measurement of $\mathcal{R}$ has yielded the result r_i. The ratio $N(r_i)/N$ expresses the probability that the result of a measurement of $\mathcal{R}$ on σ – prepared according to the fixed procedure – equals r_i (cfr. [1], p. 299). On the basis of these probabilities, the expression $\sum_i r_i N(r_i)/N$ gives the expectation value of $\mathcal{R}$ (denoted by $Exp(\mathcal{R})$).[4] Let I denote the index set of the outcome space $\{r_i\}_{i \in I}$. If there exists an $i \in I$ such that

$$\frac{N(r_{k \in I})}{N} = \begin{cases} 1, & k = i \\ 0, & \text{otherwise} \end{cases}$$

then the probability measure that can be constructed on the basis of the ratio $\frac{N(r_i)}{N}$ is a *dichotomic* probability measure (its only values are 0 and 1). Otherwise, the dispersion is positive.

According to von Neumann, the use of ensembles has two main advantages: first, for N large enough, the alteration of the system due to the interaction occurring during the measurement even of a single physical quantity has an arbitrarily small influence on the final distribution (p. 299);[5] second, if $\mathcal{R}$ and $\mathcal{S}$ are two quantities that are not simultaneously measurable on σ, the distributions obtained with the measurements on $\mathcal{R}$ and $\mathcal{S}$ separately can be obtained with arbitrary accuracy in suitable different subensembles of Σ (pp. 299-300). When explaining in details the second point, von Neumann says that "in order to measure two (or several) quantities $\mathcal{R}, \mathcal{S}$ *simultaneously*, we need two sub-ensembles [...] of such a type that the first is employed in obtaining the statistics of $\mathcal{R}$ and the second in obtaining those of $\mathcal{S}$" (our emphasis), and for what von Neumann just said about the reason why statistical ensembles are introduced, the italicized adverb is clearly unfortunate. Moreover, these two assumptions are unwarranted just because the possibility to have a joint measurement of two arbitrary quantities fails to hold in general in quantum mechanics. As pointed out by Rédei (in this volume) implicit in von Neumann's reasoning is the highly non-trivial assumption that subensembles of Σ where $\mathcal{R}$ and $\mathcal{S}$ are supposed to be measured are representative of Σ in the sense that "the relative frequency of every attribute is the same both in Σ and in the subensembles". According to Rédei, von Neumann's splitting of the ensemble undermines a relative frequency interpretation of quantum probability, since one of the basic assumptions of this interpretation is that an ensemble should not be altered by the measurement of any physical quantity performed on any element of the ensemble.

It now suffices to consider the case of a single quantity $\mathcal{R}$ measured on σ with an ensemble like Σ. The fact that in general we obtain not a definite result but a probability distribution for possible results, namely an expectation value with positive dispersion, can be interpreted according to von Neumann in two different ways:

I. The individual systems $\sigma_1, \ldots, \sigma_n$ of our ensemble can be in different states, so that the ensemble $[\sigma_1, \ldots, \sigma_n]$ is defined by their relative frequency. The fact that we do not obtain sharp values for the physical quantities in this case is caused by our lack of information: we do not know in which state we are measuring, and therefore we cannot predict the results.
II. All individual systems $\sigma_1, \ldots, \sigma_n$ are in the same state, but the laws of nature are not causal. Then the cause of the dispersions is not our lack of information, but is nature itself, which has disregarded the 'principle of sufficient cause'.[6] ([1], p. 302)

The principle of sufficient cause (i.e. of sufficient reason) is, according to von Neumann, simply another name for causality, namely the fact that "two identical objects S_1, S_2, – i.e., two replicas of the system S which are in the same state – will remain identical in all conceivable interactions [...] For if S_1, S_2 could react differently to the same intervention in their interaction (i.e., if they gave different values in the measurement of a quantity $\mathcal{R}$), then we would not have called them identical." (p. 303).

The alternative envisaged here is in fact at the heart of the controversy, since the question at stake is whether the quantum mechanical description of states may be regarded as encoding *everything* there is to know about the physical properties of systems, namely whether quantum mechanics is a *complete* theory or not. For

those who are not prepared to abandon causality, the option I is the only interpretation that is compatible with the principle of sufficient reason and, at the same time,
with the fact that there is a positive dispersion for the expectation value of certain
physical quantities. It then appears that, according to von Neumann, the notion
of causality is equivalent to the notion of determinism. For in the interpretation
I causality is not questioned and the statistical element is contingent: the state of
the system is represented by a mixture in which different probabilistic weights are
attached to the different 'real' states of the replicas $\sigma_1, \ldots, \sigma_n$ in the ensemble. In
the interpretation II, on the other hand, from an identical state different measurement results are obtained, that is we have indeterminism and then – according to
von Neumann – a violation of causality.

Every ensemble determines a functional Exp, which is supposed to characterize it completely from a statistical point of view: the problem is now to single out
the conditions that a given functional must satisfy in order to be the Exp of an
ensemble. According to von Neumann, such conditions are:

A. If the quantity $\mathcal{R}$ is identically 1 [...], then $Exp(\mathcal{R}) = 1$.
B. For each $\mathcal{R}$ and for each real number a, $Exp(a\mathcal{R}) = aExp(\mathcal{R})$.
C. If the quantity $\mathcal{R}$ is by nature non-negative, [...] then also $Exp(\mathcal{R}) \geq 0$.
D. If the quantities $\mathcal{R}, \mathcal{S}, \ldots$ are simultaneously measurable, then $Exp(\mathcal{R} + \mathcal{S} + \ldots) = Exp(\mathcal{R}) + Exp(\mathcal{S}) + \ldots$

Given a functional Exp of an ensemble, there are two alternatives: either its dispersion equals zero (in which case the functional is said to be *dispersion-free*), or
its dispersion is positive. In the latter case, it is possible to find two functionals
Exp', Exp'' such that, for any $\mathcal{R}$, we have $Exp(\mathcal{R}) \neq Exp'(\mathcal{R}) \neq Exp''(\mathcal{R})$ and

$$Exp(\mathcal{R}) = aExp'(\mathcal{R}) + bExp''(\mathcal{R}), \tag{6}$$

with $a, b > 0$, $a + b = 1$. Whenever (6) implies that $Exp(\mathcal{R}) = Exp'(\mathcal{R}) = Exp''(\mathcal{R})$, then Exp is said to be *pure*.

For a pure functional Exp of an ensemble Σ, the latter is said to be *homogeneous* (i.e., representative of a system in a pure state). It is to be stressed that the ensembles von Neumann is concerned with should be both dispersion-free *and* pure;
as Bub emphasizes, "[...] von Neumann is obviously not concerned to prove that
no quantum mechanical statistical state is dispersion-free, for this follows immediately from the limitation of the statistical algorithm to compatible magnitudes. Nor
could the absence of dispersion-free states in the theory by itself exclude the possibility of reconstructing the statistical relations on a classical probability space. [...]
It is the existence of *homogeneous ensembles with dispersion* that von Neumann
regards as significant."[7]

The condition that will turn out to be crucial to the whole von Neumann's
strategy is **D**. Von Neumann stresses initially that "its correctness depends on this
theorem on probability: the expectation value of a sum is always the sum of the
expectation values of the individual terms. [...] That we have formulated it only
for simultaneously measurable $\mathcal{R}, \mathcal{S}, \ldots$ is natural, since otherwise $\mathcal{R} + \mathcal{S} + \ldots$

is meaningless." ([1], pp. 308-9). Nevertheless, von Neumann attempts to justify the extension of **D** also to *non-simultaneously measurable* quantities with the aid of a purely mathematical property of Hermitean operators, namely by suddenly plunging the discussion into Hilbert space quantum mechanics.

But the algorithm of quantum mechanics contains still another operation, which goes beyond the one just discussed: namely, the addition of two arbitrary quantities, which are not necessarily simultaneously measurable. This operation depends on the fact that for two Hermitean operators, R, S, the sum $R + S$ is also an Hermitean operator, even if the R, S do not commute.[8]

It is this property (let us call it SUM) that, according to von Neumann, allows to extend the condition **D**, which so far was "not specialized to quantum mechanics", also to non simultaneously measurable quantities (p. 309). In fact one should say "not specialized to *Hilbert space quantum mechanics*", since SUM is a simple consequence of the fact that now von Neumann employs effectively the axiomatized formulation of quantum mechanics (i.e. quantum mechanics in Hilbert space). The extension then reads as follows:

E. If $\mathcal{R}, \mathcal{S}, \ldots$ are arbitrary quantities, then there is an additional quantity $\mathcal{R} + \mathcal{S} + \ldots$ [...] such that $Exp(\mathcal{R} + \mathcal{S} + \ldots) = Exp(\mathcal{R}) + Exp(\mathcal{S}) + \ldots$[9]

At this point von Neumann summarizes his assumptions in a compact way and renumbers them as follows: **C** is preserved identical and named **A'**, whereas **B** and **E** are reunited as follows:

> **B'.** If $\mathcal{R}, \mathcal{S}, \ldots$ are arbitrary quantities, and $a, b, \ldots$ are real numbers, then
> $$Exp(a\mathcal{R} + b\mathcal{S} + \ldots) = aExp(\mathcal{R}) + bExp(\mathcal{S}) + \ldots$$

Finally, the assumption of a one-to-one correspondence between Hermitean operators and physical quantities allows to derive the two final conditions, defined as follows:

> **I.** If the quantity $\mathcal{R}$ is represented by the Hermitean operator R, then the quantity $f(\mathcal{R})$ is represented by the operator $f(R)$.

> **II.** If $\mathcal{R}, \mathcal{S}, \ldots$ are arbitrary quantities represented by the operators $R, S, \ldots$, then the quantity $\mathcal{R} + \mathcal{S} + \ldots$ is represented by the operator $R + S + \ldots$

Von Neumann's final result is then that the assumptions **A'**, **B'**, **I.** and **II.** imply $Exp(\mathcal{R}) - Tr(UR)$. But let us summarize the argument from a logical point of view: let us call (HSQM) the assumption of the Hilbert space axiomatization of quantum mechanics, and (CORR) the assumption of a one-to-one correspondence between physical quantities and Hermitean operators. Then

> (i) (HSQM) implies (SUM), which implies **E**,

> (ii) (CORR) implies both **I** and **II**,

and we know that **A'=C**, and **B'=B+E**. Hence it follows that (HSQM), (CORR), **A'** and **B** are sufficient to derive $Exp(\mathcal{R}) = \text{Tr}(UR)$.

What does von Neumann's result really establish? Let us define *von Neumann functionals* the expectation functionals satisfying conditions **A'**, **B'**, **I.** and **II**. These functionals, associated to statistical ensembles, are completely determined and they turn out to be in one-to-one correspondence (via the functional "trace") with the class of all statistical operators. As pointed out by Varadarajan ([8]) such a correspondence preserves *convexity*. Accordingly, unidimensional projections (associated to pure von Neumann expectation functionals) are mapped onto the extreme points of the convex set determined by the class of all statistical operators. Let us then consider the restriction of a von Neumann expectation functional Exp to the set $P(\mathcal{H})$ of all projections in the Hilbert space $\mathcal{H}$. It turns out that Exp becomes a finitely additive probability measure on the orthomodular lattice $\mathcal{P}(\mathcal{H})$ induced by $P(\mathcal{H})$, for, if $P \in P(\mathcal{H})$, then $|Exp(P)| \leq \|P\|$. Conversely, given any finitely additive probability measure on $\mathcal{P}(\mathcal{H})$, there exists, by the spectral theorem, a functional Exp (on the set $\mathcal{B}(\mathcal{H})$ of all bounded self-adjoint operators) satisfying **A**, **B**, **C** and **D**, but not necessarily **E** (we will call a functional Exp satisfying **A**, **B**, **C** and **D**, a *Gleason expectation functional*).

About twenty years later, Gleason ([6]) proved[10] that every σ-additive probability measure on $\mathcal{P}(\mathcal{H})$ is determined (again via the functional trace) by a statistical operator, provided the dimension of $\mathcal{H}$ is at least 3. As a consequence, every Gleason expectation functional Exp does also satisfy **E**, provided the dimension of $\mathcal{H}$ is at least 3. Thus every σ-additive probability measure on $\mathcal{P}(\mathcal{H})$ induces a von Neumann expectation functional. Summing up: if the dimension of $\mathcal{H}$ is at least 3, then Exp is a Gleason expectation functional if and only if Exp is a von Neumann expectation functional if and only if Exp is determined by a statistical operator. Consequently, both in the Gleason and in the von Neumann case, the conclusion is the same: statistical operators are the necessary and sufficient mathematical ingredients to describe completely the quantum statistical algorithm. Differently from Gleason's theorem, however, von Neumann's result does not involve any restriction about the dimension of the Hilbert space associated to the physical system at issue. On the other hand, Gleason's theorem teaches us that the assumption **E** is not necessary, assumption **D** being sufficient to derive the statistical formula. Gleason's theorem confirms, *a posteriori*, the validity of von Neumann's result and indicates at the same time its limit: assumption **E**. Von Neumann's result is questionable not because of its alleged circularity but, as many authors have pointed out, for the unduly restrictive and unjustified character of **E**. This is confirmed by the existence (in the case of bidimensional Hilbert spaces) of Gleason expectation functionals Exp that are not von Neumann expectation functionals. Thus, in this case, there are probability measures on $\mathcal{P}(\mathbb{C}^2)$ determined by Gleason expectation functionals that are not von Neumann. This result clearly shows that there is no need for a Gleason expectation functional Exp to satisfy **E** in order to induce a probability measure on $\mathcal{P}(\mathcal{H})$. One might say that, *a posteriori*, assumption **E** is not so strong as it appears *prima facie*. As we have already mentioned, von Neumann's and

Gleason's results coincide with the exception of the pathological bidimensional case. Gleason's theorem represents, in a sense, the best generalization of von Neumann's result. It states that von Neumann's result can be achieved by using **D** instead of **E**. Moreover, this generalization cannot be further strengthened.

As to the hidden-variable problem, von Neumann's theorem implies, as we have seen, that there exists no dispersion-free von Neumann expectation functional on $\mathcal{B}(\mathcal{H})_0$. Dispersion-free expectation functionals (both Gleason and von Neumann) clearly determine $\{0, 1\}$-probability measures on $\mathcal{P}(\mathcal{H})$. Accordingly, we can conclude that there exists no dichotomic finitely additive probability measure on $\mathcal{P}(\mathcal{H})$ *determined by a dispersion-free von Neumann expectation functional*. Von Neumann's theorem, however, does not rule out the possibility of the existence of *other* dichotomic finitely additive probability measure on $\mathcal{P}(\mathcal{H})$! To do that, Gleason's theorem is needed. In this sense, as pointed out by Varadarajan ([8]), Gleason's theorem represents the strongest generalization of von Neumann impossibility proof. Indeed, if $3 \leq \dim(\mathcal{H}) < \infty$, we know, by Gleason's theorem, that every probability measure on $\mathcal{P}(\mathcal{H})$ arises from a statistical operator. But every probability measure induced by a statistical operator is the restriction to $\mathcal{P}(\mathcal{H})$ of a von Neumann expectation functional, and hence von Neumann's impossibility result already applies here. This argument can also be used to exclude the existence of dispersion-free Gleason expectation functionals for Hilbert spaces $\mathcal{H}$ of infinite dimension. It is sufficient to recall that in this case, $\mathcal{P}(\mathbb{C}^3)$ can be embedded into $\mathcal{P}(\mathcal{H})$, so that every dichotomic σ-additive probability measure on $\mathcal{P}(\mathcal{H})$ would be a dichotomic probability measure on $\mathcal{P}(\mathbb{C}^3)$, which is not the case. ([8], p. 125).

3. 'HIDDEN PARAMETERS' AND CLASSICAL STATES

Thus, as early as 1932 in the development of quantum mechanics, the aim of restoring causality in the quantum domain had been declared impossible to achieve. Von Neumann's theorem was meant to show that the ordinary formulation of quantum theory could not be simply adapted in order to comply with causal (i.e. deterministic, in von Neumann's view) requirements, and that a completely different theory, if any, had to be created in order to attempt a recovery of causality in the quantum domain.

Among the conditions required by the no hidden variables theorem, one turned out to be especially critical, namely a linearity requirement for expectation functionals *no matter what kind of states the expectation is calculated in*: if A and B arc operators representing physical quantities, a and b are real numbers, ψ is a quantum state and (ψ, ξ) denotes the 'completion' of ψ, it was assumed that

$$\langle aA + bB \rangle_{\psi,\xi} = a\langle A \rangle_{\psi,\xi} + b\langle B \rangle_{\psi,\xi},$$

a requirement that a general hidden variable theory need not satisfy when A and B represent *incompatible* physical quantities. As is well known, the lack of generality deriving from this linearity assumption for expectation functionals is the essential objection raised twenty years later by David Bohm (and others). The unwarranted restrictive character of von Neumann's theorem, however, was fully appreciated only with Bell's work, in which Bohm is credited for realizing that von Neumann's theorem was only relevant for a limited class of hidden variable theories. In his seminal paper [4] Bell, after recognizing the unduly restrictive nature of von Neumann's assumption, addressed the question of the relevance of Gleason's theorem to the hidden variables issue. He showed that von Neumann's additivity requirement, even if it is weakened, i.e. if its validity is restricted to commuting operators, leads to a contradiction with Gleason's theorem. But in addition Bell also argued that Gleason's theorem actually rules out only *non-contextual* hidden variable theories, and it is as irrelevant as von Neumann's theorem with respect to *contextual* hidden variable theories, i.e. theories in which the result of a measurement of an observable A, unlike non-contextual theories, may depend not only on the state of the system but also on the set of (compatible) observables A is measured with: Bohm's causal interpretation of quantum mechanics turns out to be contextual exactly in this sense, and the upshot of Bell's 1966 paper was then that a *consistent formulation* of hidden variable theories is perfectly possible.[11] Focusing on the linearity assumption for expectation functionals, however, may help us to better assess the significance of von Neumann's achievement. The 'impossibility' theorem clearly does not rule out the *logical* possibility of a deterministic completion of quantum mechanics, nor could von Neumann ever mean his theorem to do so: he simply says, for instance, that "an introduction of hidden parameters is certainly not possible without a basic change in the present theory" ([1], p. 210), and after completing the proof of his theorem he concludes tentatively:

The question of causality could be put to a true test only in the atom, in the elementary processes themselves, and here everything in the present state of our knowledge militates against it. The only formal theory existing at the present time which orders and summarizes our experiences in this area in a half-way statisfactory way, i.e. quantum mechanics, is in compelling contradiction with causality. *Of course it would be an exaggeration to maintain that causality has thereby been done away with*: quantum mechanics has, in its present form, several serious lacunae, and it may be that it is false, although this latter possibility is highly unlikely [...] However, mindful of such precautions, we may still say that there is at present no occasion and no reason to speak of causality in nature – because no experiment indicates its presence, since the macroscopic are unsuitable in principle, and the only known theory which is compatible with our experiences relative to elementary processes, quantum mechanics, contradicts it. ([1], pp. 327-28, our emphasis.)[12]

What the theorem *does* show, however, is that a hypothetical deterministic completion cannot be consistently formulated in *classical* terms ([10]), namely that the complete states one is supposed to average over to yield the 'right' statistics, i.e. statistics in agreement with quantum mechanical predictions, cannot be simply represented as one-point measures on the hypothetical phase space of the system under scrutiny: any deterministic completion, whatever form it might assume, will be a theory very remote from a classical theory (a statement that – by the way – David Bohm could have easily subscribed to).[13]

4. VON NEUMANN AND EINSTEIN ON THE COMPLETENESS
OF QUANTUM MECHANICS

Von Neumann's effort to give a mathematical proof of the claim, that the goal of completing quantum mechanics was unattainable, was consistent with his belief in the 'acausal' character of the laws concerning the microworld. As we have seen, the completion of quantum mechanics consisted, in von Neumann's presentation, in the possibility of defining dispersion-free and homogeneous statistical ensembles (representatives of the pure states of physical systems). Von Neumann aimed to prove that such attempts at completion were doomed to fail, and that such failure sanctioned, once and for all, the acausal character of the laws of nature ([1], p. 210.)

However, as it was briefly recalled above, the completeness issue owes a lot also to the debate on the EPR argument, of which there are three main versions. The first two are variants of the EPR argument itself, whereas the third is the incompleteness argument that Einstein formulated *on his own* in some of his contributions after 1935. The first argument takes into account a pair of quantities for each particle, in such a way that the members of each pair are mutually incompatible. This is the argument in the original EPR version. The second argument, a simplified variant of the first, deals with just one quantity for each particle. Finally, there is Einstein's own incompleteness argument which, although employing the background of the EPR argument, does not mention the controversial condition for the elements of physical reality and is grounded upon a sharp distinction between a principle of separability and a principle of locality. The importance of Einstein's own argument has been emphasized quite recently, after the discovery of a letter from Einstein to Schrödinger (the date is 19 June 1935, a month later than the appearance of the EPR paper on *Physical Review*). In this letter Einstein expresses his unhappiness about how the EPR paper failed to express his real point of view, since its main point was, according to him, "buried by the erudition".[14]

The physical situation under scrutiny in Einstein's own argument (as reconstructed in [13]) comprises a physical system S_{12}, which is a composite system made up of two noninteracting subsystems S_1 and S_2. A general condition that is assumed at the outset is the so-called *principle of separation (Trennungsprinzip)*. This condition is the conjunction of two distinct principles, the principle of *separability* and the principle of *locality*. According to separability, spatially separated systems possess *distinct real states* whereas, according to locality, the state of the system can be modified only by local influences or interactions. In Howard's words "the separability principle operates at a more basic level as, in effect, a principle of individuation for physical systems, a principle whereby we determine whether in a given situation we have only one system or two" ([13], p. 173.) The condition of *completeness* can then be formulated as follows: the quantum mechanical wave function of the system S_{12}, denoted by Ψ_{12}, is associated one-to-one with the real state of the system. Clearly, this formulation of the completeness condition differs from the completeness condition in the EPR paper: let us call EPR-completeness the latter condition and E-completeness the former.

Now suppose that S_1 consists of a single particle. Then we can choose whether to measure its position or its momentum: with reference to this choice, we will obtain for Ψ_2 [the wave function for S_2] different representations, since from the choice of the measurement performed on S_1 different statistical predictions concerning the successive measurements to be performed on S_2 can be derived. The particular physical situation under scrutiny then requires that different wave functions be ascribed to the state of the second system, depending on the type of measurement that one chooses to perform on the first system. Therefore, several wave functions $\Psi_2', \Psi_2'', \ldots$ are associated with the same real state of the second system. The condition of separation embodies the relative independence of the two systems (according to separability) and prohibits immediate influences between the two systems (according to locality). The definition of completeness just formulated, however, demands a one-to-one association between the wave function and the real state. The ideal experiment displays instead the association between one real state and a plurality of wave functions $\Psi_2', \Psi_2'', \ldots$. The quantum mechanical description 'encoded' in the wave function is then E-incomplete.

If this is the structure of Einstein's own argument for the incompleteness of quantum mechanics, then an interesting question arises: what is the conceptual relationship between Einstein's and von Neumann's notions of completeness?[15]

As we have seen, von Neumann's reasoning is straightforward on this point. The relation between a quantum-mechanical state ψ, associated with a statistical ensemble $\sigma_1, \ldots, \sigma_n$, and the values of physical quantities is in general a statistical one: since quantum mechanics rejects the universal validity of an *ignorance interpretation*, and holds that all $\sigma_1, \ldots, \sigma_n$ are in the same state, it follows that the laws of nature are not causal ([1], p. 302). To make the point even clearer, von Neumann sketches what could be the argument of the defenders of the possibility of completing quantum mechanics, and in doing so he actually formulates an *incompleteness argument* that shows a substantial agreement between von Neumann's notion of completeness and Einstein's own one, outlined above. The argument is formulated as follows:

[...] two identical objects S_1, S_2, – i.e., two replicas of the system S which are in the same state – will remain identical in all conceivable interactions [...] For if S_1, S_2 could react differently to the same intervention in their interaction (i.e., if they gave different values in the measurement of a quantity $\mathcal{R}$), then we would not have called them identical. Therefore, in an ensemble $[S_1, \ldots, S_n]$ which has dispersion relative to a quantity $\mathcal{R}$, the individual systems $S_1, \ldots, S_N$ cannot (by definition) all be in the same state. [...] Since one will obtain different values in the measurement of the same quantity $\mathcal{R}$ in several systems, *which all are in the state with the wave function ϕ* – if ϕ is not an eigenfunction of the operator R of $\mathcal{R}$ – therefore these systems are not equal to one another – i.e., the description by the wave function is not complete. ([1], p. 303, our emphasis.)

As can be seen, the condition of completeness singled out above by von Neumann is identical to the condition of completeness in Einstein's own incompleteness argument, although expressed in a somewhat involved way. Von Neumann's completeness criterion (cp. the italicized sentence above) requires a one-to-one correspondence between the real state of the systems belonging to the ensemble and the wave function, which is the mathematical element entering the statistical algorithm for determining the probability distributions of values for the physical quantities. The presence of a nonzero dispersion for the ensemble, relative to the quantity $\mathcal{R}$, determines different values of $\mathcal{R}$ for the several elements of the ensemble: but the elements of the ensemble are – by definition – replicas of a system S, that remain 'identical in all conceivable interactions'. There being a single wave function associated with the totality of the elements of the ensemble, it follows that the description provided by the wave function is not complete.

The connection between von Neumann's notion of completeness and Einstein's own notion can be further clarified by recalling a passage contained in [18]. In this passage Einstein speaks of the plurality of representations for Ψ_2, the wave function for the system S_2, corresponding to the choice of the physical quantity to be measured on the system S_1 ([18], pp. 320-24). Translating this EPR-type physical situation into von Neumann's language, let us suppose we introduce the ensemble of the replicas of S_2, where each replica is associated with one of these possible representations of Ψ_2 for S_2. The elements of this ensemble are identical, since they are by definition replicas of S_2, but they are associated with different representations of Ψ_2: this means that the description of the Ψ is not von Neumann complete. Two remarks are in order, however. First, there is an asymmetry in the way in which completeness is violated way when the Ψ-representation is associated with the real state of the system. In Einstein's argument, the real state of the system comes to be associated with a plurality of wave functions $\Psi_1, \Psi_2, \ldots$, whereas in von Neumann's argument a plurality of values in the measurement of a physical quantity comes to be associated with a single wave function ϕ describing an ensemble of replicas of the given system. Furthermore, an assumption of determinism is required for von Neumann's argument to work. However, although the differences between the two arguments are not to be overlooked, von Neumann's argumentation strategy is very close to Einstein's: the agreement between Einstein and von Neumann on the idea of completeness (a one-to-one correspondence between the real state of the system and the wave function) turns out to be substantial, although it must be emphasized that von Neumann never investigated any possible connection between his thoughts on completeness and the question of nonlocality in quantum mechanics.

5. CONCLUSION

Von Neumann's Theorem has been both under- and over-estimated. It has been overestimated when meant to show the impossibility of *any* non-contextual hidden variable reconstruction of quantum mechanics. But it has been underestimated when it has been considered only as a "no-go theorem". As we have remarked, the content of von Neumann's theorem is "positive", the impossibility of a *certain* class of hidden variable models being only one of its consequences. The foundational relevance of von Neumann's theorem can be appreciated only in a probabilistic context, namely, only if one takes into due account both the specific von Neumann interpretation of probability (i.e. the relative frequency interpretation) and the difficulties to which such interpretation leads in the framework of quantum mechanics (in particular with respect to the disturbance view of measurement that appears crucial in von Neumann's reasoning). The tension between a relative frequency interpretation and the failure of the joint measurability of all physical quantities on a quantum system turns out to be the real weak point in the von Neumann's ingenious construction. It is a tension which emerges already *within* quantum mechanics, and with respect to it the usual charge raised against von Neumann's theorem (see Section 3) does not appear as crucial as it is usually thought to be.

NOTES

1. For a more accessible presentation, we will employ the English version [1], although the translation is not always completely satisfactory.
2. See note 128. It is interesting to remark that this puzzling statement is missing in the 1955 American edition.
3. Here von Neumann stresses that his discussion is meant to be carried on without mentioning Hilbert space quantum mechanics (p. 297). However, quantum mechanics as a purely physical theory is always in the background, due to the all-pervasive circumstance of non-simultaneous measurability of all sets of quantities. According to von Neumann, the very introduction of statistical ensembles is in fact justified by this typical quantum-mechanical circumstance, although, as we will see shortly, the use of the notion of statistical ensemble in von Neumann's argument is rather ambiguous.
4. It is to be stressed that, in the sequel, von Neumann omits any notational reference to the state in which the expectation is evaluated. Clearly, from what has been said concerning the fixed preparation of each element of the statistical ensemble, such a reference is to be meant as implicit.
5. The first point presupposes adhering to a 'disturbance theory of measurement', a theory by no means endorsed by all members of the Copenhagen orthodoxy, most important of all Niels Bohr (cfr. on the issue [2], pp. 110 ff).
6. We stress that the English translation here is particularly incorrect: for the German edition has the expression *Prinzip vom hinreichenden Grunde*, namely the celebrated principle of sufficient reason of the old philosophical tradition, which the latter has long identified simply with determinism.
7. [3], p. 52; the possibility of reconstructing the statistical relations of quantum mechanics on a classical probability space has been made actual by the models exhibited by [4] and [5].
8. Moreover, as is clear from the quotation, von Neumann assumes implicitly a one-to-one correspondence between Hermitean operators and physical quantities, although few pages later he states the assumption in an explicit way ([1], p. 313).
9. The way in which von Neumann attempts to extend **D** shows moreover, once again, the tension between the non-joint measurability of all quantities in quantum mechanics and relative frequency interpretation of quantum probabilities: the ensembles determining the Exp functionals that satisfy **E** cannot be all genuine ensembles in the sense of relative frequency interpretation.

10. Actually, Gleason's theorem is stronger and refers to measures that are not necessarily probability measures on $P(\mathcal{H})$.

11. On this and related points, cfr. [9].

12. Still, von Neumann is not always straightforward on this point: as a matter of fact, two pages before this passage he argues much more strictly: "It is therefore not, as is often assumed, a question of a reinterpretation of quantum mechanics – the present system of quantum mechanics would have to be *objectively false*, in order that another description of the elementary processes than the statistical be possible." ([1], p. 325, our emphasis).

13. Moreover, the formulation of the completeness issue in analogy with the mechanical 'reduction' of thermodynamics through classical statistical mechanics may be considered rather natural in the early thirties, and especially from the perspective of von Neumann, who was working on problems in statistical physics just in the period of preparation of the 1932 book (cfr. [11], pp. 141-2).

14. The main works investigating the Einstein's incompleteness argument are [12] and [13]. On the basis of these works, it can be safely concluded that the EPR argument published in *Physical Review* in 1935 cannot be considered as a completely faithful representation of Einstein's own views on the issue.

15. To our knowledge, this question is investigated for the first time in [14]. In his [15], Caruana points out the need for such an investigation but without working out any useful analysis. Surprisingly, there seems to be no trace of any public comment on von Neumann's theorem by Einstein. However, in a later appendix to the reprinted version of his [16], Shimony tells us that Einstein, in a conversation with Valentine Bargmann in 1938, was reported to have argued that there is no reason why the linearity assumption in von Neumann's theorem should hold also for a class of states not acknowledged by quantum mechanics, which is essentially the criticism raised against the theorem twenty years later by David Bohm ([17], p. 89).

REFERENCES

[1]   von Neumann J., *Mathematical Foundations of Quantum Mechanics*, Princeton University Press, Princeton 1955 (orig. ed. *Mathematische Grundlagen der Quantenmechanik*, Springer 1932).

[2]   Folse H.J., *The Philosophy of Niels Bohr. The Framework of Complementarity*, North-Holland, Amsterdam 1985.

[3]   Bub J. *The Interpretation of Quantum Mechanics*, Reidel, Dordrecht 1974.

[4]   Bell J.S., "On the Problem of Hidden Variables in Quantum Mechanics", *Reviews of Modern Physics* **38** (1966), pp. 447-452 (reprinted in *Speakable and Unspeakable in Quantum Mechanics*, Cambridge University Press, Cambridge 1987, pp. 1-13).

[5]   Kochen S., Specker E., "The Problem of Hidden Variables in Quantum Mechanics", *Journal of Mathematics and Mechanics* **17** (1966), pp. 59-88 (reprinted in [7], pp. 293-328).

[6]   Gleason A.M, "Measures on the Closed Subspaces of a Hilbert Space", *Journal of Mathematics and Mechanics* **6** (1957), pp. 885-893 (reprinted in [7], pp. pp. 123-134).

[7]   Hooker C.A. (ed.), *The Logico-Algebraic Approach to Quantum Mechanics*, vol. I, Reidel, Dordrecht 1975.

[8]   Varadarajan V.S., *Geometry of Quantum Theory*, 2nd ed., Springer, Berlin 1985.

[9]   Laudisa F., "Contextualism and Nonlocality in the Algebra of EPR Observables", *Philosophy of Science* **64** (1997), pp. 478-496.

[10] Giuntini R., *Quantum Logic and Hidden Variables*, Bibliographische Institut, Mannheim 1991.

[11] Rédei M., *Quantum Logic in Algebraic Approach*, Kluwer, Dordrecht 1998.

[12] Fine A., "Einstein's Critique of Quantum Theory: the Roots and Significance of EPR", in P. Barker, C.G. Shugart (eds.), *After Einstein*, Memphis University Press, Memphis 1981 (reprinted in A. Fine, *The Shaky Game. Einstein Realism and the Quantum Theory*, University of Chicago Press, Chicago 1986, pp. 26-39).

[13] Howard D., "Einstein on Locality and Separability", *Studies in History and Philosophy of Science* **16** (1985), pp. 171-201.

[14] Laudisa F., "Einstein, Bell and Nonseparable Realism", *British Journal for the Philosophy of Science* **46** (1995), pp. 309-329.

[15] Caruana L., "John von Neumann's 'Impossibility Proof' in a Historical Perspective", *Physis* **32** (1995), pp. 109-124.

[16] Shimony A., "Experimental Test of Local Hidden-Variable Theories", in B. d'Espagnat (ed.) *Foundations of Quantum Mechanics*, Academic Press, New York-London 1971, pp. 182-194 (reprinted in [17], pp. 130-139).

[17] Shimony A. *Search for a Naturalistic World View*, vol. II, Cambridge University Press, Cambridge 1993.

[18] Einstein A., "Quantenmechanik und Wirklichkeit", *Dialectica* **2** (1948), pp. 320-324.

Roberto Giuntini
Dipartimento di Scienze Pedagogiche e Filosofiche
Università di Cagliari
Via Is Mirrionis 1
I-09123 Cagliari
Italy

Federico Laudisa
Dipartimento di Filosofia
Università di Firenze
Via Bolognese 52
I-50139 Firenze
Italy

PETER MITTELSTAEDT

QUANTUM MECHANICS WITHOUT PROBABILITIES

1. INTRODUCTION

Usually, quantum mechanics is considered as the prototype of a probabilistic theory. In contrast to statistical mechanics, dice throwing, and roulette game, quantum mechanical probability statements cannot be reduced to causally determined individual events, whose explicit calculation is, however, too complicated for all practical purposes. Even hypothetically, one must not assume that quantum mechanical events were determined in principle and merely computationally intractable, since that assumption would lead to probabilistic predictions which contradict quantum mechanics. Hence, the title of this article seems somewhat surprising at first glance, and in particular it seems difficult to connect a probability free quantum mechanics with the work of John von Neumann.

Although it was definitely not the intention of von Neumann to eliminate probabilities in quantum mechanics, – he formulated the first no-go theorem for causality preserving hidden variables – his work is nevertheless the origin of such a program and in addition he provided very important tools which are needed for achieving this goal. Already in his book of 1932, but more explicitly in the quantum logic paper (together with Birkhoff) of 1936 and in the fundamental mathematical article of 1938 on infinite quantum systems, there are many important ideas and motivations for a scientific development which finally lead to a formulation of quantum mechanics without probabilities. This theory will briefly be sketched in the present article.

2. HISTORICAL REMARKS

Starting from von Neumann's above mentioned contributions, there are several ways of reasoning which lead to a critical reconsideration of the rôle of probability in quantum mechanics. Here, we will briefly mention four different attempts to reconstruct the probability structure in quantum mechanics.

189

M. Rédei and M. Stöltzner (eds.),
John von Neumann and the Foundations of Quantum Physics, 189–200.
© 2001 Kluwer Academic Publishers. Printed in the Netherlands.

2.1 Quantum theory of measurement

In his book *Mathematische Grundlagen der Quantenmechanik* of 1932 [Neu 32] von Neumann provides, among many other things, the first quantum theory of measurement which also considers the apparatus as a proper quantum system. Since the objectification of the measurement results cannot be achieved by a unitary premeasurement, von Neumann added the projection postulate as an additional requirement whose justification remained an open question. Many years later, in 1957, H. Everett in his paper "Relative State formulation of Quantum Mechanics" [Eve 57] reconsidered von Neumann's measurement process. Adopting only the theory of unitary measurements and omitting additional requirements like the projection postulate, Everett arrived at an interpretation of quantum mechanics without objectification. Since the measurement process is a unitary dynamical process, after each measurement the world splits into many different worlds, each of which corresponds to a possible measurement result. Since the whole process is completely determined by unitary dynamics, the temporal development of a quantum system and its apparatus is known in principle. Obviously, there is no need and no room for probabilities which correspond to some deficiency of information.

Everett tried to explain the meaning of probabilities by calculating relative frequencies of "worlds" in repeated measurements. His arguments are not fully convincing and they were often considered to be circular. Except for several misleading formulations his approach was the first, albeit incomplete attempt to derive the probability structure of quantum mechanics from a probability free theory. The many worlds interpretation is an interpretation without objectification and without probabilities. It results from a rigorous application of von Neumann's quantum theory of measurement without any additional assumptions.

2.2 Quantum logic

The second attempt starts from the article "The Logic of Quantum Mechanics" which was written with Birkhoff in 1936. In this paper the authors showed that an orthomodular lattice L_Q of quantum mechanical yes-no propositions is a basic structure of quantum mechanics. Since this lattice L_Q is a relaxation on the well known Boolean lattice L_B of classical logic, the authors gave the impression that a relaxed "quantum logic" of yes-no propositions is the fundamental structure of quantum mechanics. This interpretation was adopted by Finkelstein in 1962 who tried to show that quantum mechanics may indeed be considered a theory of yes-no propositions without probabilities. For this reason he reformulated the probability statements of quantum mechanics as yes-no propositions that refer to an ensemble of equally prepared systems and indicated a proof of this equivalence.

Finkelstein did not refer to the measurement process. Instead, he formulated his results with respect to the preparation states of a single system and of the ensemble.

In his paper "The Logic of Quantum Physics" [Fink 62], Finkelstein considered the following problem: For a single quantum system one can formulate, as in classical physics, yes-no propositions that refer to the objective properties of the system. In addition, in quantum mechanics there are probability statements that are concerned with the nonobjective properties of the system. However, according to Finkelstein these probability statements may be considered as yes-no propositions whose referent is an ensemble of a sufficiently large number of equally prepared systems.

2.3 Individual systems

In 1968 Hartle published a paper with the title "Quantum Mechanics of Individual Systems". In this article, he presented a formulation of quantum mechanics that begins with assertions for individual systems and derives the statistical predictions of the theory. In particular Hartle succeeded in showing that quantum probabilities are almost equal to relative frequencies in an ensemble of equally prepared systems provided the number of systems is sufficiently large. As in the approach of Everett and Finkelstein, Hartle did not refer to the statistical predictions about the state of the object system after the measurement but proved the approximate equivalence of probability propositions about properties of the object with yes-no propositions (about an ensemble) for the pure state prior to the measurement. From a technical point of view Hartle's result is more rigorous than that of Everett and Finkelstein, since it deals carefully with the problem of defining the Hilbert space in the limit of infinitely many systems, making use extensively of von Neumann's results of 1938.

2.4 Operational approach to quantum mechanics

Finally, I mention the operational approach to quantum mechanics [Mi 76, 78] which tries to justify quantum mechanics on the basis of a priori reasons and which is still discussed in the present-day literature. Starting from the most general propositional language of quantum physics one arrives at a propositional system which turns out to be an orthomodular lattice. In addition, if all propositions of this lattice refer to properties of the same individual system, then the lattice in question is atomic and fulfils the covering law [Sta 84]. The gap between this lattice and the Hilbert lattice of subspaces of a Hilbert space can be closed by a more complicated property which was recently discovered by Solèr [Sol 95]. The three classical Hilbert spaces with real, complex, and hypercomplex numbers can then immediately be obtained.

Since yes-no propositions refer to objective properties only the meaning of probabilities for the values of nonobjective properties must still be explained. In the realistic interpretation of quantum mechanics [Mi 98] these probabilities refer to a situation *after* the measurement of the respective observable. Accordingly, it turns out that the relative frequency of a post-measurement value of this observable in a large ensemble of identically prepared systems approaches the value of the quantum mechanical probability if the number of systems is sufficiently large. This result can be formulated as a rigorous theorem, the proof of which once again makes extensive use of the results on infinite quantum systems which were obtained by von Neumann in his 1938 paper. [Mi 90, BLM 91, Mi 98] More details about this approach will be presented in the following sections.

3. THE PROBABILITY INTERPRETATION OF QUANTUM MECHANICS

For a systematic treatment of the problem mentioned, we apply the realistic interpretation and not the many worlds interpretation, which means that probabilities are reproduced in the statistics of the objectified system properties after the measurement. In order to avoid circularity we must begin with a probability free formulation and interpretation of quantum physics and then derive the probability structure. The mathematical methods are partly based on von Neumann's investigations [Neu 38] and partly on recent results [Gu 95]. Finally, we apply the modern version of the quantum theory of measurement which clearly distinguishes between preparation, premeasurement, objectification, and reading.

Let us discuss a simple problem in the two-dimensional Hilbert space $H_2 = C^2$. In the split beam experiment in Fig.1, which has been realised both with photons and with neutrons, the state φ of the incoming photon is split by a half-transparent mirror, beam splitter BS_1, into two components described by orthonormal states $\varphi(B)$ and $\varphi(\neg B)$. The two parts of the beam are reflected at two (fully reflecting) mirrors M_1 and M_2 and recombined with a phase shift δ at a second half-transparent mirror, beam splitter BS_2. Hence, $\varphi = 1/\sqrt{2}(\varphi(B) + e^{i\delta}\varphi(\neg B))$. In the experiment there are two mutually exclusive measuring arrangements: If the detectors D_1 and D_2 are in the positions (D_1^B, D_2^B) one observes which way (B or $\neg B$) the photon or neutron came. If the detectors are in the position (D_1^A, D_2^A) one observes the interference pattern, i.e. the intensities which depend on the phase δ.

In this experiment the object system S is prepared in the state $\varphi \in H_2$. There are two incommensurable and nonobjective observables, the path observable B with eigenstates $\varphi(B)$, $\varphi(\neg B)$ and the interference observable A with eigenstates $\varphi(A)$, $\varphi(\neg A)$. The probability for B (to register the system in D_2^B) and for $\neg B$ (to register the system in D_1^B) reads

$$p(\varphi, B) = p(\varphi, \neg B) = \tfrac{1}{2}. \tag{3.1}$$

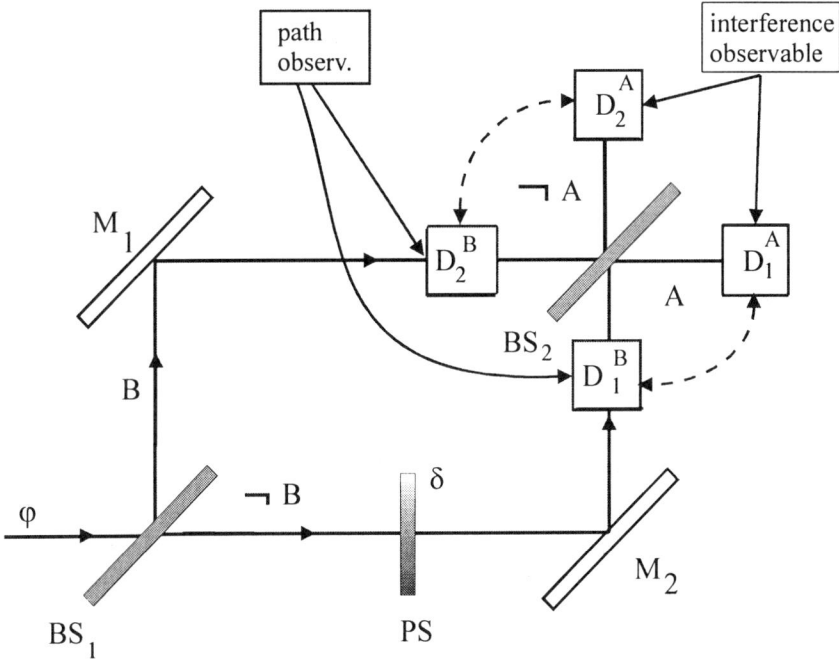

Fig. 1: Photon split-beam experiment with beam splitters BS_1 and BS_2, both half-reflecting mirrors, two fully reflecting mirrors M_1 and M_2, a phase shifter PS providing a phase shift δ, and two detectors D_1 and D_2 in mutually exclusive positions (D_1^A, D_2^A) and (D_1^B, D_2^B).

The probability for A (to register the system in D_1^A) and for $\neg A$ (to register the system in D_2^A) reads

$$p(\varphi, A) = \cos^2(\delta/2) \quad \text{and} \quad p(\varphi, \neg A) = \sin^2(\delta/2), \tag{3.2}$$

respectively. [Mi 97] This means that the relative frequency of systems arriving at D_1^A is approximately given by $\cos^2(\delta/2)$ and the relative frequency of systems arriving at D_2^A by $\sin^2(\delta/2)$.

Since the quantum mechanical probabilities refer to the state of the system *after* the measurement, we will briefly discuss the measurement process for the observable A and in particular for the observable P(A) which is given by the projection operator $P[\varphi(A)]$ of the state $\varphi(A)$. In order to describe that measurement process we consider the object system S with Hilbert space H_S and the measuring apparatus M with Hilbert space H_M. Let $\varphi \in H_S$ and $\Phi \in H_M$ be the preparations of S and M, respectively, i.e. the states prior to the measurement process. Here we consider a unitary and repeatable premeasurement of P(A), which can be described by a unitary operator U_A acting on the tensor product $\varphi \otimes \Phi$ of the compound system S+M. The operator U_A is further determined by the calibration postulate. If the object system S is in one of the two eigenstates $\varphi(A)$ or $\varphi(\neg A)$ of P(A), then the unitary and repeatable premeasurement must reproduce this state. This means that for the special preparation $\varphi(\delta=0) = \varphi(A)$ and $\varphi(\delta=\pi) = \varphi(\neg A)$ we have

$$\varphi(A) \otimes \Phi \quad \rightarrow \quad U_A(\varphi(A) \otimes \Phi) = \varphi(A) \otimes \Phi_A$$

$$\varphi(\neg A) \otimes \Phi \rightarrow \quad U_A(\varphi(\neg A) \otimes \Phi) = \varphi(\neg A) \otimes \Phi_{\neg A} \qquad (3.3)$$

where Φ_A and $\Phi_{\neg A}$ are eigenstates of a pointer observable $Z = Z_A P[\Phi_A] + Z_{\neg A} P[\Phi_{\neg A}]$ whose eigenvalues Z_A and $Z_{\neg A}$ indicate the measuring results A and $\neg A$, respectively. By means of the unitarity of U_A it follows for an arbitrary preparation

$$\varphi = (\varphi(A), \varphi) \, \varphi(A) + (\varphi(\neg A, \varphi) \, \varphi(\neg A) = \frac{1}{2}(1+e^{i\delta})\varphi(A) + \frac{1}{2}(1-e^{i\delta})\varphi(\neg A)$$

$$\varphi \otimes \Phi \rightarrow U_A(\varphi \otimes \Phi) = \frac{1}{2}(1+e^{i\delta}) \, \varphi(A) \otimes \Phi_A + \frac{1}{2}(1-e^{i\delta}) \, \varphi(\neg A) \otimes \Phi_{\neg A}. \qquad (3.4)$$

The state of the object system after the premeasurement is then given by the reduced mixed state

$$W_S(\varphi, A) = \cos^2 \delta/2 \ P[\varphi(A)] + \sin^2 \delta/2 \ P[\varphi(\neg A)]. \qquad (3.5)$$

The interpretation of this mixed state of the object system after the premeasurement is usually given by the probability reproducibility condition: The probability distribution $p(\varphi, A_i)$, $A_i \in \{A, \neg A\}$ which is induced by the preparation φ and the measured observable P(A), is reproduced in the statistics of the post-measurement values $(Z_A, Z_{\neg A})$ and states $(\Phi_A, \Phi_{\neg A})$ of the pointer. In case of repeatable measurements, i.e. in the realistic interpretation, this means that $p(\varphi, A_i)$ is also reproduced in the statistics of the states $(\varphi(A), \varphi(\neg A))$.

4. THE PROBABILITY REPRODUCIBILITY CONDITION

On the basis of these arguments we can now formulate the main problem: Given an ensemble of (before the measurement) identically prepared systems S which are in the reduced state (3.5) after the premeasurement of A, is it possible to justify that the (formal) probability $p(\varphi, A_i)$ is reproduced in the statistics of the measurement results A and $\neg A$, respectively? In order to answer this question consider a large number of identically prepared systems S_i in states φ^i which are not eigenstates of the observable P(A). Let us further assume that the unitary operator U_A which is used for a measurement of the observable P(A), fulfils the calibration postulate for repeatable measurements. Then we know that a measurement of the observable P(A) in case of the particular preparation $\varphi(A)$ leads with certainty to the states Φ_A and $\varphi(A)$ showing the result A. On the basis of this *probability free interpretation* of quantum mechanics we want to show that for arbitrary preparations $\varphi \neq \varphi(A)$, $\varphi \neq \varphi(\neg A)$ the formal probability $p(\varphi, A_i)$, which is induced by φ and P(A), is reproduced in the statistics of the measuring outcomes A_i . The *probability reproducibility condition* would then be a theorem of the probability free theory and no longer an additional postulate.

Let us consider N independent systems S_i with identical preparation φ^i as a compound system S^N in the tensor product state

$$(\varphi)^N = \varphi^1 \otimes \varphi^2 \otimes \ldots \otimes \varphi^N , \quad (\varphi)^N \in H(S^N)$$

where $H(S^N)$ is the tensor product of N Hilbert spaces $H(S_i)$. A premeasurement of A transforms the initial state φ^i of each system S_i into the mixed state

$$W^i = p(\varphi^i, A) P[\varphi^i(A)] + p(\varphi^i, \neg A) P[\varphi^i(\neg A)]$$

with eigenstates $\varphi^i(A_k)$ of A corresponding to results A_k . If A is measured on each system S_i, then the measurement result is given by a sequence $\{A_{l(1)}, \ldots A_{l(N)}\}$ of system properties $A_{l(i)}$ and states $\varphi(A_{l(i)})$, respectively, with an index sequence l: = { l(1), l(2), $\cdots$ l(N) } such that $A_{l(i)} \in \{A_k\} = \{A, \neg A\}$.

In the N-fold tensor product Hilbert space $H(S^N)$ of the compound system S^N the special states $(\varphi)_l^N = \varphi^{(1)}(A_{l(1)}) \otimes \cdots \otimes \varphi^{(N)}(A_{l(N)})$ with $\varphi^{(i)}(A_{l(i)}) \in H(S_i)$ form an orthonormal basis. The relative frequency $f^N(k, l)$ of outcomes $A_k \in \{A, \neg A\}$ in the state $(\varphi)^N_l$ is then given by $f^N(k, l) = 1/N \sum \delta_{l(i), k}$. We can now define in $H(S^N)$ an operator *"relative frequency of systems with properties A_k"* by

$$f^N_k : = \sum f^N(k, l) P[(\varphi)^N_l]$$

where the sum runs over all sequences l . The eigenvalue equation of this operator

$$f^N_k (\varphi)^N_l = f^N(k, l) (\varphi)^N_l$$

then shows that the relative frequency of the measurement result A_k is an objective property of S^N in the state $(\varphi)^N_l$ and given by $f^N(k,l)$. The eigenvalue equation can also be written in the equivalent form

$$tr\{ P[(\varphi)^N_l] (f^N_k - f^N(k, l))^2 \} = 0 .$$

After a premeasurement of P(A) a system S_i is in a mixed state W^i. If N pre-measurements of P(A) are performed, then the state of the compound system S^N is given by the N-fold tensor product state

$$(W)^N = W^1 \otimes W^2 \otimes \cdots \otimes W^N$$

of these mixed states W^i. One easily verifies that the expectation value of f^N_k in this product state is given by $p(\varphi, A_k)$. However, in general the state $(W)^N$ is not an eigenstate of the relative frequency operator f^N_k with eigenvalue $p(\varphi, A_k)$. This means that

$$T^N_k := \mathrm{tr}\{ (W)^N (f^N_k - p(\varphi, A_k))^2 \} \neq 0$$

and that the relative frequency of outcomes A_k is not an objective property of the system S^N in the state $(W)^N$.

In contrast to this somewhat unsatisfactory result one finds that for large values of N the post-measurement product state $(W)^N$ of the compound system S^N becomes an eigenstate of the operator f^N_k and the value of the relative frequency of results A_k approaches the probability $p(\varphi, A_k)$. Indeed, one finds after some tedious calculations [DeW 71], [Mi 90, 96]

$$T^N_k = 1/N\ p(\varphi, A_k) (1 - p(\varphi, A_k))$$

and thus one finally obtains the desired result

$$\lim_{N \to \infty} \mathrm{tr} \{ (W)^N (f^N_k - p(\varphi, A_k))^2 \} = 0$$

This means that in the limit of an infinite number N of systems the state $(W)^N$ is an eigenstate of the operator f^N_k of the relative frequency of results A_k and that the compound system S^N possesses the relative frequency $p(\varphi, A_k)$ of A_k as a objective property.

In order to ensure this way of reasoning against mathematical objections one has to guarantee that the overwhelming majority of index sequences $l = \{l(i)\}$ are random sequences and that the contribution of the non random sequences can be neglected. As a first orientation let us define the function $\delta(l) = \Sigma_k (f^N(k, l) - p(\varphi, A_k))^2$ in order to measure the degree to which a given sequence l deviates from a random sequence with weights $p(\varphi, A_k)$. A sequence will be called first random if $\delta(l) < \varepsilon$ for arbitrary positive ε. It can then be shown that in the limit $N \to \infty$ the contribution of the non first random sequences disappear. [Mi 90, 98] This first confirmation of our statistical results will be made more rigorous in the following section.

5. ELIMINATION OF PROBABILITIES – MATHEMATICAL CONSIDERATIONS

In classical probability theory it is well known that probabilities are not relative frequencies. According to the law of large numbers, for large N the relative frequency $f^N(k, l)$ for some index value k approaches the probability p(k) for almost all sequences l, i.e. with a probability which is equal to one. This means that probability statements cannot be replaced in general by probability-free statements, even if the number of systems is infinite. Hence it must be clarified for the above-mentioned result that in quantum mechanics probabilities can in fact be completely eliminated, provided the number of systems or of measurement outcomes is infinite.

According to a recent investigation [Gu 95] this can actually be shown to apply for the problem of the present paper. Let S^∞ be a compound system which is composed of infinitely many copies of split beam photons S_i. The system S^∞ can be described in the infinite tensor product space $H(S^\infty)$ which is a non-separable Hilbert space [Neu 38]. The eigenvalue equation of the observable P(A), say, for the individual system S_i is written here as

$$P^i(A)\,\varphi^i(A_k) = e_1^i(A_k)\,\varphi^i(A_k),\, k \in \{1, 2\},\, e_1^i(A_1) = 1,\, e_1^i(A_2) = 0.$$

For an arbitrary state $\varphi \in H(S)$ with $(\varphi,\varphi) = 1$ the tensor product state $(\varphi)^\infty$ is a state in $H(S^\infty)$. If one performs P(A) – measurements on each system S_i one obtains a sequence $s_{\{1\}}$ of P(A) – eigenvalues $e_1^i(A_{l(i)}) \in \{0, 1\}$, $l = \{l(1), l(2),\cdots\}$ and $l(i) \in \{1,2\}$. Let $\Sigma = \{s_{\{1\}}\}$ be the nondenumerable set of sequences of this kind. Any subset $\Sigma^{(\alpha)} \subseteq \Sigma$ describes a property. E.g. the set $\Sigma^{(p)}$ of sequences $s_{\{1\}}$ with the "probability p law of large number property" (with respect to $e_1^i = 1$) reads

$$\Sigma^{(p)} = \left\{(s_{\{1\}}) : \lim_{N \to \infty} 1/N \sum_1^N e_1^i(A_{l(i)}) = p \right\}$$

For any subset we introduce an indicator function $F^{(\alpha)}(s_{\{1\}})$ by $F^{(\alpha)}(s_{\{1\}}) = 1$ if $s_{\{1\}} \in \Sigma^{(\alpha)}$ and $F^{(\alpha)} = 0$ otherwise.

According to the spectral theorem by an indicator function $F^{(\alpha)}$ a projection operator $P^{(\alpha)}$ is uniquely defined in $H(S^\infty)$. Hence, for the product state $(\varphi)^\infty_1$ we obtain

$$P^{(\alpha)}(\varphi)^\infty_1 = F^{(\alpha)}(s_{\{1\}})(\varphi)^\infty_1.$$

In the nonseparable Hilbert space $H(S^\infty)$ there are many product states $(\psi)_k^\infty = \psi_{k(1)}^1 \otimes \psi_{k(2)}^2 \cdots$ which are not superpositions of the states $(\varphi)_l^\infty$. For these states one obtains

$$| P^{(\alpha)}(\psi)_k^\infty |^2 = \int F^{(\alpha)}(s_{\{1\}})d\mu \qquad (5.1)$$

where the measure μ depends on the state $(\psi)_k^\infty$ and on the measured observable. [Gu 95]

Let $P^{(1/2)}$ be the projection operator of the "probability $p = \frac{1}{2}$ law of large number property" of a sequence $s_{\{1\}}$:

$$F^{(\frac{1}{2})}(s_{\{1\}}) = \begin{cases} 1 & \text{if } \lim_{N\to\infty} 1/N \sum e_1^i(A_{1(i)}) = \frac{1}{2} \\ 0 & \text{otherwise.} \end{cases}$$

This situation is realised, for example, if all systems S_n are prepared in states

$$\psi^n = \frac{1}{2}(1+e^{i\pi/2})\,\varphi^n(A) + \frac{1}{2}(1-e^{i\pi/2})\,\varphi^n(\neg A)$$

and $P^n(A)$ is measured. The post-premeasurement mixed states read in this case

$$W^n = \frac{1}{2}P[\varphi^{(n)}(A)] + \frac{1}{2}P[\varphi^{(n)}(\neg A)]$$

and the probability p to find the value $e_1^n(A_n) = 1$ is then given by $p = |< \varphi^n (A_1)|\psi^n >|^2 = \frac{1}{2}$ for each n. The classical law of large numbers asserts in this case that the probability for the relative frequency $\frac{1}{2}$ for the value "1" in a sequence $s_{\{1\}}$ is equal to 1 , i.e.

$$\int F^{(1/2)}(s_{\{1\}})\, d\mu(s_{\{1\}}) = 1$$

This means that the relative frequency of "1" for almost all sequences amounts $\frac{1}{2}$. Together with equation (5.1) we get the relation $|P^{(1/2)}(\psi)^\infty|^2 = 1$ and thus

$$P^{(1/2)}(\psi)^\infty = 1 \cdot (\psi)^\infty. \qquad (5.2)$$

According to the realistic interpretation of quantum mechanics this eigenvalue equation (5.2) means that the compound system S^∞ *possesses* the property given by $P^{(1/2)}$. Hence this property, or the relative frequency value $\frac{1}{2}$ property, pertains to the system without any reference to probability.

It can be shown [Gu 95] that this way of reasoning can be applied to any "tail-property" with probability 1, e.g. to randomness. Within the context of the present problem this means, that in quantum mechanics the probabilistic way of speaking can be replaced by statements which *do not* refer to probability. This result justifies the interpretation of the approximate results of Section 4. Whereas in classical probability theory probabilistic statements like p = ½ can never be reduced to statements which, for an arbitrary large ensemble, hold with certainty but only to statements which are almost true, in quantum mechanics probability statements can be replaced by propositions which hold without reference to probability.

6. CONCLUDING REMARKS

In the short period from 1932 to 1938 John von Neumann initiated three important scientific developments. He formulated the first quantum mechanical theory of the measurement process, he discovered the logical structure underlying quantum mechanics, and he formulated the theory of infinite tensor products as a means for dealing with infinite quantum systems. During the last decades these contributions have led to the contemporary quantum theory of measurement, the fully elaborated quantum logic and its operational foundation, and to the theory of infinite quantum ensembles, respectively.

These results allow for a reformulation of quantum mechanics as a theory without probabilities. The operational approach to quantum logic and the quantum theory of measurement lead to a quantum theory of yes-no observables and to an interpretation which is based exclusively on the calibration postulate. This rudimentary quantum mechanics is a theory without probabilities. However, it can be shown that within this probability-free theory, the probability reproducibility condition can be proved as a theorem which holds for infinite quantum ensembles. This result – which is based essentially on the theory of infinite tensor products – shows that quantum mechanics is first of all a theory without probabilities and that probability statements emerge from this theory as yes-no propositions about infinite quantum ensembles.

REFERENCES

[BLM 91]   Busch, P., P. Lahti and P. Mittelstaedt (1991), *The Quantum Theory of Measurement,* Springer, Heidelberg (2nd edition 1996).

[DeW 71]   DeWitt, B.S. (1971), "The Many-Universes Interpretation of Quantum Mechanics", in: *Foundations of Quantum Mechanics*, IL Corso, B. d'Espagnat, ed., Academic Press, New York, pp. 167-218.

[Eve 57]   Everett, H. (1957), "Relative State Formulation of Quantum Mechanics", *Review of Modern Physics,* **29**, pp. 454-62.

[Fink 62]   Finkelstein, D. (1962), "The logic of quantum physics", *Trans. New York Acad. Sci.* **25**, pp. 621-37.

[Gu 95]   Gutmann, S. (1995), "Using Classical Probability to Guarantee Properties of Infinite Quantum Sequences", quant-ph/ 9506016.

[Hart 68]   Hartle, J. B. (1968), "Quantum mechanics of individual systems", *Am. Journ. Phys.* **36**, pp. 704-12.

[Mi 76]   Mittelstaedt, P. (1976), *Philosphical Problems of Modern Physics*, D. Reidel, Dordrecht.

[Mi 78]   Mittelstaedt, P. (1978), *Quantum Logic*, D. Reidel, Dordrecht.

[Mi 90]   Mittelstaedt, P. (1990), "The objectification in the measuring process and the many worlds interpretation", in: *Symposium on the Foundations of Modern Physics 1990*, World Scientific, Singapore, pp. 261-279.

[Mi 98]   Mittelstaedt, P. (1998), *The Interpretation of Quantum Mechanics and the Measurement Process*, Cambridge, University Press.

[Neu 32]   von Neumann, J. (1932), *Mathematische Grundlagen der Quantenmechanik*, Springer Verlag, Berlin.

[Neu 36]   Birkhoff, G. and J.v. Neumann, (1936), "The Logic of Quantum Mechanics", *Annals of Mathematics* **37**, pp. 823-43.

[Neu 38]   von Neumann, J. (1938), "On infinite direct products", *Compositio Mathematica* **6**, pp. 1-77.

[Sta 84]   Stachow, E.-W. (1984), "Structures of a Quantum Language for Individual Systems", in: *Recent Developments in Quantum Logic*, eds. P. Mittelstaedt and E.-W.Stachow, BI-Wissenschaftsverlag, Mannheim.

[Sol 95]   Solèr, M.P. (1995), "Characterisation of Hilbert Spaces by Orthomodular Lattices", *Communications in Algebra,* **23**(1), pp. 219-243.

Institut für Theoretische Physik
Universität zu Köln
Zülpicher Straße 77
D-50937 Köln
Germany

LÁSZLÓ E. SZABÓ

CRITICAL REFLECTIONS ON QUANTUM PROBABILITY THEORY[*]

1. INTRODUCTION

The story of *quantum probability theory* and *quantum logic* begins with von Neumann's recognition[1], that quantum mechanics *can be regarded* as a kind of "probability theory", if the subspace lattice $L(H)$ of the system's Hilbert space H plays the role of event algebra and the '$tr(WE)$'-s play the role of probability distributions over these events. This idea had been completed in the *Gleason theorem*[2]:

Definition 1 *A non-negative real function μ on $L(H)$ is called (quantum) probability measure if $\mu(H) = 1$ and if whenever $E_1, E_2, \ldots$ are pairwise orthogonal subspaces, and $E = \bigvee_{i=1}^{\infty} E_i$, then $\mu(E) = \sum_{i=1}^{\infty} \mu(E_i)$.*

Theorem 1 (Gleason 1957) *If H is a real or complex Hilbert space of dimension greater than 2, and μ is a probability measure on $L(H)$, then there exists a density operator W on H, such that $(\forall E \in L(H)) [\mu(E) = tr(WE)]$.*[3]

Formally, on the basis of mathematical analogy, the intersection and the (closed) linear union of subspaces are called 'conjunction' and 'disjunction' in the underlying 'event' lattice of *quantum probability theory*.

So far so good, but many think that we can go beyond this simple mathematical analogy, and regard quantum probability theory as a real probability theory replacing the classical one in its role in describing our world. As if the 'change of meaning' were such an easy matter[4], we are suggested to use the quantum logical connectives completely incompatible with the logical connectives of the metalanguage, at least when we are talking about the microphysical reality.

The quantum probability/quantum logic approach is based on the conviction that there are phenomena of quantum physics which cannot be accommodated in a world describable by the classical Kolmogorov theory of probability alone. The majority of experts share this conviction and, due to Feynmann[5], this opinion is also quite common among physicists whose field of interest is not foundations of physics.

There has been serious criticism of this approach, too. The first paper pointing out the contradictions which may appear if we assume that the event algebra is isomorphic with the subspace lattice of a Hilbert space was published by Strauss[6] a year after the famous Birkhoff and Neumann paper. It seems, however, that

M. Rédei and M. Stöltzner (eds.),
John von Neumann and the Foundations of Quantum Physics, 201–219.
© 2001 *Kluwer Academic Publishers. Printed in the Netherlands.*

quantum probabilists and quantum logicians completely ignore the serious pitfalls pointed out by these authors. Bell expressed quite a similar disappointment[7]:

Why did such serious people take so seriously axioms which now seem so arbitrary? I suspect that they were misled by the pernicious misuse of the word 'measurement' in contemporary theory. This word very strongly suggests the ascertaining of some preexisting property of some thing, any instrument involved playing a purely passive role. Quantum experiments are just not like that, as we learned especially from Bohr. The results have to be regarded as the joint product of 'system' and 'apparatus,' the complete experimental set-up. But the misuse of the word 'measurement' makes it easy to forget this and then to expect that the 'results of measurements' should obey some simple logic in which the apparatus is not mentioned. The resulting difficulties soon show that any such logic is not ordinary logic. It is my impression that the whole vast subject of 'Quantum Logic' has arisen in this way from the misuse of a word. I am convinced that the word 'measurement' has now been so abused that the field would be significantly advanced by banning its use altogether, in favor for example of the word 'experiment'.

My aim is to show in this paper that, beyond its counterintuitiveness, quantum probability theory is *inadequate* and *unnecessary*. It is inadequate because there cannot exist events in reality the *relative frequencies* of which would be equal to quantum probabilities. And it is unnecessary too, because there is no need in quantum mechanics to supersede the Kolmogorov theory of probability; we will see how quantum phenomena can, in general, be accommodated in the classical Kolmogorov theory of probability.

2. NO FREQUENCY INTERPRETATION FOR QUANTUM PROBABILITIES

2.1 Nonsensical probabilities for non-commuting elements

The following theorem illustrates that for non-commuting elements of $L(H)$ quantum probability theory predicts "probabilities" which are not interpretable as relative frequencies.

Theorem 1 Let E_1 and E_2 be two non-commuting elements of $L(H)$. There exists a pure state Ψ for which the probabilities violate inequality

$$p(E_1) + p(E_2) - p(E_1 \wedge E_2) \leq 1 \qquad (1)$$

Proof Arbitrary E_1 and E_2 can be written in the following form:

$$E_1 = (E_1 \wedge E_2) \vee A,$$
$$E_2 = (E_1 \wedge E_2) \vee B, \qquad (2)$$

such that $A \perp E_1 \wedge E_2$ and $B \perp E_1 \wedge E_2$.

First we prove the following statements:

(a) $A \neq \emptyset$ and $B \neq \emptyset$ and $A \neq B$.

(b) $A \not\!\!\perp B$.

Indeed, if $A = \emptyset$ or $B = \emptyset$ or $A = B$ would hold then either $E_1 < E_2$ or $E_2 < E_1$, that would contradict to the assumed non-commutativity of E_1 and E_2. For proving (b) we show that from $A \perp B$ also the commutativity of E_1 and E_2 would follow. Commutativity is equivalent with $E_1 = (E_1 \wedge E_2) \vee (E_1 \wedge E_2^\perp)$. Using (2) we have

$$
\begin{aligned}
&(E_1 \wedge E_2) \vee \left[((E_1 \wedge E_2) \vee A) \wedge ((E_1 \wedge E_2) \vee B)^\perp\right] \\
&= (E_1 \wedge E_2) \vee \left[((E_1 \wedge E_2) \vee A) \wedge ((E_1 \wedge E_2)^\perp \wedge B^\perp)\right] \qquad (3)\\
&= (E_1 \wedge E_2) \vee \left[((E_1 \wedge E_2) \vee A) \wedge (E_1 \wedge E_2)^\perp\right] \wedge B^\perp
\end{aligned}
$$

Since $A \perp (E_1 \wedge E_2)$, the distributivity holds in the square brackets. Therefore we can continue (3) as follows:

$$
\begin{aligned}
&= (E_1 \wedge E_2) \vee (A \wedge B^\perp) \\
&= (E_1 \wedge E_2) \vee A = E_1
\end{aligned}
$$

which proves (b).

Now, from (a) and (b) it follows that there exists at least one normalized vector $\Psi \in A$ such that $\Psi \notin B$. Such a Ψ is a state vector for which the inequality

$$
\underbrace{\langle \Psi, E_1 \Psi \rangle}_{1} + \underbrace{\langle \Psi, E_2 \Psi \rangle}_{>0} - \underbrace{\langle \Psi, (E_1 \wedge E_2 \Psi) \rangle}_{0} > 1 \qquad (4)
$$

holds, which is a violation of (1).

The strange meaning of (4) is obvious! If E_1 happens with certainty, how can E_2 occur without E_1? It is remarkable that equation

$$
p(E_1 \vee E_2) = p(E_1) + p(E_2) - p(E_1 \wedge E_2) \qquad (5)
$$

implies inequality (1). Von Neumann regarded (5) as a fundamental property of a probability measure, on which we should insist in the quantum case, too.[8]

2.2 The Laboratory Record Argument

While I am completely convinced that Theorem 1 indicates a serious problem, a quantum probabilist might argue that this problem is a fictitious one because E_1

and E_2 do not commute and therefore they are not measurable simultaneously. Consequently, there is no experimental situation which would give rise to the nonsensical probabilities (4).

My second argument does not, however, appeal to conjunctions of noncommuting elements of $L(H)$. Neither does this argument appeal to a clear concept of 'event' in quantum mechanics. The only thing it appeals to is that quantum mechanics must be applicable to the everyday laboratory situations.

Example I don't know what is "quantum event", the probability of which is a number like $tr(WE)$, but anyone who knows should be able to tell a laboratory assistant when does such an "event" occur. According to the instruction (s)he makes a record like this:

Run	A_1	A_2	A_3	A_4	$A_1 A_3$	$A_1 A_4$	$A_2 A_3$	$A_2 A_4$
1	0	0	1	0	0	0	0	0
2	1	0	1	0	1	0	0	0
3	1	0	0	0	0	0	0	0
4	0	1	0	1	0	0	0	1
5	1	0	0	1	0	1	0	0
6	1	0	1	0	1	0	0	0
7	1	0	1	0	1	0	0	0
8	0	0	1	0	0	0	0	0
$\vdots$	$\vdots$	$\vdots$	$\vdots$	$\vdots$	$\vdots$	$\vdots$	$\vdots$	$\vdots$
99998	1	0	0	0	0	0	0	0
99999	0	0	1	0	0	0	0	0
100000	1	0	0	1	0	1	0	0
$N = 100000$	N_1	N_2	N_3	N_4	N_{13}	N_{14}	N_{23}	N_{24}

where, assume, (A_1, A_3), (A_1, A_4), (A_2, A_3) and (A_2, A_4) are pairs of "quantum events" corresponding to commuting projectors. "0" stands for the case if an event does not happen and "1" if it does. The relative frequencies can be computed from this table:

$$\nu_1 = \frac{N_1}{N}, \ \nu_2 = \frac{N_2}{N}, \ \dots \nu_{24} = \frac{N_{24}}{N}$$

Notice that each row of this table corresponds to one of the 2^4 possible classical truth-value functions over propositions A_1 happened, A_2 happened, A_3 happened, and A_4 happened. In Pitowsky's language (see Appendix I) we could say that each row corresponds to a vertex, $\mathbf{u}^\varepsilon$, of the corresponding classical polytope $\mathcal{C}(4, S)$, where $S = \{\{i, j\} \mid i = 1, 2; \ j = 3, 4\}$.

We can sum up our observation in the following

Stipulation *If the components of a (finite) correlation vector* $\mathbf{p} = (\nu_i, \nu_{ij})$ *can be interpreted as relative frequencies of events, computed from a laboratory report,* *then*

$$(\forall i)\,(\forall j) \left[\begin{array}{c} \nu_i = \sum_{\varepsilon \in \{0,1\}^n} \lambda_\varepsilon u_i^\varepsilon \\ \nu_{ij} = \sum_{\varepsilon \in \{0,1\}^n} \lambda_\varepsilon u_{ij}^\varepsilon \end{array} \;\middle|\; \lambda_\varepsilon \geq 0; \; \sum_{\varepsilon \in \{0,1\}^n} \lambda_\varepsilon = 1 \right] \quad (6)$$

$\lambda_\varepsilon = \frac{N_\varepsilon}{N}$, *where N_ε is the number of type-u^ε rows in the record.*

In other words, relative frequencies are weighted averages of the classical truth-values. By virtue of Pitowsky theorem (see Appendix I) (6) implies that relative frequencies must satisfy condition $\mathbf{p} \in \mathcal{C}(n, S)$ and, consequently, the corresponding Bell-type inequalities.

Now, consider the well known EPR-Aspect experiment (Appendix II). Just as in the above example, we have four events and four conjunctions. Each of the conjunctions belongs to *commuting* projectors! In case of a particular choice of directions along which the spin components are measured, the quantum probabilities form the following correlation vector:

$$\mathbf{p} = (q\,(A_1)\,,\, q\,(A_2)\,,\, q\,(A_3) \,\ldots\, q\,(A_2 A_4)) = \left(\frac{1}{2}, \frac{1}{2}, \frac{1}{2}, \frac{1}{2}, \frac{3}{8}, \frac{3}{8}, 0, \frac{3}{8} \right)$$

But these numbers do not satisfy the Clauser-Horne inequalities! That is, *quantum probabilities cannot be, in general, interpreted as relative frequencies of events.*
In order to clarify the importance of this result, I need to make a few remarks:

1. What we have actually recognized here is that "quantum probability" must not be interpreted as probability, if we insist on the frequentists' understanding of the term. The quantum mechanical $tr\,(WE)$ is not the (absolute) probability of a real event, but it is a conditional probability $p(A\,|a)$, which means the probability of the outcome-event A, given that the measurement-preparation a has happened. $tr\,(WE)$ no doubt does have such a meaning, I believe, in accordance with the everyday laboratory practice. The controversial question is, whether it means something more: whether there exists an event $\widetilde{E}$ in reality, such that $tr\,(WE) = p\left(\widetilde{E}\right)$ holds. As we could see, if probability means relative frequency, then the probability function on an event algebra can be nothing else but the weighted average of the classical truth-value functions, therefore, according to Pitowsky's theorem, it must be Kolmogorovian, consequently, it satisfies the Bell inequalities. Since the $tr\,(WE)$-type quantities violate Bell inequalities, they are not interpretable as relative frequencies. In other words, *there are no events, in general, that would happen with probabilities like $tr\,(WE)$.*

2. And accordingly, as I pointed out in an earlier paper[9], such a question, for example, whether there exists a common cause explanation for the EPR correlations, in its original form, is meaningless, because it is about the existence of a common cause for correlations among *non-existing* events. *What*

the violation of the Bell inequalities indicates is not that the EPR correla-
tions do not have a common cause, but rather that there are no events which
would have such correlations.

3. The same holds for the question whether the classical Kolmogorov theory of
   probability can or cannot describe the probabilities of events observed in the
   quantum world. If we insist on the frequency interpretation of probability,
   the answer is clear: yes, it can, because relative frequencies *are* Kolmogorovian. If "quantum probabilities" do not satisfy the Kolmogorov axioms, then
   they are not interpretable as relative frequencies, or, if they are thought as
   relative frequencies, then *there are no events in reality which would happen*
   with these frequencies.

4. The fact that "non-Kolmogorovian probabilities" are not interpretable as relative frequencies explains why von Neumann's program to create a "non-
   commuting version of probability theory" necessarily failed[10]. In his new
   mathematical model he wanted to *reproduce* the same *values* of "quantum
   probabilities"!

$$\text{something new} \overset{1}{=} tr(WE) \overset{2}{=} \text{relative frequency of event}$$

Solving the problem of equation 1 does not resolve the contradiction at equation 2.

5. Some people claim that "a quantum probabilist is, of course, not a frequentist". Then what is (s)he? What is a tenable interpretation of probability, different from the frequency interpretation, applicable to quantum mechanics?
   In addition to the problem, that even the subjective probabilities must, in
   final analysis, satisfy the Kolmogorov axioms, we have no freedom in interpreting quantum mechanical probabilities, at all. It is because the laboratory
   practice singles out the frequency interpretation. If quantum probabilist's
   quantum mechanics is identical with the experimental physicist's one, then
   there is no other meaning of probability than relative frequency. Only in this
   sense we can say that quantum mechanics is an experimentally confirmed
   physical theory.

3. DO WE REALLY NEED QUANTUM PROBABILITY THEORY?

3.1 The double slit experiment

The conclusion we can draw from the previous section is that $L(H)$ can hardly play
the role of an "algebra of events" for a probability theory and the numbers $tr(WE)$
cannot be interpreted as relative frequencies of events. Sometimes it is claimed that

in spite of the above "difficulties" – which is of course an understatement – we must figure out something, we must solve these interpretational problems of "quantum probability theory" because there are phenomena in quantum physics which are not describable with the classical theory of probability. In this section I want to show that this is not true, we don't need quantum probability theory.

Our next example is the double slit experiment which is often quoted in order to justify why we need quantum probability theory (Fig. 1).

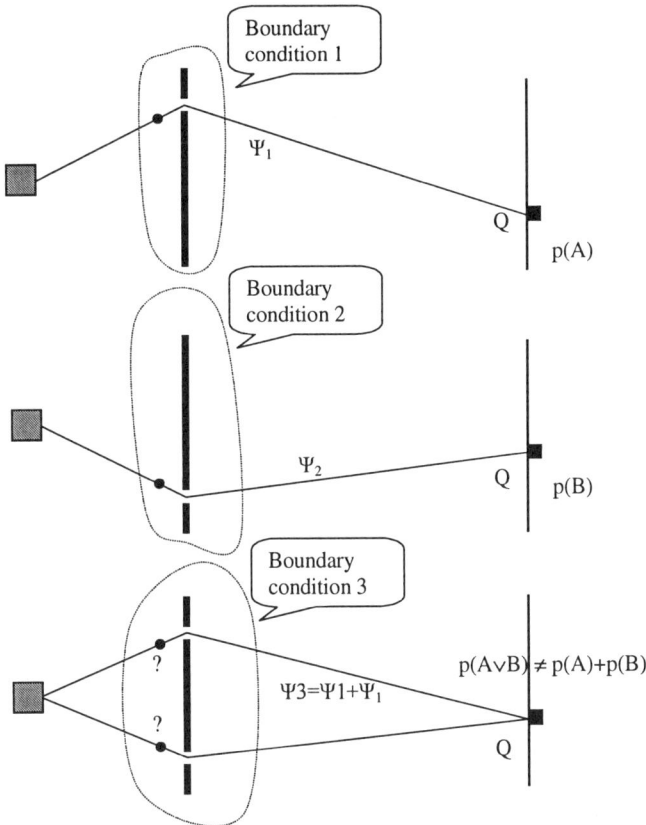

Figure 1: According to the usual interpretation the double slit experiment indicates that the rules of computing probabilities in quantum mechanics must be different from that of classical probability theory.

Denote $p(A)$ the probability of that "the particle arrives at a given point of the screen, Q, when only slit 1 is open". $p(B)$ denotes the similar probability for slit 2. In the experiment one finds that

$$p(A) + p(B) \neq p(A \vee B) \qquad (7)$$

where $p(A \vee B)$ stands for the probability of "the particle arrives at point Q, when both slits are open'. According to the usual interpretation the double slit experiment shows that "*the method of computing probabilities involving subatomic particles is different from that of classical probability theory*"[11]. Therefore we must, as the usual conclusion says, 1) change probabilities for complex amplitudes (Feynmann) or 2) give up the Boolean event lattice and classical probability theory (quantum probability theory). There is, however, a bad mistake in this standard claim, namely a misinterpretation of the superposition principle. While it is true, approximately, that the solution of the linear Schrödinger equation with boundary condition 3 (both slits are open) is a superposition of the solutions belonging to boundary conditions 1 and 2 (see Fig. 1), that is, $\Psi_3 = \Psi_1 + \Psi_2$, it doesn't mean, on the other hand, that the relation of the three situations could be described as a logical disjunction.

Let me show at least two different ways in which we can, contrary to the standard view, correctly describe the double slit experiment within the framework of classical probability theory. In both cases, the precise usage of notions "event" and "disjunction" is what makes the classical probability theory satisfactory, while the formula (7) is, as we will see soon, based on the misuse of these notions.

Version I

We must precisely distinguish the following events:

A: "Slit 1 is open and slit 2 is closed and the particle is detected at Q"

B: "Slit 1 is closed and slit 2 is open and the particle is detected at Q"

C: "Slit 1 is open and slit 2 is open and the particle is detected at Q"

Obviously,

$$A \vee B \neq C$$

Consequently, we are not surprised that

$$p(A \vee B) = p(A) + p(B) \neq p(C)$$

That is, formula (7) is incorrect, consequently there is no violation of classical rules of probability calculation.

Version II

There is only one event:

D: "The particle is detected at Q"

There are, however, different *conditions* under which the probabilities are understood. But the Kolmogorov axioms are meant to apply to probabilities belonging to *one common* system of conditions! Consequently, it does not mean a violation of the Kolmogorov axioms if

$$p_{1 \text{ is open; 2 is closed}}(D) + p_{1 \text{ is closed; 2 is open}}(D) \neq p_{1 \text{ is open; 2 is open}}(D) \qquad (8)$$

It is to be mentioned here that the conditions written in the indexes of probabilities in (8) are sometimes meant as conditioning events:

$$\alpha: \quad \text{"1 is open; 2 is closed"}$$
$$\beta: \quad \text{"1 is closed; 2 is open"}$$
$$\gamma: \quad \text{"1 is open; 2 is open"}$$

If, falsely, γ is taken as the disjunction of α and β

$$\gamma = \alpha \vee \beta \qquad (9)$$

then the conditional probabilities should satisfy the following inequality:

$$min\,(p(D\,|\alpha), p(D\,|\beta)) \leq p(D\,|\alpha \vee \beta) \leq max\,(p(D\,|\alpha), p(D\,|\beta)) \qquad (10)$$

In the experiment, the interference pattern shows many violations of this condition. And again, we arrive at an "argument" for rejecting the probability theory and/or logic involved in the deduction of (10). The other reaction is, as van Fraassen rightly remarks[12], to reject (9). However I disagree with his explanation: *"To put it in other words: we reject the idea that the electron must have a definite position (in slit 1 or in slit 2) at time of its passing the barrier."* Nothing forces us to jump to such a conclusion. (9) fails simply because sentence γ is not the disjunction of α and β.

3.2 The EPR experiment

We have seen that the double slit experiment does not prove the nonapplicability of Kolmogorov's classical theory of probability. It is true, however, that this example is not regarded as a serious one: it is rather used in quantum mechanics text books only. But now we are going to analyze the Einstein-Podolsky-Rosen experiment which is regarded as a crucial, empirically tested situation providing probabilities which do not conform with the Kolmogorovian theory. You can find the description of the experiment in Appendix II.

The question we would like to answer is whether the probabilities (17), measured in the Aspect experiment, can be accommodated in a Kolmogorovian probability model, or not. Let me first recall the standard argumentation which yields to a negative answer. Consider

$$p_1 = tr(WA_1), \ p_2 = tr(WA_2), \ p_3 = tr(WA_3), \ p_4 = tr(WA_4)$$

$$p_{13} = tr(WA_1A_3), \ p_{14} = tr(WA_1A_4), \ p_{23} = tr(WA_2A_3), \ p_{24} = tr(WA_2A_4)$$

the values of which are given in (18). Substituting these values into the last Clauser-Horne inequality of (16) we find that

$$\mathbf{p} = \left(\frac{1}{2}, \frac{1}{2}, \frac{1}{2}, \frac{1}{2}, \frac{3}{8}, \frac{3}{8}, 0, \frac{3}{8}\right) \notin C(n, S).$$

According to the Pitowsky theorem, as the standard argument goes on, the probabilities observed in the Aspect experiment have no Kolmogorovian representation.

We must recognize, however, that this is just the same problem we discussed in section 2.2. What the violation of the Clauser-Horne inequalities indicates is not that probabilities (17) cannot be represented in a Kolmogorov probability space, but *there are no events in reality which would happen with these frequencies*. The probabilities of the real physical events observed in the experiment, $p(A_1), p(A_2), p(A_3), p(A_4), p(a_1), p(a_2), p(a_3), p(a_4)$, do not violate Kolmogorovity, as it must be the case, since they are relative frequencies counting down from a laboratory record of a real experiment:

$$
\begin{aligned}
\mathbf{p} &= (p(A_1), p(A_2), p(A_3), p(A_4), p(a_1), p(a_2), p(a_3), p(a_4), \\
&\quad p(A_1 \wedge A_2), \ldots p(A_4 \wedge a_4), \ldots p(a_3 \wedge a_4)) \\
&= \left(\frac{1}{4}, \frac{1}{4}, \frac{1}{4}, \frac{1}{4}, \frac{1}{2}, \frac{1}{2}, \frac{1}{2}, \frac{1}{2}, 0, \frac{3}{32}, \frac{3}{32}, \frac{1}{4}, 0, \frac{1}{8}, \frac{1}{8}, 0, \frac{3}{32}, 0, \frac{1}{4}, \frac{1}{8}, \right. \quad (11) \\
&\quad \left. \frac{1}{8}, 0, \frac{1}{8}, \frac{1}{8}, \frac{1}{4}, 0, \frac{1}{8}, \frac{1}{8}, 0, \frac{1}{4}, 0, \frac{1}{4}, \frac{1}{4}, \frac{1}{4}, \frac{1}{4}, 0\right) \\
&\in C(8, S_{\max})
\end{aligned}
$$

Of course, there are no derived Bell-type inequalities for 36-dimensional correlation vectors, therefore condition (11) is tested numerically, by computer[13]. However, it is quite plausible if you take into account that the correlation vector made of the measured relative frequencies of the outcome events only,

$$
\begin{aligned}
\mathbf{p} &= (p(A_1), p(A_2), p(A_3), p(A_4), p(A_1 \wedge A_3), p(A_1 \wedge A_4), \\
&\quad p(A_2 \wedge A_3), p(A_2 \wedge A_4)) \\
&= \left(\frac{1}{4}, \frac{1}{4}, \frac{1}{4}, \frac{1}{4}, \frac{3}{32}, \frac{3}{32}, 0, \frac{3}{32}\right)
\end{aligned}
$$

satisfies the Clauser-Horne inequalities.

What we can observe here is nothing else but what I formulated in my "Kolmogorovian Censorship" hypothesis[14]. We never encounter "naked" quantum probabilities in reality. A correlation vector consisting of empirically observed probabilities is always of the following form:

$$\mathbf{p} = (q_1 \widetilde{p}_1, q_2 \widetilde{p}_2, \ldots \widetilde{p}_1, \widetilde{p}_2, \ldots q_{ij} \widetilde{p}_{ij}, \ldots \widetilde{p}_{kl}, \ldots)$$

where $(q_1 \ldots q_n \ldots q_{ij} \ldots)$ are quantum probabilities and $(\widetilde{p}_1 \ldots \widetilde{p}_n \ldots \widetilde{p}_{ij} \ldots)$ are classical probabilities with which the corresponding measurement preparations occur. The hypothesis says that such a $\mathbf{p}$ is always classical. (In general, a product of quantum and classical probabilities does not necessarily form a classical correlation vector.) Bana and Durt proved such a theorem for finite number of measurements[15]. In the next section I shall give a more simple proof which is also valid for the infinite (but countable) case.

4. THE MEANING OF QUANTUM PROBABILITY

We have seen that quantum probability is not probability, because it cannot be, in general, the relative frequency of an event. What is then the correct interpretation of the $tr(WE)$-type quantitites? As I have mentioned already, it is a *conditional* probability, $tr(WE) = p(E \mid e)$, that is, the probabilitiy of the measurement oucome E, given that the measurement preparation e happened. There is nothing new in this interpretation; whenever a statistical prediction of quantum mechanics is experimentally tested, $tr(WE)$ is identified with $p(E \mid e)$.

In order to make this picture complete, we want to see a big Kolmogorovian probability space where all these conditional probabilities are, jointly, represented. First, let me give an example, how we can solve a similar problem in the classical theory of probability.

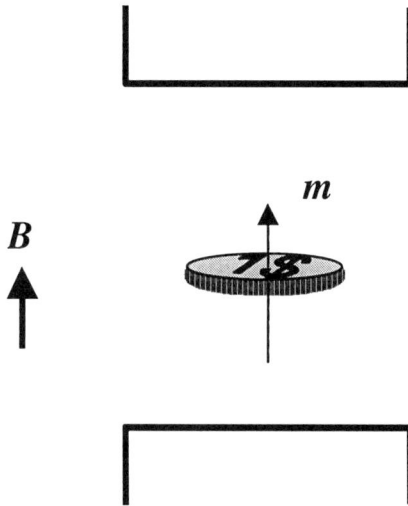

Figure 2: We are tossing a coin which has a little magnetic moment. The probabilities of Heads (H) and Tails (T) are modified if the magnetic field is on.

We are tossing a coin which has a little magnetic momentum (Fig. 2). If the magnetic field is off, the probabilities are

$$p_{\text{off}}(\text{H}) \quad = \quad 0.5$$
$$p_{\text{off}}(\text{T}) \quad = \quad 0.5$$

If the magnetic field is on, the probabilities are different:

$$p_{\text{on}}(\text{H}) \quad = \quad 0.2$$
$$p_{\text{on}}(\text{T}) \quad = \quad 0.8$$

The event algebra $\mathcal{A}$ is shown in Figure 3. For the two different physical conditions we have two separate probability models: $(\mathcal{A}, p_{\text{off}})$ and $(\mathcal{A}, p_{\text{on}})$, which are, separately, Kolmogorovian. For example, they satisfy the simplest Bell-type inequality (1):

$$p_{\text{off}}(\text{H}) + p_{\text{off}}(\text{T}) - p_{\text{off}}(\text{H} \wedge \text{T}) \leq 1$$

and separately,

$$p_{\text{on}}(H) + p_{\text{on}}(T) - p_{\text{on}}(H \wedge T) \leq 1$$

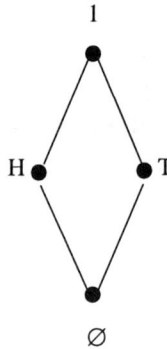

Figure 3: The original event algebra $\mathcal{A}$.

Now, if we "forgot" that these probabilities belong to different physical conditions, and put them together into one formula prescribed for a Kolmogorovian probability theory, then we would find the "violation of the rules of classical probability theory":

$$p_{\text{off}}(\text{H}) + p_{\text{on}}(\text{T})) - p_{\text{off}}(\text{H} \wedge \text{T}) = 0.5 + 0.8 > 1$$

or

$$p_{\text{on}}(\text{H}) + p_{\text{off}}(\text{T}) = 0.2 + 0.5 \neq 1 = p_{\text{off}}(1) = p_{\text{off}}(\text{H} \vee \text{T})$$

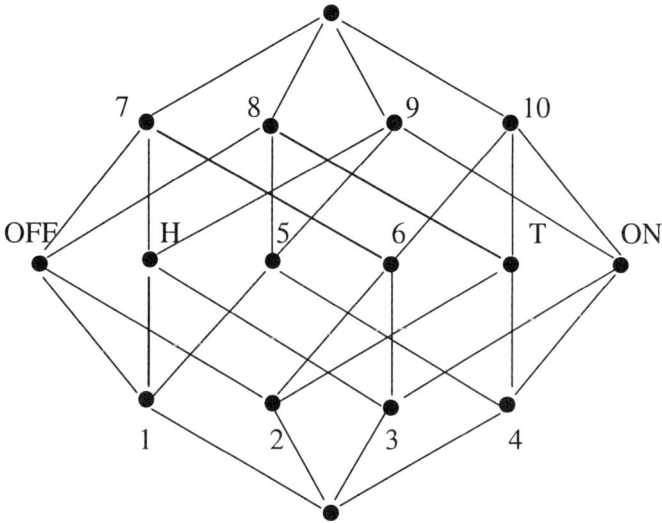

Figure 4: The extended event algebra $\mathcal{A}'$.

If we wish to see these probabilities together in one probability model, then it is necessary 1) to extend the original event algebra (Fig. 4) in order to include new events corresponding to the different conditions, and, of course, 2) we must be able to tell the probabilities of the conditioning events. In the example in question assume that $p(\text{OFF}) = 0.5$ and $p(\text{ON}) = 0.5$. So, the unified probability model is $(\mathcal{A}', p)$, where

$$
\begin{aligned}
p(1) = p(2) = p(9) = p(10) &= 0.25 \\
p(3) = p(8) &= 0.1 \\
p(4) = p(7) &= 0.4 \\
p(\text{OFF}) = p(\text{ON}) &= 0.5 \qquad\qquad (12) \\
p(\text{H}) = p(6) &= 0.35 \\
p(\text{T}) = p(5) &= 0.65
\end{aligned}
$$

The original probabilities are represented as *conditional* probabilities (defined by the Bayes law):

$$
\begin{aligned}
p_{\text{on}}(\text{H}) &= \frac{p(\text{H} \wedge \text{ON})}{p(\text{ON})} = \frac{p(3)}{p(\text{ON})} = \frac{0.1}{0.5} = 0.2 \\
p_{\text{on}}(\text{T}) &= \frac{p(\text{T} \wedge \text{ON})}{p(\text{ON})} = \frac{p(4)}{p(\text{ON})} = \frac{0.4}{0.5} = 0.8
\end{aligned}
$$

$$p_{\text{off}}(\text{H}) \quad = \quad \frac{p(\text{H} \wedge \text{OFF})}{p(\text{OFF})} = \frac{p(1)}{p(\text{OFF})} = \frac{0.25}{0.5} = 0.5 \qquad (13)$$

$$p_{\text{off}}(\text{T}) \quad = \quad \frac{p(\text{T} \wedge \text{OFF})}{p(\text{OFF})} = \frac{p(2)}{p(\text{OFF})} = \frac{0.25}{0.5} = 0.5$$

To come back to the quantum case, the method can be the same as in the above classical example. For sake of simplicity, assume we consider a countable set of physically different measurement setups, denoted by $\mathcal{M}$. A Boolean σ-algebra of outcomes $\mathcal{A}_m$ belongs to each $m \in \mathcal{M}$. A good picture to have in mind is that each $\mathcal{A}_m$ is isomorphic with one of the maximal Boolean sublattices of $L(H)$. It is a fact of quantum mechanics that quantum probability q_W, belonging to a given state operator W, forms a Kolmogorovian probability measure on $\mathcal{A}_m$. Let us denote it by p_m.

Again, if we wish to create a joint Kolmogorovian representation for all these p_m-s, we must embed algebras $\mathcal{A}_m$ into one larger algebra $\mathcal{A}$ and specify the probability distribution ϱ over the different measurement setups. To specify such a probability distribution is, of course, a delicate question, but not less delicate than to specify how frequently is the magnetic field on and off in the above example. The only thing we assume about this probability distribution is that $(\mathcal{M}, \varrho)$ is a discrete Kolmogorov probability space, i.e., $\varrho : \mathcal{M} \to [0, 1]$ and $\sum_{m \in \mathcal{M}} \varrho(m) = 1$.

Let the extended algebra $\mathcal{A}$ consist of the sections of bundle $\bigcup_{m \in \mathcal{M}} \mathcal{A}_m$:

$$\mathcal{A} = \left\{ \alpha \,\middle|\, \alpha : \mathcal{M} \to \bigcup_{m \in \mathcal{M}} \mathcal{A}_m \text{ and } (\forall m)\, [\alpha(m) \in \mathcal{A}_m] \right\}$$

$\mathcal{A}$ is a Boolean σ-algebra with the following operations:

$$(\alpha \wedge \beta)(m) \quad = \quad \alpha(m) \wedge \beta(m)$$
$$(\alpha \vee \beta)(m) \quad = \quad \alpha(m) \vee \beta(m)$$
$$(\alpha^{\perp})(m) \quad = \quad \alpha(m)^{\perp}$$

The minimal and maximal elements of $\mathcal{A}$ are

$$\emptyset, \mathbb{I} \quad : \quad \mathcal{M} \to \bigcup_{m \in \mathcal{M}} \mathcal{A}_m$$
$$\emptyset(m) \quad = \quad \emptyset_m$$
$$\mathbb{I}(m) \quad = \quad \mathbb{I}_m$$

It is easy to check that the following map defines a probability measure on $\mathcal{A}$:

$$p \quad : \quad \mathcal{A} \to [0, 1]$$
$$p(\alpha) \quad = \quad \sum_{m \in \mathcal{M}} p_m(\alpha(m))\, \varrho(m)$$

Elements $A_m \in \mathcal{A}_m$ and $m \in \mathcal{M}$ can be represented as follows:

$$A_m \quad \leftrightarrow \quad \gamma_{A_m} \in \mathcal{A}$$

$$\gamma_{A_m}(i) \;=\; \begin{cases} \emptyset_i \in \mathcal{A}_i & \text{if } i \neq m \\ A_m & \text{if } i = m \end{cases}$$

and

$$m \quad \leftrightarrow \quad \gamma_m \in \mathcal{A}$$

$$\gamma_m(i) \;=\; \begin{cases} \emptyset_i \in \mathcal{A}_i & \text{if } i \neq m \\ \mathbb{I}_m \in \mathcal{A}_m & \text{if } i = m \end{cases}$$

Moreover,

$$p(\gamma_m) \;=\; \sum_{i \in \mathcal{M}} p_i\left(\gamma_m(i)\right) \varrho(i) = \varrho(m) \tag{14}$$

$$p(\gamma_{A_m}) \;=\; \sum_{i \in \mathcal{M}} p_i\left(\gamma_{A_m}(i)\right) \varrho(i) = p_m(A_m)\,\varrho(m) = tr\,(W A_m)\,\varrho(m) \tag{15}$$

Therefore, "quantum probability" $tr\,(W A_m)$ obtains its interpretation in conditional probability of getting outcome A_m given that measurement m is performed:

$$p\left(\gamma_{A_m} \mid \gamma_m\right) = \frac{p\left(\gamma_{A_m} \wedge \gamma_m\right)}{p\left(\gamma_m\right)} = \frac{p\left(\gamma_{A_m}\right)}{p\left(\gamma_m\right)} = tr\,(W A_m)$$

The existence of representations (14) and (15) proves the Kolmogorovian Censorship Hypothesis for countable number of measurements.

Remarks

- Since it was supposed in the above construction that $\mathcal{M}$ is countable, projectors $\{A_m\}_{m \in \mathcal{M}}$ do not cover the whole continuous $L(H)$. To cover the whole lattice with a continuous $\mathcal{M}$ needs further technical elaborations, which I want to publish elsewhere. It is true, however, that any countable subset of $L(H)$ can be covered in this way.

- The above representation of quantum probabilities is consistent in the sense that if a projector $P \in L(H)$ appears in more that one maximal Boolean sublattices in question, for example, $A_m \in \mathcal{A}_m$ and $A'_{m'} \in \mathcal{A}_{m'}$ correspond to the same P, then

$$p\left(\gamma_{A_m} \mid \gamma_m\right) = p\left(\gamma_{A'_{m'}} \mid \gamma_{m'}\right) = tr(W P)$$

It is also consistent in the sense that the corresponding conditional probabilities reproduce quantum probabilities independently of how $(\mathcal{M}, \varrho)$ is chosen.

5. APPENDIX I.

Pitowsky elaborated a convenient geometric language for the discussion of the problem whether empirically given probabilities are Kolmogorovian or not[16].

Let S be a set of pairs of integers $S \subseteq \{\{i,j\} \mid 1 \le i < j \le n\}$. Denote by $R(n,S)$ the linear space of real vectors having a form like $(f_1, f_2, \ldots f_{ij}, \ldots)$. For each $\varepsilon \in \{0,1\}^n$, let u^ε be the following vector in $R(n,S)$:

Definition 2 *The* classical correlation polytope $\mathcal{C}(n,S)$ *is the closed convex hull in* $R(n,S)$ *of vectors* $\{u^\varepsilon\}_{\varepsilon \in \{0,1\}^n}$:

$$\mathcal{C}(n,S) := \left\{ a \,\middle|\, a \in R(n,S) \text{ and } a = \sum_{\varepsilon \in \{0,1\}^n} \lambda_\varepsilon u^\varepsilon, \text{ where } \lambda_\varepsilon \ge 0 \text{ and } \sum_{\varepsilon \in \{0,1\}^n} \lambda_\varepsilon = 1 \right\}$$

Consider now events $A_1, A_2, \ldots A_n$ and some of their conjunctions $A_i \wedge A_j$: $(\{i,j\} \in S)$. Assume that we know their probabilities from which we can form a so-called *correlation vector*:

$$\begin{aligned} \mathbf{p} &= (p_1, p_2, \ldots p_n, \ldots p_{ij}, \ldots) \\ &= (p(A_1), p(A_2), \ldots, p(A_n), \ldots p(A_i \wedge A_j), \ldots) \in R(n,S) \end{aligned}$$

Definition 3 *We will then say that* $\mathbf{p}$ *has a Kolmogorovian representation if there exist a Kolmogorovian probability space* (Ω, Σ, μ) *and measurable subsets*

$$X_{A_1}, X_{A_2}, \ldots X_{A_n} \in \Sigma$$

such that Pitowsky's theorem tells us the necessary and sufficient condition a correlation vector must satisfy in order to be Kolmogorovian.

Theorem 2 (Pitowsky, 1989) *A correlation vector*

$$\mathbf{p} = (p_1, p_2, \ldots p_n, \ldots p_{ij}, \ldots)$$

has a Kolmogorovian representation if and only if $\mathbf{p} \in \mathcal{C}(n,S)$.

In case $n = 4$ and $S = S_4 = \{1,3\}, \{1,4\}, \{2,3\}, \{2,4\}\}$ the condition $\mathbf{p} \in \mathcal{C}(n,S)$ is equivalent with the following inequalities:

$$\begin{aligned} & 0 \le p_{ij} \le p_i \le 1, \\ & 0 \le p_{ij} \le p_j \le 1, \qquad\qquad i = 1, 2 \ \ j = 3, 4 \\ & p_i + p_j - p_{ij} \le 1, \\ & -1 \le p_{13} + p_{14} + p_{24} - p_{23} - p_1 - p_4 \le 0, \qquad\qquad (16) \\ & -1 \le p_{23} + p_{24} + p_{14} - p_{13} - p_2 - p_4 \le 0, \\ & -1 \le p_{14} + p_{13} + p_{23} - p_{24} - p_1 - p_3 \le 0, \\ & -1 \le p_{24} + p_{23} + p_{13} - p_{14} - p_2 - p_3 \le 0. \end{aligned}$$

(16) reminds us the well known *Clauser-Horne inequalities*[17]

6. APPENDIX II.

Consider an Aspect-type EPR experiment with spin-$\frac{1}{2}$ particles (Fig. 5). The four detectors detect the spin-up events. The two switches are making choice from sending the particles to the Stern-Gerlach magnets directed into different directions. The *observed* events are the followings:

A_1 : The "left particle has spin 'up' along direction **a**" detector beeps
A_2 : The "left particle has spin 'up' along direction **a'**" detector beeps
A_3 : The "right particle has spin 'up' along direction **b**" detector beeps
A_4 : The "right particle has spin 'up' along direction **b'**" detector beeps
a_1 : The left switch selects direction **a**
a_2 : The left switch selects direction **a'**
a_3 : The right switch selects direction **b**
a_4 : The right switch selects direction **b'**

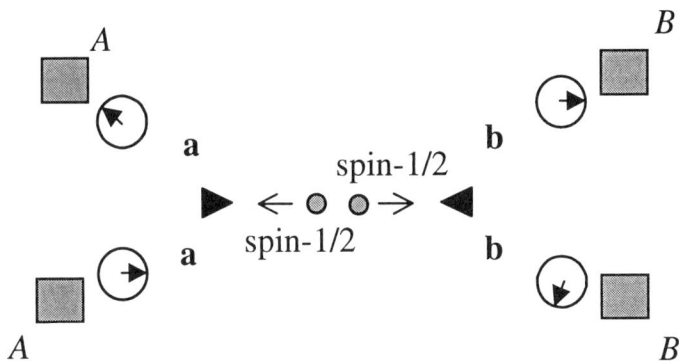

Figure 5: The Aspect experiment with spin-$\frac{1}{2}$ particles.

For the probabilities of these events, in case of $\angle\,(\mathbf{a},\mathbf{a}') = \angle\,(\mathbf{a}',\mathbf{b}) = \angle\,(\mathbf{a},\mathbf{b}') = 120°$ and $\angle\,(\mathbf{b},\mathbf{a}') = 0$, we have

$$p(A_1) = p(A_2) = p(A_3) = p(A_4) \quad = \quad \frac{1}{4}$$

$$p(a_1) = p(a_2) = p(a_3) = p(a_4) \quad = \quad \frac{1}{2}$$

$$p(A_1 \wedge a_1) = p(A_1) \quad = \quad \frac{1}{4}$$

$$p(A_2 \wedge a_2) = p(A_2) \quad = \quad \frac{1}{4}$$

$$p(A_3 \wedge a_3) = p(A_3) \quad = \quad \frac{1}{4}$$

$$p(A_4 \wedge a_4) = p(A_4) \quad = \quad \frac{1}{4}$$

$$p(A_1 \wedge a_2) = p(A_2 \wedge a_1) = p(A_3 \wedge a_4) = p(A_4 \wedge a_3) \quad = \quad 0 \qquad (17)$$

$$p(A_1 \wedge A_3) = p(A_1 \wedge A_4) = p(A_2 \wedge A_4) \quad = \quad \frac{3}{32}$$

$$p(A_2 \wedge A_3) \quad = \quad 0$$

$$p(a_1 \wedge a_2) = p(a_3 \wedge a_4) \quad = \quad 0$$

$$p(a_1 \wedge a_3) = p(a_1 \wedge a_4) = p(a_2 \wedge a_3) = p(a_2 \wedge a_4) \quad = \quad \frac{1}{4}$$

$$p(A_1 \wedge a_3) = p(A_1 \wedge a_4) = p(A_2 \wedge a_3) = p(A_2 \wedge a_4)$$

$$= p(A_3 \wedge a_1) = p(A_3 \wedge a_2) = p(A_4 \wedge a_1) = p(A_4 \wedge a_2) \quad = \quad \frac{1}{8}$$

These statistical data *agree* with quantum mechanical results, in the following sense:

$$\frac{p(A_1 \wedge a_1)}{p(a_1)} = tr(W A_1) = \frac{p(A_2 \wedge a_2)}{p(a_2)} = tr(W A_2)$$

$$= \frac{p(A_3 \wedge a_3)}{p(a_3)} = tr(W A_3) = \frac{p(A_4 \wedge a_4)}{p(a_4)} = tr(W A_4) \quad = \quad \frac{1}{2}$$

$$\frac{p(A_1 \wedge A_3 \wedge a_1 \wedge a_3)}{p(a_1 \wedge a_3)} = \frac{p(A_1 \wedge A_3)}{p(a_1 \wedge a_3)} = tr(W A_1 A_3) = \frac{1}{2}\sin^2 \angle(\mathbf{a}, \mathbf{b}) \quad = \quad \frac{3}{8}$$

$$\frac{p(A_1 \wedge A_4 \wedge a_1 \wedge a_4)}{p(a_1 \wedge a_4)} = \frac{p(A_1 \wedge A_4)}{p(a_1 \wedge a_4)} = tr(W A_1 A_4) = \frac{1}{2}\sin^2 \angle(\mathbf{a}, \mathbf{b}') \quad = \quad \frac{3}{8} \qquad (18)$$

$$\frac{p(A_2 \wedge A_3 \wedge a_2 \wedge a_3)}{p(a_2 \wedge a_3)} = \frac{p(A_2 \wedge A_3)}{p(a_2 \wedge a_3)} = tr(W A_2 A_3) = \frac{1}{2}\sin^2 \angle(\mathbf{a}', \mathbf{b}) \quad = \quad 0,$$

$$\frac{p(A_2 \wedge A_4 \wedge a_2 \wedge a_4)}{p(a_2 \wedge a_4)} = \frac{p(A_2 \wedge A_4)}{p(a_2 \wedge a_4)} = tr(W A_2 A_4) = \frac{1}{2}\sin^2 \angle(\mathbf{a}', \mathbf{b}') \quad = \quad \frac{3}{8}$$

where the outcomes are identified with the following projectors

$$A_1 \quad = \quad P_{span\left\{\psi_{+\mathbf{a}} \otimes \psi_{+\mathbf{a}}, \psi_{+\mathbf{a}} \otimes \psi_{-\mathbf{a}}\right\}}$$

$$A_2 \quad = \quad P_{span\left\{\psi_{+\mathbf{a}'} \otimes \psi_{+\mathbf{a}'}, \psi_{+\mathbf{a}'} \otimes \psi_{-\mathbf{a}'}\right\}}$$

$$A_3 \quad = \quad P_{span\left\{\psi_{-\mathbf{b}} \otimes \psi_{+\mathbf{b}}, \psi_{+\mathbf{b}} \otimes \psi_{+\mathbf{b}}\right\}}$$

$$A_4 \quad = \quad P_{span\left\{\psi_{-\mathbf{b}'} \otimes \psi_{+\mathbf{b}'}, \psi_{+\mathbf{b}'} \otimes \psi_{+\mathbf{b}'}\right\}}$$

of the Hilbert space $H^2 \otimes H^2$. The state of the system is assumed to be represented by $W = P_{\Psi_s}$, where $\Psi_s = \frac{1}{\sqrt{2}}\left(\psi_{+\mathbf{a}} \otimes \psi_{-\mathbf{a}} - \psi_{-\mathbf{a}} \otimes \psi_{+\mathbf{a}}\right)$.

NOTES

* Supported by OTKA Foundation (T 015606, T 025841, T 032771).

1. This idea appeared in J. von Neumann, *Mathematische Grundlagen der Quantenmechanik*, (Berlin: Springer, 1932). One can find it in a more explicit and somewhat different form in G. Birkhoff and J. von Neumann, "The logic of quantum mechanics", *Ann. Math.* **37** (1936), 823-843.
2. A. M. Gleason, "Measures on the closed subspaces of a Hilbert space", *J. math. Phys.* **6** (1957), 8855-8893.
3. The subspaces, the corresponding projectors and the corresponding events are denoted by the same letter.
4. Cf. M. Dummett, *The logical basis of metaphysics*, (London: Duckworth, 1995).
5. R. Feynmann and A. Hibbs, *Quantum Mechanics and Path Integrals*, (New York: McGraw-Hill, 1965).
6. M. Strauss, "Mathematics as logical syntax – A method to formalize the language of a physical theory", *Erkenntnis*, **7** (1937), 147-153.
7. J. S. Bell, *Speakable and unspeakable in quantum mechanics*, (Cambridge: Cambridge University Press, 1987), p. 166.
8. Cf. M. Rédei, "Why John von Neumann did not like the Hilbert space formalism of quantum mechanics (and what he liked instead)", *Studies in the History and Philosophy of Modern Physics*, **27** (1996) 493-510.
9. L. E. Szabó, "On an attempt to resolve the EPR-Bell paradox via Reichenbachian concept of common cause", forthcoming in *Int. J. of Theor. Phys.*
10. Cf. M. Rédei, Von Neumann's concept of quantum logic and quantum probability, in this volume.
11. S. P. Gudder, *Quantum probability*, Academic Press Inc., San Diego, 1988, p. 57.
12. B. Van Fraassen, *Quantum Mechanics – An Empiricist View*, (Oxford: Clarendon Press, 1991), p. 111.
13. L. E. Szabó, "Is quantum mechanics compatible with a deterministic universe? Two interpretations of quantum probabilities", *Foundations of Physics Letters*, **8** (1995), 421-440.
14. L. E. Szabó, *Op. cit.*
15. Bana, G. and Durt, T., Proof of Kolmogorovian Censorship, *Foundations of Physics*, **27**, 1355-1373. (1997)
16. *Op. cit.*
17. See J. F. Clauser and A. Shimony, "Bell's theorem: experimental tests and implications", *Rep. Prog. Phys.* **41** (1978), 1881-1927. There is, however, an important conceptual disagreement between (16) and the original Clauser-Horne inequalities, see L. E. Szabó, "Quantum mechanics in an entirely deterministic universe", *Int. J. Theor. Phys.*, **34** (1995), 1751-1766.

Department of Theoretical Physics
Department of History and Philosophy of Science
Eötvös University of Budapest
Pf. 32
H-1518 Budapest
Hungary

Unpublished Letters and Lectures
by John von Neumann

Editors' notes

The documents selected for publication in this book have never been published before. All are deposited in the von Neumann Archive of the Manuscript Division of the Library of Congress (Washington, D.C., USA). The rich von Neumann Archive contains numerous documents that have never been published in any form. The documents included in this volume have been selected on the basis of their direct relevance to von Neumann's work on the foundations of quantum physics, the main topic of this volume. Three kinds of documents are published here: (i) letters by and to von Neumann, (ii) von Neumann's 1954 address to the International Congress of Mathematicians in Amsterdam, and (iii) the unpublished manuscript "Quantum mechanics of infinite systems" authored by von Neumann. The comments below put these documents in context.

(i)

Sometime around October 1944, Dr. F.B. Silsbee, the president of the Washington Philosophical Society invited von Neumann to give its Fourteenth Joseph Henry Lecture. In a letter to Silsbee, dated November 3, 1944, von Neumann accepts the invitation together with Silsbee's suggestion concerning the topic of the lecture (application of mathematics to quantum mechanics). On March 17, 1945, von Neumann did deliver the lecture that discussed quantum mechanics and its relation to logic and probability. As one can infer from a letter to Silsbee (June 11, 1945), von Neumann also promised to submit a paper entitled "Logic of quantum mechanics", a paper that was to be based on his talk. But the promised paper was never written. In the letter to Silsbee published here, von Neumann explains why. This letter is partly a monologue of a man exhausted and torn-apart by war-related work, disrupted from his "pure" scientific activity, a personal document of the human side of a genius. It is interesting, however, what it was that von Neumann found so difficult to accomplish: to give a satisfactory account of the unified theory of logic, probability and quantum mechanics, a web of problems that is still with us and which von Neumann also selects as the main topic of his Amsterdam talk.

M. Rédei and M. Stöltzner (eds.),
John von Neumann and the Foundations of Quantum Physics, 221–224.
© *Marina von Neumann Whitman. Printed in the Netherlands.*

The exchange of letters between H.D. Kloosterman, the chairman of the Program Committee of the International Congress of Mathematicians held in Amsterdam in 1954, needs no detailed comments. The invitation is testimony to von Neumann's exceptional standing in the mathematics community in the early fifties, while von Neumann's reply shows his down-to-earth attitude and modesty.

(ii)

The typescript of von Neumann's lecture in Amsterdam is published here in exactly the form in which it can be found in the Library of Congress, no corrections of any sort having been made to this document, not even corrections of the obvious misprints. We left this document untouched partly because we want the readers to be aware that this typescript is an unrevised work containing not only misprints and grammatically problematic sentences but also some sentences that do not seem to have any meaning at all. It is unclear who prepared the typescript and there is no evidence that von Neumann had ever seen it. Yet it is an important historical document because it is the only detailed and systematic formulation of von Neumann's post-1932 views on quantum mechanics, which is intimately related to the theory of von Neumann algebras, and which differ considerably from the view he had held in the period of creating abstract Hilbert space quantum mechanics (1926-1932). The handwritten sketch of the lecture shows that von Neumann wanted to address much more in his talk than he actually did. He must have realized that some of the topics he planned to present could not even be stated clearly within the time he had available. On p. 246 we reproduce in facsimile the second page of this sketch. After the introductory re- marks which correspond to the first two paragraphs of the typescript, he gives a classification of the problems to be treated that apparently had been left out in the talk. The page is followed by the detailed plan provided on p. 247f.

(iii)

Von Neumann's "Quantum mechanics of infinite systems" is a neatly typed manuscript in the von Neumann Archive, which emerged from a lecture von Neumann had given in Wolfgang Pauli's seminar "The Theory of the Positron and Related Topics" at the Institute for Advanced Study in Princeton in the winter of 1935/6. Further talks were given by Maurice Pryce, Gregory Breit, Maurice Rose, and Pauli himself. Although all lectures had been typed by Banesh Hoffmann and were later mimeographed, we found it important to make the text available to all those interested in von Neumann's work – for the following reasons. First, it represents the only paper known to the editors in which von Neumann explicitly develops quantum field theory and he does this in terms of operator algebras. In view of the importance of von Neumann's work for a rigorous foundation of quantum field theory, his position should not be without interest. Second, this work and his later mathematical work on the infinite tensor product had surprisingly little impact on the subsequent developments, at least until the algebraic view regained importance from the 1960s on. Third, Pauli took von Neumann's ideas about navigating around infinities quite seriously. On October 26, 1936, he writes to Heisenberg[1]:

Perhaps all the haunting [of infinite eigenvalues for the Hamiltonian] results from observing states about which one actually cannot know anything. – I am at present considering again *Neumann's* idea that for systems with infinitely many degrees of freedom one should consider only such *mixtures* for which one knows something only about finitely many states, while for 'almost all' one has to assume complete ignorance. (458)

Heisenberg subsequently embarks on a study of von Neumann's mathematics. In a postcard of November 4, 1936, Pauli writes he is mailing Heisenberg a copy of von Neumann's lecture[2] emphasizing the

mathematics of density matrices in particular with respect to the limit of the composition of infinitely many independent systems (= eigenvibrations) To be sure, it is doubtful whether in this way already something physically useful results. (This even he himself did *not* claim.) But it is quite possible that one can learn something from Neumann's particular way of taking the limit to infinitely many degrees of freedom. (463)

On November 16, 1936, Heisenberg is more skeptical:

I have read Neumann's manuscript with great interest. Yet my impression is that in searching for a quantum theory of waves one should, to be sure, always keep in mind Neumann's idea, but that this idea alone has little to do with real physics. I am actually quite convinced that the true wave quantum theory can be found only after introducing a universal length, and in Neumann for the time being there is no other place for this than in any other theory. (471)

In his answer of November 18, 1936, Pauli agrees with Heisenberg on the importance of the fundamental length. Here the discussion of von Neumann's lecture ends. As is well-known, Heisenberg's faith in the fundamental length was never borne out, so that one may wonder whether really all perspectives of the approach had been exploited at the time. Together with the above-mentioned question how von Neumann's personal ideas about quantum field theory relate to the tradition so much indebted to his mathematics, there is still some detailed historical work to be done.

NOTES

1. Wolfgang Pauli. *Scientific Correspondence with Bohr, Einstein, Heisenberg, a.o. Volume II: 1930-1939*, edited by Karl von Meyenn with the cooperation of Armin Herrmann and Victor F. Weisskopf, Heidelberg-Berlin-New York: Springer, 1985. All page numbers refer to this volume. See also von Meyenn's very illuminating editorial remarks for the context of this short discussion within the work of both Heisenberg and Pauli. Quotations from the letters have been translated by the editors.
2. This suggests that the typescript had just been completed by this time, which would also explain why it was dated 1937 in the von Neumann Archive.

Unpublished Correspondence

Institute of Advanced Study
School of Mathematics
Princeton, New Jersey

July 2, 1945

Dear Doctor Silsbee,

It is with great regret that I am writing these lines to you, but I simply cannot help myself. In spite of very serious attempts to write the article on the "Logics of quantum mechanics" I find it completely impossible to do it at this time. As you may know, I wrote a paper on this subject with Garrett Birkhoff in 1936 ("Annals of Mathematics", vol. 37, pp. 823-843), and I have thought a good deal on the subject since. My work on continuous geometries, on which I gave the Amer. Math. Soc. Colloqium lectures of 1937, comes to a considerable extent from this source. Also a good deal concerning the relationship between strict- and probability logics (upon which I touched briefly in the Henry Joseph Lecture) and on the extension of this "Propositional calculus" work to "logics with quantifiers" (which I never so far discussed in public). All these things should be presented as a connected whole (I mean the propositional and the "quantifier" strict logics, the probability logics, plus a short indication of the ideas of "continuous" projective geometry), and I have been mainly interrupted in this (as well as in writing a book on continuous geometries, which I still owe the Amer. Math. Soc. Colloqium Series) by the war. To do it properly would require a good deal of work, since the subjects that have to be correlated are a very heterogenous collection – although I think that I can show how they belong together.

When I offered to give the Henry Joseph Lecture on this subject, I thought (and I hope that I was not too far wrong in this) that I could give a reasonable general survey of at least part of the subject in a talk, which might have some interest to the audience. I did not realize the importance nor the difficulties of reducing this to writing.

I have now learned – after a considerable number of serious but very unsuccessful efforts – that they are exceedingly great. I must, of course, accept a good part of the responsibility for my method of writing – I write rather freely and fast if a

225

M. Rédei and M. Stöltzner (eds.),
John von Neumann and the Foundations of Quantum Physics, 225–229.
© *Marina von Neumann Whitman. Printed in the Netherlands.*

subject is "mature" in my mind, but I develop the worst traits of pedantism and inefficiency if I attempt to give a preliminary account of a subject which I do not have yet in what I can believe to be its final form.

I have tried to live up to my promise and to force myself to write this article, and spent much more time on it than on many comparable ones which I wrote with no difficulty at all – and it just didn't work. Perhaps if I were not continually interrupted by journeys and other obligations arising from still surviving war work, I might have been able to do it – although I am not even sure of this. As things are my work in all respects has suffered, I have not succeded in producing any paper I would like to offer to you for publication and I was not able to live up to my other duties as I would have liked to. (I owe the NDRC two long reports which I am unable to start while this "struggle" goes on – I realize that this is an irrational inhibition, but I cannot help it.)

Also, I would like to stress it once more, that apart from my unwillingness to publish something I do not like, and my inhibitions to finish "at any rate" a manuscript which does not satisfy me, it is my respect for the WPS and my appreciation of the honor done to me in inviting me to give the Joseph Henry Lecture, which prevent me from turning in a manuscript. In peace time I would try to stick to a decision of doing nothing else until I have thought out this subject until it is mature for publication as a whole and then written a satisfactory account – this might succeed, although it is a rather painful operation which I did certainly not foresee when I suggested the topic. At present, or in the immediate postwar period, however, I would not enforce such a decision upon myself.

I must therefore ask you to excuse me for not being able to fulfill my promise. I will certainly sometime write up this subject, probably in much more detail. If the Washington Philosophical Society or the Washington Academy should still be interested in it, I will be only too glad to offer it to you for publication. Hoping that you do not think too badly of me for having caused all these complications and hoping to see you again before too long,

I am sincerely yours

John von Neumann

P.S. I would like to tell you once more what an unmixed pleasure it was for me to talk before the Washington Philosophical Society, and how I appreciate this privilege. On the other hand, I feel that I should not have accepted a honorarium, particularly since I am not able to fulfill in a reasonable time the only really onerous obligation: To write the article. Would you let me know, in which way I can correct this most conveniently for the Washington Philosophical Society?

From:
Professor H. D. Kloosterman
van Oldenbarnevelstraat 52
Leiden, Holland

November 27, 1952

Professor John von Neumann
Institute for Advanced Study
Princeton, N.J., U.S.A.

Dear Professor von Neumann:

In the committee which is planning the program for the International Congress of Mathematicians, to be held in Amsterdam, 1954, September 2-9, a proposal has been made to consider the possibility of an address of the same nature as Hilbert's famous address in 1900 in Paris about those unsolved problems of outstanding importance in mathematics, from the solution of which a more or less important progress of mathematical science can reasonably be expected.

The committee realizes that the recent development of mathematics has an ever increasing tendency towards specialization and that for this reason the task of preparing an address of the character mentioned above might prove to be a too heavy one for one person. It has therefore taken the following three possibilities (stated in order of preference) into consideration:

1. an address on unsolved problems is prepared and delivered by one mathematician;
2. a team of (say: three) mathematicians prepare together a list of unsolved problems and one of them reports to the congress;
3. a team of (say: three) mathematicians prepare together a list of unsolved problems and each of them reports to the congress.

As I mentioned already the program committee has a preference for the first of these suggestions. On the other hand the committee's opinion is that you are probably the only active mathematician in the world who is master of the whole of mathematics to such a degree as to be able to deliver an address of the character as expressed above.

For this reason you would oblige me very much to communicate to me if you would kindly accept an invitation to deliver before the International Congress of Mathematicians in Amsterdam an address on unsolved problems in mathematics.

In any case your opinion about the three suggestions stated would be most valuable to our committee.

Yours most sincerely,

H. D. Kloosterman

March 25, 1953

Prof. Dr. H. D. Kloosterman
International Congress of Mathematicians
2d Boerhaavestraat 49
Amsterdam-0
The Netherlands

Dear Professor Kloosterman:

I have just received your letter of March 20, and the copy of your letter of November 27 which was attached to it. I am extremely sorry that what appears to have been a piece of exceptionally bad luck has delayed your work and that of the Program Committee, and has caused me to be unpunctual. A very thorough search of my memory and of that of our files shows your letter of November 27 did not reach me. Needless to say, I would otherwise have answered immediately.

I am deeply appreciative of the great distinction that the invitation and the considerations contained in your letter imply. As to which of the three alternatives that you mention would be best, this is a very difficult problem. I must admit that the task implied in alternative (1) is a staggering one. In view of the exceptional confidence that your invitation expresses, I do not see how I can do otherwise than accept your invitation. May I, however, ask you to permit me to consider the matter further for about a week, and then to write you again, as to whether I can undertake the task entirely in the spirit of your alternative (1), or whether some compromise like alternative (2) or some latitude in interpretation might not be worth considering.

I am,
Yours most sincerely,

John von Neumann

April 10, 1953

Prof. Dr. H. D. Kloosterman
International Congress of Mathematicians
2d Boerhaavestraat 49
Amsterdam-0
The Netherlands

Dear Professor Kloosterman:

Since I wrote you on March 25, I have thought a great deal about the possibilities described in your letter of November 27. The conclusion that I have reached is as follows.

If this is the preference of your Committee, and if it is also otherwise acceptable to you, I will give an address on the basis of alternative (1) that you mentioned – that is, an individual address "On Unsolved Problems in Mathematics."

The total subject of mathematics is clearly too broad for any one of us. I do not think that any mathematician since Gauss has covered it uniformly and fully, even Hilbert did not, and all of us are of considerably lesser width (quite apart from the question of depth) than Hilbert. It would, therefore, be quite unrealistic not to admit, that any address I could possibly give would not be biased towards some areas in mathematics in which I have had experience, to the detriment of others which may be equally or more important. To be specific, I could not avoid a bias towards those parts of analysis, logics, and certain border areas of the applications of mathematics to other sciences, in which I have worked. If your Committee feels that an address which is affected by such imperfections still fits into the program of the Congress, and if the very generous confidence in my ability to deliver continues, I shall be glad to undertake it. The task represents a very interesting and inspiring challenge, and I would certainly try to make the limitations that I have described above as palatable to the audience as I can.

I shall be very much interested in your own and the Committee's views and comments concerning these matters.

I am,
Yours most sincerely,

John von Neumann

JOHN VON NEUMANN

UNSOLVED PROBLEMS IN MATHEMATICS

Typescript of von Neumann's address to the International Congress of Mathematicians, Amsterdam, September 2-9, 1954

Professor Schouten, Members of the Congress.

The invitation of the Organizing Committee for me to speak about "Unsolved problems in mathematics" fills me as it should with considerable trepidation and a prevailing feeling of personal inadequacy. Hilbert gave a talk on this subject at the similar congress about 50 years ago and this is a very formidable precedent. He stated about a dozen unsolved problems in another widely separated areas of mathematics, and they proved to be prototypical for much of the development that followed in the next decades. It would be absolutely foolish, if I tried to emulate this quite singular feat. In addition I do not know the future and the future at any rate can only be predicted ex post with any degree of reliability. I will, therefore, define what I am trying to do in a much more narrow way, hoping that in this manner I have a better chance of not failing. I will limit myself to a particular area of mathematics which I think I know and I will talk about it and about what its open ends appear to be, particularly in some directions which are not the ones that the evolution so far has mainly emphasized and which are, I think, quite important.

I will speak about operator theory and about its connections with various areas and quite particularly about how it hangs together with a number of open questions in physics and how I think it hangs together or ought to hang together with a number of questions in logics and probability theory and questions of the foundations of these and certain reformulations of these which I think it puts into a quite different light from the one with which we usually look at these subjects. Some of the things that I will say have been said before, but – at any rate at this point, I think, I can follow one precept of Hilbert, that one should not be afraid in saying the same thing over and over again, if one really believes that it matters. So, as I say, if I will talk about operator theory and about its connections with various other things where a large number of quite open unsolved problems exist, I will mention a number of unsolved problems quite specifically.

I will, however, take the additional liberty, and I hope that you will forgive it to me, of actually emphasizing mainly the kind of unsolved problem where one cannot be absolutely specific as to what the problem is, but rather sees that there is a group of problems in an area and that the whole area ought to be looked at much more closely or should be looked at from a different point of view. So beyond that

231

M. Rédei and M. Stöltzner (eds.),
John von Neumann and the Foundations of Quantum Physics, 231–246.
© *Marina von Neumann Whitman. Printed in the Netherlands.*

point the formulations will be, that a certain general complex of questions should be looked at, rather than that a particular one which can be answered yes or no, should be so answered.

Let me first say a few things about the general character and the origins of operator theory. As you know, it deals with linear operations applied to various things, usually interpreted as functions, equally possibly interpreted as vectors, at any rate in a Hilbert space. This means that it is a vectorial space, and this means that it is further specialized, the specialization being not only that it is metric, in which case it would be a Banach space, but that there is an inner product, which essentially means that the concept of an angle is defined in the conventional manner. And, geometrically speaking, it would be quite sufficient to say that what one has postulated is that the concept of an angle should apply there. The elegant way to state it is that one talks not about the angle but about the cosine and actually not about the cosine but about the inner product of two vectors. So a Hilbert space is defined by the existence of the inner product and I will not go into the details of that which are well-known. It is also defined by some plausible topological properties which, however, are not terribly decisive, since what one normally postulates is first of all completeness which, if it were not present, one could achieve by simple operation of completing, and also separability which in some cases has a very seriously narrowing effect but in many cases simply means that parts of the space which could be handled separately anyhow, are singled out.

I will not go into this but just say in this generality, that operator theory deals with linear operations in a Hilbert space which means essentially that it is a space with an inner product, that is with a concept of an angle. This fact that the concept of an angle is central will come up again and again in what I will say later and you will see that it is quite crucial. A linear operator is simply a linear operation in this space, in other words, an operation which has this property:

$$T(f + g) = Tf + Tg, \qquad T(af) = aTf.$$

What I have written is, of course, the familiar statement; f and g are symbols for points in that space or for vectorial entities, the small letters like a, b, c are numbers, T is the operator, and I stated that the operator is linear. This concept is, of course, due to Hilbert and Hilbert considered it in conjunction with the important property that it is continuous or, in the terminology used by Hilbert, bounded.

At this point one of the remarkable properties of this whole area manifests itself, namely that almost all topological assumptions which one usually introduces in order to make things simpler, become equivalent and that one does not have to consider all the distinctions which one normally runs into in real function theory. Specifically being bounded in a particular bounded set, say in a sphere, in any particular sphere is equivalent, no matter which sphere you take, and is also equivalent to ordinary continuity. Furthermore, if you analyse the theory of Hilbert space, you find that topology can be defined in it in various ways, first of all with the metric which one is using and secondly also in others, and continuity with respect to all of these happens to be the same thing. So, so far, the kind of simplifications, which is quite familiar in Hilbert space theory occurs, as soon as you have uttered the topological term; it means the same thing no matter which of half a dozen of possible interpretations you attach to it. This is true quite long on the way, although not absolutely always.

Hilbert considered bounded operators, derived for them a number of properties including the decisive one that when the operator is symmetric it has a spectrum and has all the properties which go with the possession of a spectrum, which I will not discuss, but which were quite fundamental in its uses, for instance, in quantum theory and also in other applications.

Another observation is that, while it is quite plausible that in a simple theory at least at first track the basic functions should be continuous, so that it is perfectly reasonable to consider operators only when they are bounded, it turns out very quickly that a number of the most important operators are not bounded. In fact the operators for the sake of which the theory was evolved by Hilbert are only about 50% bounded, because the plausible integral operators usually are bounded, but no differential operator ever is. The earliest theory of the subject is somewhat obscured by the fact that, if you handle differential operators properly and especially if you exclude some cases which can be excluded or circumvented, differential operators are not bounded, whereas integral ones are.

And since you are in a non compact space, continuity of the function and continuity of its inverse are not the same thing. However, you can get practically everything that you get from the continuity of the function, equally well from the continuity of its inverse. It is very characteristic of the terminology of the early epoch of operator theory that many of the concepts are so worded, that it was always slightly ambiguous whether they referred to the operator or to its inverse, the reason being that one always wanted to apply them to that one of the two which is continuous.

The [missing word – *the editors*][1] of quantum theory and of quantum mechanics and the wide use of operators, the quite fundamental rôle of operators in quantum mechanics, changes the situation, inasmuch as it there became absolutely clear that one had to use non-bounded operators and there could be no two ways about it. The quite decisive phenomenon is that the two operators which play a fundamental rôle in quantum mechanics, namely those which stand for the basic mechanical concepts for a coordinate and for its conjugate momentum, had to satisfy a certain algebraical condition of which it is quite easy to show that it can never be satisfied by bounded operators. I mean the Heisenberg commutational relation which I will write down

$$PQ - QP = i\mathcal{J}.$$

P and Q are the two operators in question, i is a number: it is the imaginary unit and $\mathcal{J}$ stands for the unit operator. As you know this is usually stated with another numerical factor there, the Planck constant $h : 2\pi$, but since that is only due to dimensional reasons and can be removed by changing units, and in any case does not alter the algebraical structure, I will omit it. The relation, of course, expresses the non-cummutativity of P and Q but expresses a good deal more. It is not at all difficult to show that bounded operators can never satisfy this relation.

I will come back in a few words later on how one gets round the difficulty that these operations are usually not even clearly defined, but at any rate it makes it quite clear that the domain of operators has to be extended including non-bounded operators. This was done and at the time of the introduction of quantum mechanics it was done immediately intuitively, but mathematically speaking one immediately notices very great difficulties. They are of the following character. First of all it is clear, as soon as one analyses the Heisenberg relation, that boundedness or rather continuity cannot be insisted on. The next thing that one can notice is that because the space involved is not compact, a number of properties which are equivalent in compact spaces are not equivalent any longer. And the certain things which for a compact space would be the same as continuity now mean something else and can still be used and simulate some at least of the properties of continuity which one needs in order to have a reasonably regulated calculus. To be specific, the following property survives.

There are any number of equivalent topological formulations of what one wants when one mentions continuity (as I mentioned Hilbert actually defined it differently), but one of the perfectly equivalent ones is that if a sequence of points f_n converges to a point f, and the operator T applies to all f_n's, then it also applies to f, and the Tf_n converge to Tf.

Now another property which in a compact space would be equivalent to this, is the following:

$$\text{If} \quad f_n \to f \quad \text{and if} \quad Tf_n \to f^* \quad \text{then} \quad Tf = f^*.$$

This says that if both the f_n and the Tf_n converge then T can be applied to the limit of the f_n and its value is the limit of the Tf_n.

It is perfectly clear which argument in the case of compactness will make these properties both to be equivalent to continuity. In the Hilbert space the second one is not equivalent to continuity; it has been called the property of being closed. This property can be insisted on in all operators, which occur in quantum theory, and one gets a quite satisfactory domain of operators by insisting on this, but not on boundedness.

It must be said that the whole thing does not make sense if one is too liberal about how far the operators are defined, specifically whether they are defined everywhere in Hilbert space or only at some special f's. If one does not insist that they are defined everywhere in Hilbert space then the statement of the second theorem must be like this: *If the f_n converge to an f and all Tf_n exist and converge to an f^*, then Tf exists and is equal to f^*.* Actually the statement "then Tf exists" is the critical one. With some of the assumptions which one makes about operators, one can usually derive from this that then Tf is equal to f^*. Specifically the property of a Hermitean symmetry implies that. But the decisive thing is that then Tf exists.

How much this question of existence is central can be seen from the fact, that one of the classical theorems of operator theory, which is a good deal older than these discussions, namely a theorem of Toeplitz and Banach, shows in this terminology: *If a closed operator is everywhere defined, then it is automatically continuous.*

In other words, if one is interested in generalizations beyond the continuous domain, then one cannot insist on the operator being everywhere defined. This, is, of course, not surprising, since the classical examples of non-bounded, non-continuous operators, the ones which are most important, and which introduced non-boundedness into quantum theory, namely the differential operators, are obviously not everywhere defined, since a function in Hilbert space need not be differentiable, or have its derivative in Hilbert space. Also it is quite clear by looking at them that the reason why they are not bounded (so that the derivative can have a large square integral while the function has a small square integral) is closely allied to the fact that the functions may not be differentiable at all. However, the theorem of Toeplitz and Banach gives the complete formalistic completion of this heuristic notion, namely it shows that any closed operator is continuous if and only if it is everywhere defined. As soon as one observes that non-bounded operators are necessarily not everywhere defined operators, one immediately gets into a host of difficulties and one immediately runs into a number of wide open problems, all of them of this variety.

If you deal with operators and deal with their algebra, evidently operations like adding two operators or multiplying two operators are the first ones with which you somehow have to determine your relationship. At the same time it is perfectly clear that these immediately tie you up in various questions of where the domain of the operators lies. Specifically $T + S$ is obviously defined at the points where both T and S are defined. TS is defined at the points where S is defined and gives a value at which T is defined.

It is immediately clear therefore that you can not introduce these concepts unless you have a well-regulated set of requirements and a theory which is matched to these and works for domains of operators. This produces at once considerable difficulties. The character of the difficulties can be most easily examplified by these observations.

All through operator theory the representation of operators by matrices plays a great role. This goes to the point that in quantum theory the operatorial aspect was only brought in via the matrix aspect. I mean, the dependence on matrices was discovered first, and the abstract operatorial formulation followed.

A matrix actually expresses a very simple property of an operator, namely, if you have a complete coordinatisation of the Hilbert space, in other words, a complete normalized orthogonal set, say a sequence $\phi_1, \phi_2, \phi_3, \ldots$, then, forming the operator with each ϕ_i, the expansion coefficients of this in the ϕ's, so the inner products $(T\phi_i, \phi_j)$ are matrix elements a_{ij}. It is clear, therefore, that to specify the matrix elements is no more and no less than to specify what T does to a complete orthogonal set ϕ_i and, since by linear combinations a complete orthogonal set can be carried into any countable everywhere dense set and v.v., this does not mean more and does not mean less as knowing the operator on a dense set. Now, knowing the operator on a dense set would be, of course, as good as knowing it everywhere if it were continuous. But it is not. Since it is, however, closed, some vestigial properties of continuity survive and, therefore, knowing it on a dense set means knowing a great deal. It is, nevertheless, not as much as knowing the operator and produces all sorts of difficulties. Actually various problems of the continuation of operators, of the boundary conditions which are natural for an operator, and the questions whether an operator has any or one or more spectra, are all connected with this. It is quite important furthermore whether it ever happens for two operators that their definitional domains have a dense set in common on which they both agree, or whether – to say something which sometimes, although not always, is more – two operators are such that everywhere where both are defined they agree and this set is dense. But they may nevertheless be not identical, in other words, both may then further be defined at other places, at each of which only one is defined and not the other.

If[2] you call two operators which agree everywhere where both are defined and which are jointly defined at least on one dense set and are both closed, if you call two such operators say partially equivalent, then one would expect that that one can only expect (and I have reason) to exist in this domain, that the concept of partial equivalence for one can be used which in some manner can be extended to a genuine equivalence which conveys some significant information.

It is not hard to construct examples which show that the opposite of this is true. The opposite is true because the following thing is true. Not only is partial equivalence not transitive, not only is it true that A can be partially equivalent to B and B to C without A being so to C: A can agree on a dense set with B, B can agree on a dense set with C, but these two dense sets may have nothing in common, and A and C's domains may intersect on a third dense set, which has nothing in common with the two former and where they are different. But even more is true, specifically it is true, if you take any two bounded operators, then by a limited number of steps which is certainly not more than nine, you can pass from A to a partially equivalent A_1, from there to a partially equivalent A_2 and so on and get in no more than nine steps to B. Nine is surely not the smallest number by the way. Consequently it is completely hopeless to try to get to any sensible concept of equivalence in this manner.

There is plenty of other pathology in this area, but the example that I have mentioned is characteristic of the rest of it.

On the other hand it is quite clear that the interesting applications, quite particularly in quantum theory, absolutely call for a settlement of these questions: to tell how to operate on non-bounded operators, how to form sums and products, what to do if their domains have nothing in common, i.e. if there is no point where both are defined, what to do if the points where they are both defined, exist and are dense, but one somehow suspects that the set is not large enough (and this can be given a sharper meaning), generally speaking, how to introduce an algebra. This is not a particularly fertile area to speak on unsolved problems, however, except to state the general malaise, because several of the most painful questions were settled. For instance the question of how to replace the relation $PQ - QP = J$ by something to which one can attach a definite meaning, has been solved in the sense that by various algebraical transformations, which I will not describe here, one can pass to a relation which holds not between P and Q, but between certain functions of them, which are clearly bounded and for which one can derive that they determine the operators involved uniquely and actually determine them as the operators which Heisenberg used in the first place.

However, one is still left without any satisfactory calculus of operators. There are various things which one can attempt to do and I will now pursue a biased course in one particular direction.

One possible procedure for them is to say that after all the assemblage of all operators does make all these troubles. One knows that certain subsets of it will not make such troubles. For instance, perfectly satisfactory calculi have been derived for families of commuting operators, even though there can be any number of unbounded ones among them.

One can therefore ask whether there are subsystems of operators in which the above pathology is avoided. In talking of subsystems of operators it would be of course very interesting whether one can find subsystems, which as far as internal properties are concerned, behave like the system of all operators, in other words, which have the same algebraical properties as all operators. Well, in that case one evidently should talk of rings, in other words, of systems which are closed under addition, multiplication and subtraction.

It is also plausible that one will have to require some topological closure. It is again true that topology for operators of Hilbert space, just as for Hilbert space itself, has these properties. First of all it is a somewhat ambiguous object inasmuch as one can topologize this space, just like the underlying Hilbert space, in several inequivalent ways for every one of which I can use some arguments. One can also show that, like in the critical properties of Hilbert space, it is largely indifferent how one defines closure, because for the critical entities one gets the same concepts. For instance, Hilbert space has several topologies, and closure under them are different things, but for a linear set the closure under all of them is the same thing.

Again Hilbert space has several topologies and at least for two of them which are the best known ones, the so-called strong and weak, which are not equivalent, the closure of a ring is the same thing, although this situation is a bit less satisfactory than before, because it matters for instance whether one asks for closure with respect to the topology or only for closure with respect to convergent sequences in the Fréchet sense. However, if one defines closure as closure in either strong or weak topology, one gets an unambigious concept. That the concept is reasonable can also be seen in this way.

If S is a set of operators, I designate by S' the set of all operators that commute with S. It is clear that this concept can be iterated. S'' are all operators which commute with every operator that commutes with S. It is clear that this definition gets verbally more awkward as you repeat the primes on S. Actually the situation is not so complicated. By very simple arguments one can show the reason.

One has a heuristic feeling that S, if it is at all reasonably closed in some reasonable sense, ought to be the same thing as S''. In other words, any sensible concept of algebraical topological closure should have this property that, if an operator A commutes with everything that conmutes with all S, then A should be obtainable from S algebraically, in other words, if S is a closed ring, then A should belong to it: S should be equal to S''.

This has peculiar properties, for instance, it is quite easy to see that there are great parts in algebra, where one can always judge whether one has a reasonable concept by checking whether this criterium holds. It is easy to see for instance that S''' agrees with S' and that as soon as you attach at least one prime to S, an even number of primes does not matter. It is not trivial whether S is equal to S'', however. One can show that "S equal to S''''" is equivalent to closure in the sense which I have mentioned, so that the ring S is equal to the commutator of its commutator, if and only if it is topologically closed in the strong or in the weak topology of the Hilbert space.

So it is not unreasonable, at any rate in a first experiment, to define things like this. However, then the next step that one is also likely to take, is to consider such systems of operators, and to ask that they should be further similar to the system of all operators of Hilbert space, by those internal properties that one can easily name and the most conspicuous one of which is a commutativity property, namely the statement that nothing in the ring except the unity and its constant multiples commute with everything in the ring, so that the ring is so uncommutative that its centre consists only of the numerical multiples of unity. Of course, it is an inner property.

So I am not talking of S' but of the part of S' which lies in S, of the centre of S, the intersection of S and S'. If you ask that this set should consist only of the numerical multiples of unity, you get the concept which in ordinary algebra is equivalent to the fundamental property of simplicity. An ordinary ring of matrices in a finite number of dimensions is actually simple, if its centre consists of the numerical multiples of unity, the ring not necessarily being the ring of all matrices.

If you again call a ring simple, if it has this property, in other words, if the only elements in it which commute with all the rest, are the numerical multiples of unity, then a quite simple theory of these entities exists. This was developed years ago by the mathematician Murray and myself and is based on the following idea.

You analyse the ring by doing to it what is one of the standard techniques in analysing ordinary finite base rings in algebra, namely you consider the idempotent elements. You consider those elements which are equal to their own squares, so those elements from which nothing new can be obtained by the standard algebraic operations, specifically by multiplication.

It is quite easy to see that in a system of Hermitean operators in Hilbert space these are identical with the projections, so these are the operators which you get if you take any closed linear set in that space, and decompose each vector into that part of it which lies in that space and that part which is orthogonal to that space. This is, of course, an operation, it is called the projection of that linear space, and there is a one to one correspondence between operators that are their own squares, idempotents, and closed linear subsets of Hilbert space. Namely, each closed linear subset is a projection and that projection is idempotent and every idempotent can be obtained in this manner with precisely one closed linear space. So the usually algebraical discussion of all idempotents of the ring is equivalent in this case to a discussion of all closed linear sets whose projections belong to the ring.

It is also quite easy to see that, if you have a ring S, and if you fix your attention not at S, but at S', the set of all operators which commmute with it, or even only at the unitary part of S', so the set of all these rotations of the space which leave the operators of S fixed, so all those rotations of the space which leave the elements of the ring in question invariant, then those linear subsets of Hilbert space, which have projections belonging to the ring, are precisely those linear subsets which are invariant under all these rotations. In other words, you can define a ring, if you wish, not only directly, but also indirectly by telling which rotations of the space leave its elements fixed.

It can always be defined as a set of fixed points, operatorially, of a family of rotations. And then it is perfectly natural to look also at all assemblages of other entities, which are associated with the Hilbert space and for which it makes sense what the rotation does to them and whether they are or not invariant under such a rotation. And specifically one can ask about the linear subsets of all sub-spaces of the space which are so invariant. Now for this, one can do the following thing. One can define for them a concept of dimensionality which is built in exactly the same way like the concept of aleph, of power, in set theory à la Cantor, but one limits it in the following manner. Instead of doing it for all sets, the way Cantor did it, one does it only for the linear subspaces of Hilbert space, and instead of calling two equivalent, if there is any one to one mapping of the first on the second, you call them only equivalent, if there is a rotation in the family of rotations, that I mentioned, which maps one of them on the other.

Let me be careful about this, I was talking first about the rotations not of S, but of S', all those rotations which leave every element of S, and therefore all the closed linear sets whose projections are in S, invariant. It is clear, of course, that these rotations will never map any of them on anything except on itself. However, you can now look at the rotations of S. S is not commutative, so the rotations of S will not commute with all elements of S and will not in general belong to S'. Therefore, a rotation of S can perfectly well transform a projection of S into another projection of S. It can perfectly well map a closed linear set in S on something else.

If you do this, then you get a concept of equivalence. You call two closed linear subsets which belong to S equivalent, if a rotational S carries one into the other. This is very much like the set theoretical concept of equivalence, but of course it is strongly limited, because the set is not general. It is a closed linear set and the mapping, the one to one mapping, is not general. It is a rotation from the ring S.

In spite of this the whole algorithm of Cantor theory is such that most of it goes over on this case. One can prove various theorems on the additivity of equivalence and the transitivity of equivalence, which one would normally expect, so that one can introduce a theory of alephs here, just as in set theory, and that it has the normal properties of being able to define an equality and being able to add. I may call this dimension since for all matrices of the ordinary space, it is nothing else but dimension.

But one can also derive, very parallelly to Cantor, things which are less obvious, for instance the equivalence theorem that, if a linear set is equivalent to a subset of another linear set and v.v., then the two are equivalent to each other.

One can, however, go a great deal further. Specifically in the case of Hilbert space, one can define finiteness and infiniteness in the same way as Cantor did by equivalence to a proper subset. One can prove most of the Cantoreal properties of finite and infinite, and, finally, one can prove that given a Hilbert space and a ring in it, a simple ring in it, either all linear sets except the null set are infinite (in which case this concept of alephs gives you nothing new), or else the dimensions, the equivalence classes, behave exactly like numbers and there are two qualitatively different cases. The dimensions either behave like integers, or else they behave like all real numbers. There are two subcases, namely there is either a finite top or there is not. So, when they behave like integers, they either behave like all integers from one to a finite n, or like all integers to infinity plus a symbol infinity. When they are continuous, they either behave like all real numbers from null to a finite number a, inclusive, or else like all real numbers up to infinity with a symbolic top at infinity.

In total there are therefore five classes, I mean like the integers which may have a finite top or not, like all real numbers which may have a finite top or not, and, finally, the case where only the infinite dimensions exist, apart from the dimension null.

The case which is entirely finite, where all you have are the dimensions which are integers that have a finite ceiling, is always isomorphic to the matrices of the Euclidean space. The case where you have integers going to infinity, is isomorphic to all matrices of Hilbert space, so there nothing is gained. About the infinite cases very little is known.

There is one residual case, however, where the dimensionality is like real numbers with a finite ceiling a. And, since the real numbers from null to a are isomorphic under addition as a group to the real numbers from 0 to 1, one might as well put the ceiling at 1. This case has a number of singular properties and this is not new either. It is very well behaved: although there are plenty of unbounded operators here, one can show that any finite number of them, in fact any countable number of them, are simultaneously defined on an everywhere dense set; one can prove that one can indulge in operations like adding and multiplying operators and one never gets into any difficulty whatever. The whole symbolic calculus goes through.

One can further show that such systems of operators are in many ways very similar to certain operator systems used in quantum theory. I will not attempt to go into detail at this occasion, but it is true that actually the so-called method of second quantization, which introduces the operators of quantum theory depending on certain processes of counting of states, permits a very plausible generalization which leads exactly into this kind of operator ring, and which is therefore immune to the usual pathology of operator rings. And there is a quite relevant question there, whether the use of a ring like this gives acceptable results which are different from the usual ones.

But various operator theoretical niceties of how one introduces operators of a second quantization matter can be seen from very other things. For instance, it is possible to show that the Dirac theory of electrons in its ordinary form and in the form in which it introduces positrons, can be formulated so that the difference between them is an operatorial nicety of this variety.

I will, however, not go into this now and also not discuss to what intuitive concepts this introduction would correspond. To say it briefly, it corresponds to a view where one does not assume what the ordinary quantum-mechanical assumption is, that almost all states are unoccupied, nor what the positron theory assumption is, that they fall into two classes, where in one almost everything is void and in the other almost everything is occupied, but this corresponds to something which can intuitively be described as saying that the presumption is that almost all states are 50% occupied.

But I will not go into this. What I would like to say, however, is this, that it is possible to operate with this last category of operator rings in a somewhat more abstract manner, namely it is quite clear from the connection with quantum theory, that an operator stands for a physical quantity and that a projection, an idempotent, an operator which is its own square, stands for a physical quantity, which has only such values as are their own squares, which means 0 and 1. In other words, the idempotents, the projection operators, represent those physical quantities which have only two values, 0 and 1.

A physical quantity with two values which have been standardized in any particular way, for instance to 0 and 1, is clearly the equivalent of a logical assertion. A logical assertion is, if you wish, a physical quantity with two truth values, yes and no. And anybody who ever tried to arithmetize this system, usually replaces yes and no by mathematical entities, and the conventional replacement is 0 for no and 1 for yes. It is the same operation as replacing the set by its characteristic function.

Therefore, it is clear that projections in closed linear sets for that matter in this representation correspond to logical propositions. This means that from the mathematical point of view the more desirable system to treat is not operator theory, but that part of it which deals with idempotents, because that corresponds to logics, whereas the whole system corresponds to a somewhat unpleasant extension of logics, namely where you deal with quantities which can have any number of numerical values, in other words, physical quantities.

So, of course, it is well-known that it is quite sufficient to limit oneself to two value things: all numerical statements can always be replaced by superposition of yes and no alternatives.

It is worth noticing that this transformation, which is plausible from the point of view of the whole tradition of logics, has a perfectly good mathematical operational meaning and justification. It just corresponds to the insight, that it is better to analyse an algebra by looking at the idempotents in it. It furthermore indicates that as long as you are tied to Hilbert space and do not proceed abstractly, it is probably better not to look at operator rings but to look at families of closed linear sub-spaces.

We know that all operator rings which matter can always be defined, not only as rings (which drags in algebraical properties that one can hardly carry over in closed linear sets, I mean that the operations for addition and multiplication are completely washed out when you limit yourself to the projections), but also by saying that we are looking at all those operators which are invariant under the given group of rotations. Well, this definition carries over to closed linear sets.

You can see that you study only those closed linear sets of a Hilbert space which are invariant under a pre-assigned group of rotations. This must in some way correspond to logics, as it gets itself transformed when you pass to quantum theory. And furthermore one would expect that the fact, that quantum theory first hits this correspondence in a Hilbert space and in operator context, is probably an accident and mainly an accident of one's being familiar to this particular mathematical machinery.

But there is probably an abstract background behind this which one ought to abstract, in finding which one ought to be guided by the fact that clearly projections of a Hilbert space seem to have had many of these traits which mattered, and also that the projections of the finite dimensional Euclidean space did not seem to have these. Therefore, this thing must look very much like projective geometry, but there must be some inherent reason for abandoning the finite dimensionality.

Furthermore, all the troubles of operator theory in Hilbert space and all the troubles of the last two decades in quantum theory with the well-known divergences indicate that the transition to infinite dimensionality has not been performed quite as happily as it might have been. In other words, in getting to something, there must be an inherent reason for going to infinite dimension, but Hilbert space, which apparently did this to nearly 100%, clearly did not do it perfectly (or else one would not have the divergence difficulties of all the more elaborate forms of quantum mechanics, in which one is).

The question, therefore, arises, and it is quite characteristic, how the physical difficulties indicate where the mathematical bird is hidden. It indicates strongly, that one ought to find an abstract machinery which imitates the machinery of Hilbert space rings quite a way, so far that it was possible to be led to this system. But at some point it must bench off.

Now there exist some attempts in this direction, and I would like to say a few words about these attempts and also point out where I think they look healthy and at which point they are clearly diseased.

For one thing, you can go to various levels of abstraction in Hilbert space operator rings. The most naive is that you use all the operators. The next higher level is that where you get the theory of rings and everything that follows from it. The next level of abstraction to which you can go is that you limit yourself to those operators which occur in quantum mechanics, which means that they have to be Hermitean. This already produces some difficulties, since the simplest operation, namely multiplication, destroys the Hermitean character.

One can go further by limiting oneself to the projections which, as I said, obviously corresponds to the transition of logics in the normal sense.

Finally one can take this, which should be a theory of certain sub-spaces of a Hilbert space, namely all those which are invariant under some given group, but free oneself of the Hilbert space and look for a set of entities which have all of the formal properties of sub-spaces of this space, but are *not* tied to Hilbert space. This operation can be performed and has been performed in the past successfully. I am looking to the so-called lattice theory. Here the basic property that you use is, that you deal with a number of entities where a relation of being a subset is defined, and you have the usual basic property of a lattice that, given any number or subset of elements, they have a least upper bound and a greatest lower bound.

It is a very deep observation, but after you have made it, it sounds very simple. The properties which among all lattices characterize those which behave like the linear subspaces of a linear space, so behave like the projective geometry, were discovered by Garrett Birkhoff quite some time ago. Their very simple algebraic property, the so-called modular law, I will not describe.

I would like to say only that the formal systems, which characterize the sets with a partial ordering, for which the partial ordering, if certain finiteness conditions are satisfied, makes them the sets of all linear subsets of a linear space, so makes them the projective geometry, are known. And the modular law is a very simple weakening of the distributive law.

However, the question arises how to proceed when you do not have the finiteness assumptions that normally go with this. A theory which does this also exists and was also developed more than a decade ago. You can, by suitable continuity assumptions, get so far in lattice theory, that you can again define equivalence and define it completely in the classical way of projective geometry, in other words, by the usual uses of concepts like perspectivity and projectivity, and get to the transitive concept of equivalence. You can prove that it behaves like dimensionality, prove that the dimensions are isomorphic to real numbers, and prove in fact that the equivalence classes you get this way behave precisely like the real numbers between 0 and 1, and that there is one and only one way to establish this correspondence.

One can also state the axioms, which bring us still closer to quantum theory and permit us to introduce those concepts of quantum theory which correspond to the inner product, so various probabilistic statements on linear manifolds. At this point one comes to the question which is not being axiomatically understood to this day and where an axiomatic treatment is certainly desired, and I would rather think would be very rewarding. Namely it is quite clear that one has here in hands the tools, with which one gets to the system of logics which immediately also contains the probability theory. It is very characteristic that in quantum theory logics and probability theory go very closely together. Let me perhaps point this out.

If you take a classical mechanism of logics, and if you exclude all those traits of logics which are difficult and where all the deep questions of the foundations come in, so if you limit yourself to logics referred to a finite set, it is perfectly clear that logics in that range is equivalent to the theory of all sub-sets of that finite set, and that probability means that you have attributed weights to single points, that you can attribute a probability to each event, which means essentially that the logical treatment corresponds to set theory in that domain and that a probabilistic treatment corresponds to introducing measures. I am, of course, taking both things now in the completely trivialized finite case.

But it is quite possible to extend this to the usual infinite sets. And one also has this parallelism that logics corresponds to set theory and probability theory corresponds to measure theory and that given a system of logics, so given a system of sets, if all is right, you can introduce measures, you can introduce probability and you can always do it in very many different ways.

In the quantum mechanical machinery the situation is quite different. Namely instead of the sets use the linear sub-sets of a suitable space, say of a Hilbert space. The set theoretical situation of logics is replaced by the machinery of projective geometry, which in itself is quite simple.

However, all quantum mechanical probabilities are defined by inner products of vectors. Essentially if a state of a system is given by one vector, the transition probability in another state is the inner product of the two which is the square of the cosine of the angle between them. In other words, probability corresponds precisely to introducing the angles geometrically. Furthermore, there is only one way to introduce it. The more so because in the quantum mechanical machinery the negation of a statement, so the negation of a statement which is represented by a linear set of vectors, corresponds to the orthogonal complement of this linear space.

And therefore, as soon as you have introduced into the projective geometry the ordinary machinery of logics, you must have introduced the concept of orthogonality. This actually is rigorously true and any axiomatic elaboration of the subject bears it out. So in order to have logics you need in this set a projective geometry with a concept of orthogonality in it.

In order to have probability all you need is a concept of all angles, I mean angles other than 90°. Now it is perfectly quite true that in a geometry, as soon as you can define the right angle, you can define all angles. Another way to put it is that if you take the case of an orthogonal space, those mappings of this space on itself, which leave orthogonality intact, leave all angles intact, in other words, in those systems which can be used as models of the logical background for quantum theory, it is true that as soon as all the ordinary concepts of logics are fixed under some isomorphic transformation, all of probability theory is already fixed.

What I now say is not more profound than saying that the concept of a priori probability in quantum mechanics is uniquely given from the start. You can derive it by counting states and all the ambiguities which are attached to it in classical theories have disappeared. This means, however, that one has a formal mechanism, in which logics and probability theory arise simultaneously and are derived simultaneously. I think that it is quite important and will probably shade a great deal of new light on logics and probably alter the whole formal structure of logics considerably, if one succeeds in deriving this system from first principles, in other words from a suitable set of axioms. All the existing axiomatisations of this system are unsatisfactory in this sense, that they bring in quite arbitrarily algebraical laws which are not clearly related to anything that one believes to be true or that one has observed in quantum theory to be true. So, while one has very satisfactorily formalistic foundations of projective geometry of some infinite generalizations of it, of generalizations of it including orthogonality, including angles, none of them are derived from intuitively plausible first principles in the manner in which axiomatisations in other areas are.

Now I think that at this point lies a very important complex of open problems, about which one does not know well of how to formulate them now, but which are likely to give logics and the whole dependent system of probability a new slam.

I want to thank you for your attention.

EDITORS' NOTES

1. On the margin of the original copy there is a word – unfortunately undecipherable to the editors – followed by a "?".
2. On the original copy, there is a huge question mark on the margin of the present paragraph. It is unclear who put the question mark there. It indicates that this paragraph appeared incomprehensible for some reader, and indeed the paragraph is very confused.

2./ afraid of saying the same thing over and over and over again! if one really believes in it, and really believes that  it matters a great deal.

The talk deals with problems in a particular area of mathematics — operator theory, viewed in its connection with certain other subject, specifically in its algebraical aspects and in its relationship to quantum theory, and through this to logic and to the theory of probability.

The unsolved problems to which attention is called fall into three groups.

① Problems involving the algebraical structure of rings of operators.

② The role and meaning of these in view of the present difficulties and uncertainties of quantum theory.

③ Problems of reformulation and unification in logic and in probability theory based on this approach.

Facsimile copy from von Neumann's notes for the Amsterdam address.

AMSTERDAM TALK ABOUT "PROBLEMS IN MATHEMATICS" SEPTEMBER 2, 1954

John von Neumann's handwritten sketch

Problems in mathematics

1. *2m* I will not try this on the broad basis of Hilbert in 1900. Reasons.

2. *1m* Instead: A special area that I know well and that exemplifies a principle: Operator theory. "Problems": More vague definition of these. "Outlook".

3. *1m* Guidance from physics as a principle of mathematics.

4. *5m* The origin of operator theory. Hilbert. Bounded operators. Unbounded operators: Differential operators.

5. *2m* Obscuration of the contrast by the possibility of using reciprocals.

6. *4m* Quantum theory. Clear need for unbounded operators.

7. *4m* Theory of unbounded operators. Their domain, Toeplitz-Banach theorem. $Q - P$ algebra, operator algebra.

8. *4m* The problem. The various counter examples.

9. *4m* Operator systems other than the totality, operator rings. "Inner" requirements: Minimum center. Algebraic meaning: "Simplicity".

10. *5m* The vN-Murray theory. Dimensionality. Analogy to alephs. The classification.

11. *3m* Finite-infinite. No pathology for finite.

12. *5m* Isomorphism problems of the finite case.

13. *4m* Characterisation problems of the infinite case.

14. *5m* More direct connection with physics: Origin of infinity: Primary: $Q - P$. Secondary: Spins. In the secondary case: Infinite direct products.

15. *3m* Theory of the infinite direct product. The "sub-products".

M. Rédei and M. Stöltzner (eds.),
John von Neumann and the Foundations of Quantum Physics, 247–248.
© *Marina von Neumann Whitman. Printed in the Netherlands.*

16. *5m* Physical meaning: E.g. the "hole" theory of Dirac. Continuous finite cases. What is the meaning of the latter? Particle theories, interpretation.

17. *2m* Divergencies in quantum theory especially electrodynamics.

18. *5m* Relationship to logics. The various algebraic levels of discussion: a) Spacial ("outer"). b) Operatorial ("inner"). c) Lattice (projections). ("Projective geometry")

19. *3m* Continuous geometries.

20. *4m* Relationship to logic. The role of probability.

21. *6m* Logic vs. probability: Classical – set theory vs. measure theory – separate. Quantum – projective geometry with orthogonality vs. same with all angles – closely connected. What is the meaning of this?

22. *4m* Integration of logics and probability theory. "Quantitative" aspects of logic. Simplest case: Information theory.

23. *3m* Classical: Measure – volume estimates. Quantum: What is it then?

24. *5m* Inherent value of this integration. A "thermodynamical" theory of logics-probability. Its value as a heuristic guide in mathematics, etc. What will the role of the "quantum" approach be?

89m

JOHN VON NEUMANN

QUANTUM MECHANICS OF INFINITE SYSTEMS

*Mimeographed version of a lecture given at Pauli's seminar
held at the Institute for Advanced Study in 1935/6*

I wish to discuss some rather incomplete ideas concerning difficulties that arise in some parts of quantum mechanics. In general there have been no serious difficulties when we are dealing with a finite number of particles, but very essential difficulties arise as soon as we treat a system having an infinite number of degrees of freedom; for example, the theory of holes, which, because of the pair generation, requires an indefinite number of particles; also the *Dirac* non-relativistic theory of light and the *Pauli-Heisenberg* relativistic quantum electro-dynamics, these being equivalent to systems consisting of an infinite number of particles.

In dealing with a continuum we find two types of infinity. One arises from the fact that we have an infinite space, but this does not lead to serious difficulties and can be avoided by considering a finite box, or, better, by assuming periodicity in space. The second type of infinity is much more serious. It comes from the fact that in a continuum a field quantity has an infinite number of proper values. The assumption of periodicity does not remove this difficulty at all. Neither does the assumption that space is discrete solve the essential difficulty since if we pass to the limit "lattice $\rightarrow$ continuum" we get just the continuum result, which diverges.

It is the fact that we have an infinite number of degrees of freedom that causes the difficulties, and we shall therefore discuss how we can change the formal part of the theory in some way so that we can treat a system having an infinite number of degrees of freedom in a less divergent way.

The thing that seems to be wrong with the usual theory is the use of wave functions, or, in the language of *Dirac*, of maximum observations. For, suppose we wish to find out experimentally, what state a given system is in. The normal method, according to the quantum theory, is to make the maximum number of compatible measurements; from these we can find not only the state of the system but also, by means of the *Schrödinger* time-dependent equation, what happens to this state afterwards. However, if our system has an infinite number of degrees of freedom, we require an infinite number of observations before we can determine its state, and this seems unreasonable. The difficulty enters in field theories as well as in particle theories since in the former the wave functions depend on all those field components that are simultaneously measurable and these are, in the *Pauli-Heisenberg* theory, for example, infinite in number.

249

M. Rédei and M. Stöltzner (eds.),
John von Neumann and the Foundations of Quantum Physics, 249–268.
© *Marina von Neumann Whitman. Printed in the Netherlands.*

We take the attitude that this is the real cause of the trouble and we shall show that we can perhaps avoid the divergences by using a model which avoids the wave functions. We shall first discuss not the electromagnetic field but the case of a system consisting of an infinite number of *Fermi-Dirac* particles, i.e. a system in which it is essential that we do not know the number of particles, as is always the case when we have pair generation. If we assume that the universe is large but finite, the energy levels will form a discrete set and we can describe the situation by telling which are the states in which a particle is present. Let us consider the question of how many particles are in a certain box. We can find a complete set of standing waves in the box and a complete set of states outside. Then any particle definitely in the box will be represented by a wave-function consisting of a linear combination of the standing waves alone and there is thus a definite meaning to the number of particles in the box. We can actually make one simple observation of this number by merely weighing the box. But in fact the weighing will not give the number of particles in the box unless the walls are impermeable, and, moreover, impermeable not only to the particles in the box but to all types of particle or energy. Experience shows that such a box cannot be realized in practice. There is an upper limit to the molecular weight of the substance of the box and thus an upper limit to its impermeability. Even if there were a sharply defined energy at which the box changes from complete impermeability to relative impermeability we could not say that energy would be conserved within the box; for particles of very high energy could enter the box and give up part of their energy by collision with particles inside the box, thus increasing the total energy. They could also cause a decrease in the the total energy by giving to particles in the box enough energy to allow them to pass through the walls.

Another question of fundamental importance is this: Can we assume that particles of extremely high energy essentially do not interfere with particles of low energy? We must certainly assume that particles of very high energy exist since with every relevant advance in experimental technic particles of higher energy are discovered and we cannot limit the number of such particles *a priori* since the number observed by us is dependent on how short wave lengths we are able to measure. The fact that we exist and do not feel any ill effects from the high energy particles seems to show that the extremely large interaction demanded by electrodynamical theory is incorrect and that high energy particles do not interact with those of low energy. The known deviations from the *Klein-Nishina* law of absorption also point to such a conclusion. We shall therefore assume that this conclusion is correct. The situation is similar to that which arises in the Newtonian theory of gravitation. In that theory, although matter distributed uniformly over a spherical shell $(r, r + dr)$ will have no resultant attraction at the center, if fluctuations occur there will be a resultant attraction and in an infinite universe approximately uniformly filled with matter the effect of the distant masses will lead to divergent results. In the present electrodynamics the particles of high energy lead to an analogous divergence. One might say: The distant regions of the momentum space have the same divergence-generating effect in (Maxwellian) electrodynamics as the distant regions of common space have in Newtonian gravitation theory. Thus in both theories the fundamental postulate which requires the existence of "closed systems" is violated.

We shall look for a theory in which closed systems exist in momentum space. To discuss mathematically what sort of change we must expect, we must consider how we can describe systems of which we do not know the maximum information. We must also consider how we put two systems together since we get infinite systems by applying this process repeatedly.

A single particle of the infinite set of particles constituting the system will, if left alone, have certain stationary states, say $\sum_1, \sum_2, \sum_3, \ldots$. In the *Bose-Einstein* case we have a complete description of the system if we know the numbers $r_1, r_2, r_3, \ldots$ of particles in the respective states $\sum_1, \sum_2, \sum_3, \ldots$. These r's are treated as coordinates and there is an infinite number of them, the system behaving as if it were the sum of the separate states $\sum_1, \sum_2, \sum_3, \ldots$, corresponding to the equivalent oscillators of the classical theory. In the *Fermi-Dirac* case the r's can have only the values 0, 1, and the states $\sum$ behave like spins rather than oscillators. Thus a *Bose-Einstein* assembly of indistinguishable particles can be considered as a classical assembly of distinguishable oscillators, while a *Fermi-Dirac* assembly of indistinguishable particles can be regarded as a classical assembly of distinguishable "spins". It is the putting together of such "spin" systems that we wish to consider. The subdivision into $\sum$'s is not unique, but this causes no difficulties. In order to avoid unessential infinities we make the following idealization; in quantum theory any state φ can be expanded in terms of an infinite number of eigen-states φ_r as

$$\varphi = \sum_{r=1}^{\infty} c_r \varphi_r \qquad (1)$$

but we shall assume here that only a finite number of eigen-states are necessary to form a basis for this type of expansion. Thus the state of system is now to be regarded as given by a finite number of c_r's. This is the same as if we had taken a lattice instead of the continuum (in a finite volume). This simplification is further justified because for the "spin" systems which we shall ultimately consider these numbers are really finite. We shall see later that the infinity we got is quite independent of what we take to characterize eigen-states – energy or some other quantity. Let the number of eigen-states of a given system γ be N and let the indices i, j go from 1 to N. Then any state σ of the system will be described by a vector x_i such that

$$\sum_{i=1}^{N} | x_i | = 1, \qquad (2)$$

and an observable $\mathcal{A}$ will be represented by a Hermitian matrix a_{ij}. The expectation value of $\mathcal{A}$ for the state σ is given by

$$Exp(\mathcal{A}, \sigma) = \sum_{i,j} a_{ij} x_i x_j^* \qquad (3)$$

[The usual notation is $\int A\varphi(x)\varphi^*(x)\mathrm{d}x$ and here $\int \mathrm{d}x \to \sum_i$, $A\varphi \to \sum_j a_{ij}x_i$ and $\varphi^* \to x_j^*$.]

If $\mathcal{A}$ and $\mathcal{B}$ are two simultaneously observable observables, their matrices must commute.

The product of two observables is given by

$$\mathcal{L} = \mathcal{AB}; \qquad c_{ij} = \sum_l a_{il}b_{lj} \qquad (4)$$

and their sum by

$$\mathcal{L} = \mathcal{A} + \mathcal{B}; \qquad c_{ij} = a_{ij} + b_{ij}. \qquad (5)$$

The product has a meaning only when $\mathcal{A}, \mathcal{B}$ are simultaneously observable – that is, when the matrices commute. The sum however is always defined. The observable $\mathcal{A}^n$ has the matrix $(a_{ij})^n$ and we can thus compute $Exp(\mathcal{A}^n, \sigma)$ for all n. It follows that the possible values of $\mathcal{A}$ are the eigenvalues of a_{ij} and that their probabilities are the squares of the corresponding expansion coefficients of the vector $x_1, \ldots, x_N$.

All of this describes a system given by a maximum observation. But we must consider how we can describe a system whose state might be any one of the several states $\sigma_\nu \sim x_{i,\nu}$, $\nu = 1, \ldots, n$. Let the respective probabilities be p_ν ($p_\nu \geq 0$). Then, if the totality of the states σ_ν is denoted by σ', we certainly want

$$Exp(\mathcal{A}, \sigma') = \sum_\nu p_\nu Exp(\mathcal{A}, \sigma_\nu). \qquad (6)$$

We can write this as

$$Exp(\mathcal{A}, \sigma') = \sum_\nu p_\nu \sum_{i,j} a_{ij} x_{i\nu} x_{j\nu}^* = \qquad (7)$$

$$= \sum_{i,j} a_{ij} \left(\sum_\nu p_\nu x_{i\nu} x_{j\nu}^* \right). \qquad (8)$$

The parenthesis on the right depends only on the states and their probabilities and is independent of the observable $\mathcal{A}$. We have thus broken up $Exp(\mathcal{A}, \sigma')$ into two parts, the one depending only on $\mathcal{A}$ and the other only on the states and their probabilities. We therefore regard the latter as the correct description of these states for our present purposes and denote it by u_{ji}:

$$u_{ji} = \sum_\nu p_\nu x_{i\nu} x_{j\nu}^*, \qquad (9)$$

$$Exp(\mathcal{A}, \sigma') = \sum_{i,j} a_{ij} u_{ji}. \qquad (10)$$

We shall call a state in which a maximal observation is not known a *mixture*, and the corresponding u_{ji} the *statistical matrix* of the mixture. If we denote the matrix of a_{ij} by a and that of u_{ji} by u we can write (10) as

$$Exp(A, \sigma') = Trace(au). \qquad (11)$$

These matrices a, u are not arbitrary. The matrix a must be Hermitean while, since $\sum_\nu p_\nu = 1$ and the x's satisfy (2), it follows that u must be a positive definite Hermitean matrix having the trace unity.

There is a disadvantage in the u_{ji} in that it has two indices instead of the single index of a state x_i. However, since for a single state we have

$$u_{ij} = x_i^* x_j, \qquad (12)$$

we see that the statistical matrix is independent of the phase of the wave function. Thus the u's are in one-to-one correspondence with the states, whereas the x's are not. Again for states we have the "linear superposition"

$$\varphi = c_1 \varphi_1 + c_2 \varphi_2, \qquad (13)$$

which must not be mixed up with the somewhat analogous operation

$$\varphi = |c_1|^2 u_1 + |c_2|^2 u_2, \qquad (14)$$

on statistical matrices. This operation, "uniting", as contrasted to the previously mentioned "linear superposition", deals directly with the probabilities, which are the physical quantities.

Having discussed how we shall represent a mixture, we must now consider how we can put two systems together. Let us assume there are two systems γ and $\mathcal{J}$ and see what can be said concerning the system consisting of these two together. Let γ have N different states $(i, j = 1, \ldots, N)$ and $\mathcal{J}$ have M $(k, l = 1, \ldots, M)$. The sum $(\gamma + \mathcal{J})$ will have NM states and we shall use the pairs of indices $(ik), (jl)$ to denote them. Let the observable $\mathcal{A}$ in γ have the matrix a_{ij} in γ. We can look for $\mathcal{A}$ in $(\gamma + \mathcal{J})$ and so we must have a matrix for it in $(\gamma + \mathcal{J})$. We write this matrix as $a_{ik/jl}$ and make the customary assumption that

$$\bar{a}_{ik/jl} = a_{ij} \delta_{kl}. \qquad (15)$$

This means that we do to the variable that characterizes γ the same as we do to it in γ alone, and that we do nothing to $\mathcal{J}$. Again, for $\mathcal{B}$ in $\mathcal{J}$ we take for $(\gamma + \mathcal{J})$ the matrix

$$\bar{b}_{ik/jl} = \delta_{ij} b_{kl}. \qquad (16)$$

the $\bar{a}$ and $\bar{b}$ so defined commute with each other, and this is necessary on physical grounds since the corresponding observations are evidently compatible.

There will be a statistical matrix $u_{ik/jl}$ for the mixture $(\gamma + \mathcal{J})$. What can we say about γ alone? If there is a statistical matrix u_{ij} that refers to γ alone and we make an observation of $\mathcal{A}$ on γ we must find the same result as if we had made the observation on $(\gamma + \mathcal{J})$. Thus we want

$$\sum_{i,j} a_{ij} u_{ji} = \sum_{ik/jl} a_{ij} \delta_{kl} u_{jl/ik} = \tag{17}$$

$$= \sum_{i,j} a_{ij} u_{jk/ik}, \tag{18}$$

and therefore

$$u_{ij} = \sum_{k} u_{ik/jk} \tag{19}$$

We have here an operation that carries us from the statistical matrix of $(\gamma + \mathcal{J})$ to the statistical matrix of γ. It is analogous to the operation of forming the trace but only partially carried through, and corresponds to the operation of contraction in the tensor calculus. If we forget γ we have for $\mathcal{J}$ the statistical matrix

$$v_{kl} = \sum_{i} u_{ik/il}. \tag{20}$$

We now ask, if we know u_{ij} about γ and v_{kl} about $\mathcal{J}$, how can we find $u_{ik/jl}$ for $(\gamma + \mathcal{J})$? Actually we have not enough information about $(\gamma + \mathcal{J})$ since there are $(N^2 + M^2)$ components of u_{ij} and v_{kl} while there are $N^2 M^2$ components of $u_{ik/jl}$. However we can obtain the statistical matrix for $(\gamma + \mathcal{J})$ in which there are no correlations between γ and $\mathcal{J}$. If there are no correlations we will have, if $\mathcal{A}$ is in γ alone and $\mathcal{B}$ in $\mathcal{J}$ alone,

$$Exp(\mathcal{A})Exp(\mathcal{B}) = Exp(\mathcal{A}\mathcal{B}). \tag{21}$$

This leads by simple computations to the result that

$$u_{ik/jl} = u_{ij} v_{kl} \tag{22}$$

If we start with a general $u_{ik/jl}$, form the corresponding u_{ij} and v_{kl}, and then form a $u_{ik/jl}$ in accordance with (22), we will not obtain the $u_{ik/jl}$ with which we started, but this $u_{ik/jl}$ without the correlations.

We are particularly interested in putting together a certain amount of information with no information at all. The case in which we have no information is

$$u_{ij} = \frac{1}{N} \delta_{ij}. \tag{23}$$

(It is easy to see that this is the mixture of all states of a complete orthogonal set of wave functions, each one having the probability $\frac{1}{N}$. This is true for every choice of the complete orthogonal set.) Thus if we know u_{ij} about the system γ and nothing at all about the system $\mathcal{J}$, we can form the corresponding statistical matrix for $(\gamma + \mathcal{J})$ by the formula

$$u_{ij} \rightarrow u_{ik/jl} = \frac{1}{N} u_{ij} \delta_{kl}. \tag{24}$$

This process we shall refer to as *expanding u.*

Again, if we know $u_{ik/jl}$ concerning $(\gamma + \mathcal{J})$ we have already seen (eq. (19)) that we know u_{ij} about γ where

$$u_{ik/jl} \to u_{ij} = \sum_k u_{ik/jl}. \tag{25}$$

This process will be called *contracting u.*

The operation of expanding followed by contracting gives the identity operation, but if we contract first and expand afterwards we no longer get what we started with.

At this point we shall introduce a change in normalization. We alter the normalization so that instead of the trace of u_{ij} being unity, it is equal to N:

$$\frac{1}{N}Trace(u) = 1 \tag{26}$$

With this normalization, (9) becomes

$$u_{ji} = N \sum_{\nu=1}^{N} p_\nu x_{i\nu} x_{j\nu}^*, \tag{27}$$

while (10) becomes

$$Exp(\mathcal{A}, \sigma) = \frac{1}{N}Trace(au). \tag{28}$$

The operations of expanding and contracting take the form

$$u_{ij} \to u_{ik/jl} = u_{ij}\delta_{kl} \tag{29}$$

$$u_{ik/jl} \to u_{ij} = \frac{1}{M} \sum_{k=1}^{M} u_{ik/jk}. \tag{30}$$

The operations of expanding and contracting can be applied also to observables. If we have an observable $\mathcal{A}$ with the matrix a_{ij} in γ we can call it an observable in $(\gamma + \mathcal{J})$:

$$a_{ij} \text{ in } \gamma \to c_{ik/jl} \text{ in } (\gamma + \mathcal{J}); \qquad c_{ik/jl} = a_{ij}\delta_{kl}. \tag{31}$$

here we have expanded the observable $\mathcal{A}$. The formula for contracting an observable is

$$c_{ik/jl} \text{ in } (\gamma + \mathcal{J}) \to a_{ij} \text{ in } \gamma; \qquad a_{ij} = \frac{1}{M} \sum_k c_{ik/jk}, \tag{32}$$

and arises in the following manner. We assume that the observable $c_{ik/jl}$ in $(\gamma+\mathcal{J})$ is being observed by an observer who can see only γ. This observer will describe the state of things by a statistical matrix of σ: u_{ij}. If we use this u_{ij} for $(\gamma + \mathcal{J})$

we must expand it to $u_{ik/jl} = u_{ij}\delta_{kl}$ (see (28)). Thus the expectation value of this observable in $(\gamma + \mathcal{J})$ for our observer is

$$\frac{1}{MN} \sum_{i,j,k,l} (u_{ji}\delta_{kl})c_{ik/jl},\tag{33}$$

which we may write as

$$\frac{1}{N} \sum_{i,j} u_{ji}(\frac{1}{M} \sum_{k} c_{ik/jk}),\tag{34}$$

showing that $(\frac{1}{M} \sum_k c_{ik/jk})$ is, from the point of view of our observer, indistinguishable from what he would call a_{ij}. We therefore call it a_{ij} and thus obtain formula (32).

Let us now consider the putting together of a sequence of systems $\sum_1$, $\sum_2$, $\sum_3$, - - -, $\sum_\nu$, $\sum_{\nu+1}$, - - -. We shall take these systems as referring to *Fermi-Dirac* particles so that $N = 2$ for each system. (*Bose-Einstein* assemblies will be discussed later.) Let us assume that the first ν states have been put together and that the $(\nu + 1)st$ has not yet been added. The total number of states at this stage will be 2^ν and we shall use as indices running from 1 to 2^ν the sets of numbers $(i_1, i_2, \ldots, i_\nu)$ and $(j_1, j_2, \ldots, j_\nu)$. Then when we wish to describe the mixture of the first $(\nu + 1)$ states we merely add an extra number to each set, as $(i_1, i_2, \ldots, i_\nu, i_{\nu+1})$, etc. With this notation we must write a statistical matrix for the first ν states as

$$u_{(i_1,\ldots,i_\nu)(j_1,\ldots,j_\nu)}\tag{35}$$

and an observable as

$$a_{(i_1,\ldots,i_\nu)(j_1,\ldots,j_\nu)}\tag{36}$$

while for the first $(\nu + 1)$ states we must write correspondingly

$$w_{(i_1,\ldots,i_\nu,i_{\nu+1})(j_1,\ldots,j_\nu,j_{\nu+1})}\tag{37}$$

and

$$c_{(i_1,\ldots,i_\nu,i_{\nu+1})(j_1,\ldots,j_\nu,j_{\nu+1})}.\tag{38}$$

We expand and contract between the first ν and the first $(\nu + 1)$ states by the following formulas:
expanding:

$$w_{(i_1,\ldots,i_{\nu+1})(j_1,\ldots,j_{\nu+1})} = u_{(i_1,\ldots,i_\nu)(j_1,\ldots,j_\nu)}\delta_{i_{\nu+1},j_{\nu+1}},\tag{39}$$

$$c_{(i_1,\ldots,i_{\nu+1})(j_1,\ldots,j_{\nu+1})} = a_{(i_1,\ldots,i_\nu)(j_1,\ldots,j_\nu)}\delta_{i_{\nu+1},j_{\nu+1}};\tag{40}$$

contracting:

$$u_{(i_1,\ldots,i_\nu)(j_1,\ldots,j_\nu)} = \frac{1}{2} \sum_{k=0,1} w_{(i_1,\ldots,i_\nu,k)(j_1,\ldots,j_\nu,k)}, \qquad (41)$$

$$a_{(i_1,\ldots,i_\nu)(j_1,\ldots,j_\nu)} = \frac{1}{2} \sum_{k=0,1} c_{(i_1,\ldots,i_\nu,k)(j_1,\ldots,j_\nu,k)}. \qquad (42)$$

Also, in the present notation, we have for the first ν states,

$$Exp(\mathcal{A},\sigma) = \frac{1}{2^\nu} \sum_{\substack{i_1,\ldots,i_\nu \\ j_1,\ldots,j_\nu}=0,1} a_{(i_1,\ldots,i_\nu)(j_1,\ldots,j_\nu)} u_{(i_1,\ldots,i_\nu)(j_1,\ldots,j_\nu)}. \qquad (43)$$

As we have already pointed out, it is suspicious that when dealing with the properties of systems having an infinite number of degrees of freedom, i.e. systems in which each particle has an infinite number of states, we use descriptions in which it is assumed that we are able to make the infinite number of measurements necessary to obtain the maximum information. We therefore want to restrict ourselves to descriptions in which an infinite number of observations is not implied. An observer who cannot measure more states than the first ν is to be considered as in the system corresponding to the states $(\sum_1 + \sum_2 + \ldots + \sum_\nu)$. We do not want an absolute upper limit for ν; we want to deal with observers for whom ν can become arbitrarily high, but we want their measurements to be such that they approach a limit as ν increases indefinitely.

There are measurements that are not of this type. For example, if we have a box and try to measure the energy inside, the measurement of the energy will not be approximated by any measurement restricting itself to a finite number of states. The same is true for the total number of particles in the box. The measurements we wish to consider are those that can be approximated by taking ν arbitrarily high.

Let $\mathcal{A}$ be an observable referring to the sum of any number of systems. Then we can always take an observer who sees only, say, the first ν systems, and his observations of $\mathcal{A}$ will give us an observable $\mathcal{A}^\nu$ in $(\sum_1 + \ldots + \sum_\nu)$. Thus we have a series of observables corresponding to observers who see only the first, the first two, the first three, ... systems:

$$\mathcal{A}^{(1)}, \mathcal{A}^{(2)}, \mathcal{A}^{(3)}, - - -. \qquad (44)$$

To these we may add the $\mathcal{A}^{(0)}$ belonging to an observer who can make no observations within the systems Σ. The matrix of $\mathcal{A}^{(0)}$ will be an ordinary number expressing the *a priori* expectation value of $\mathcal{A}$ (corresponding to "complete ignorance" of the state). Since $\mathcal{A}^{(\nu)}$ is what an observer who can only see the first ν systems would see when he tried to measure $\mathcal{A}^{(\nu+1)}$, it follows that $\mathcal{A}^{(\nu)}$ must be a contraction of $\mathcal{A}^{(\nu+1)}$. Thus we have for all ν

$$a^{(\nu)}_{(i_1,\ldots,i_\nu)(j_1,\ldots,j_\nu)} = \frac{1}{2} \sum_{k=0,1} a^{(\nu+1)}_{(i_1,\ldots,i_\nu,k)(j_1,\ldots,j_\nu,k)}. \qquad (45)$$

For $\nu = 0$ we have merely a number $a^{(0)}$.

For $\nu = 1$ we have the matrix

$$\mathcal{A}^{(1)} \sim \begin{vmatrix} a_{00}^{(1)} & a_{01}^{(1)} \\ a_{10}^{(1)} & a_{11}^{(1)} \end{vmatrix}, \tag{46}$$

and (45) says that $a^{(0)}$ is the diagonal mean of this matrix:

$$a^{(0)} = \frac{1}{2}(a_{00}^{(1)} + a_{11}^{(1)}) \tag{47}$$

For $\nu = 2$ we have a four-rowed square matrix indicated below:

$$\mathcal{A}^{(2)} \sim \qquad\qquad\qquad\qquad\qquad\qquad\qquad\qquad , \tag{48}$$

and the means of the elements joined by dotted lines are respectively equal to the corresponding elements of the matrix of $\mathcal{A}^{(1)}$. For $\nu = 3$ we have an eight-rowed square matrix related to the matrix for $\mathcal{A}^{(2)}$ in the same general manner, and so on for all ν.

Thus, if we look at an observable very roughly it looks like a number, if we look more closely it turns out to be a two-rowed square matrix, if more closely still, a four-rowed square matrix, and so on.

For the states we use a similar description. Each mixture will be described by giving its statistical matrix, as seen by an observer who is restricted to the states $\sum_1, ---, \sum_\nu$, this being done for every $\nu = 1, 2, \ldots$. So the description will consist of a sequence of statistical matrices

$$U^{(1)}, U^{(2)}, \ldots, U^{(\nu)}, \ldots, \tag{49}$$

where $U^{(\nu)}$ refers to the first ν states. We require that $U^{(\nu)}$ shall be a contraction of $U^{(\nu+1)}$:

$$u_{(i_1,\ldots,i_\nu)(j_1,\ldots,j_\nu)}^{(\nu)} = \frac{1}{2} \sum_{k=0,1} u_{(i_1,\ldots,i_\nu,k)(j_1,\ldots,j_\nu,k)}^{(\nu+1)} \tag{50}$$

This formula is of the same type as (45) for the observables.

Although the theory presented here seems to be the same as the usual theory, there is nevertheless a real difference. Consider, for example, what happens when we deal with an observable whose nature is such that we do not wish to include it in a valid theory. Let us know that there is exactly one particle in the state $\sum_1$ and none in any other of the states $\sum_2, \sum_3$ - - -, $\sum_\nu$, - - -. This is a statement about all the systems and if we contract to the first ν systems this means that we know that there is no particle in the state $\sum_\nu$ ($\nu \neq 1$), and this is true for any large ν. But if we know this for arbitrarily large ν we can say further that the total number of particles is unity, which we could not say for finite ν. Thus as $\nu \to \infty$ we obtain more information than we could for large but finite ν and we wish to avoid such a situation. For the statement that we have only one particle in a given box has no meaning, as we have already pointed out, because it implies a knowledge of an infinite number of systems.

We can look at this question mathematically. Let the state $\sum_1$ be given by the indices $i_1 = i_2 = \ldots = i_\nu = 0$. Then the state vector is given by

$$x_{(i_1,\ldots,i_\nu)} = \begin{cases} 1 & \text{if } i_1 = i_2 = \ldots = i_\nu = 0, \\ 0 & \text{otherwise}, \end{cases} \tag{51}$$

and the statistical matrix is given by

$$u_{(i_1,\ldots,i_\nu)(j_1,\ldots,j_\nu)} = \begin{cases} 2^\nu & \text{if } \left.\begin{matrix} i_1,\ldots,i_\nu \\ j_1,\ldots,j_\nu \end{matrix}\right\} = 0, \\ 0 & \text{otherwise}. \end{cases} \tag{52}$$

Thus we have the following sequence of statistical matrices:

$$u^{(1)} \sim 1$$

$$u^{(2)} \sim \begin{vmatrix} 2 & 0 \\ 0 & 0 \end{vmatrix}$$

$$u^{(3)} \sim \begin{vmatrix} 4 & 0 & 0 & 0 \\ 0 & 0 & 0 & 0 \\ 0 & 0 & 0 & 0 \\ 0 & 0 & 0 & 0 \end{vmatrix}$$

$$u^{(4)} \sim \begin{vmatrix} 8 & 0 & 0 & 0 & 0 & 0 & 0 & 0 \\ 0 & 0 & 0 & 0 & 0 & 0 & 0 & 0 \\ 0 & 0 & 0 & 0 & 0 & 0 & 0 & 0 \\ 0 & 0 & 0 & 0 & 0 & 0 & 0 & 0 \\ 0 & 0 & 0 & 0 & 0 & 0 & 0 & 0 \\ 0 & 0 & 0 & 0 & 0 & 0 & 0 & 0 \\ 0 & 0 & 0 & 0 & 0 & 0 & 0 & 0 \\ 0 & 0 & 0 & 0 & 0 & 0 & 0 & 0 \end{vmatrix}$$

and so on. And as $\nu \to \infty$, we see that the top diagonal element becomes arbitrarily large. We shall later make a restriction that will have the effect of preventing the occurrence of such divergencies in the statistical matrices.

We shall now describe the dual procedure. We have been concerned with a state; we now discuss an observable. We want to consider a meaningless observable, such as the total energy in a box, and for simplicity we take the observable that is defined to be unity if there is one particle in the state $\sum_1$ and no particle in any other state, and to be zero if this is not the case. Thus our observable represents the decision whether or not the total number of particles is one or not. Let us consider the corresponding observable referring to the first ν states, and let us order the states as in (51). Then the observable will have the matrix

$$a_{(i_1,\ldots,i_\nu)(j_1,\ldots,j_\nu)} = \left\{ \begin{array}{ll} 1 & \text{if} \quad \left. \begin{array}{c} i_1,\ldots,i_\nu \\ j_1,\ldots,j_\nu \end{array} \right\} = 0, \\ 0 & \text{otherwise.} \end{array} \right. \tag{53}$$

Consider how this would be described by an observer who can see only the first μ states ($\mu < \nu$). Since

$$a^{(\nu)} = \begin{vmatrix} 1 & 0 & 0 & . & . & . & 0 \\ 0 & 0 & 0 & . & . & . & 0 \\ 0 & 0 & 0 & . & . & . & 0 \\ . & . & . & . & . & . & . \\ 0 & 0 & 0 & . & . & . & 0 \\ 0 & 0 & 0 & . & . & . & 0 \\ 0 & 0 & 0 & . & . & . & 0 \end{vmatrix} \updownarrow 2^\nu, \tag{54}$$

we have, by (45),

$$a^{(\nu-1)} = \begin{vmatrix} \frac{1}{2} & 0 & 0 & . & . & . & 0 \\ 0 & 0 & 0 & . & . & . & 0 \\ 0 & 0 & 0 & . & . & . & 0 \\ . & . & . & . & . & . & . \\ 0 & 0 & 0 & . & . & . & 0 \\ 0 & 0 & 0 & . & . & . & 0 \\ 0 & 0 & 0 & . & . & . & 0 \end{vmatrix} \updownarrow 2^{\nu-1}, \tag{55}$$

and so on, and ultimately we find that

$$a^{(\mu)} = \begin{vmatrix} \frac{2^\mu}{2^\nu} & 0 & 0 & . & . & . & 0 \\ 0 & 0 & 0 & . & . & . & 0 \\ 0 & 0 & 0 & . & . & . & 0 \\ . & . & . & . & . & . & . \\ 0 & 0 & 0 & . & . & . & 0 \\ 0 & 0 & 0 & . & . & . & 0 \\ 0 & 0 & 0 & . & . & . & 0 \end{vmatrix} \updownarrow 2^\mu. \tag{56}$$

If we now consider a fixed μ, and if $\nu \to \infty$, then this matrix tends to the zero matrix, which means that if we know we have a particle in $\sum_1$ and no particles in

$\sum_2, - - -, \sum_\mu$, then the *a priori* probability that this is still true for $\sum_2, - - -,$ $\sum_\nu$ as $\nu \to \infty$ is zero.

We need some regularity assumption. It turns out to be best to make this assumption refer to the observables. An instrument cannot measure over an infinite range and we therefore do not actually measure unbounded observables; only observables that are like them over a certain range but are in fact bounded. We therefore impose the condition that only bounded observables are permitted in the theory, where by a "bounded" observable we mean one whose range of values in $\sum_1, \sum_2 - - -, \sum_\nu$ is uniformly bounded for all ν. In symbols, our regularity condition can be written as

$$\overline{\lim_{\nu \to \infty}} \|a_{ij}^{(\nu)}\| \text{ shall be finite} \tag{57}$$

(By $\|b_{ij}\|$ we mean the "absolute value" of the matrix b_{ij}: The maximum length of the vector (y_i), $y_i = \sum_j b_j x_j$, if the vector x_i is of length 1. This is, for Hermitean (b_{ij}), the greatest absolute proper value.) Let us consider the statistical matrices. We desire the normalization

$$\frac{1}{2^\nu} Trace\, U^{(\nu)} = 1 \tag{58}$$

Since $U^{(\nu)}$ is a contraction of $U^{(\nu+1)}$ it is evident that the left hand side is at any rate independent of ν. We must therefore require only that $u^{(0)} = 1$ and then be careful that $U^{(\nu)}$ remain definite for $\nu > 0$. We define $Exp(\mathcal{A}, \sigma)$ as

$$Exp(\mathcal{A}, \sigma) = \lim_{\nu \to \infty} Exp_\nu(\mathcal{A}, \sigma) = \tag{59}$$

$$= \lim_{\nu \to \infty} \frac{1}{2^\nu} Trace(\mathcal{A}^{(\nu)} U^{(\nu)}), \tag{60}$$

and therefore we require that for allowed $\mathcal{A}$'s the limit

$$\lim_{\nu \to \infty} \frac{1}{2^\nu} Trace(\mathcal{A}^{(\nu)} U^{(\nu)}) \tag{61}$$

shall exist. This limitation is of the following nature: if we write $U'^{(\mu)}$ for the expansion of $U^{(\mu)}$ to the same size matrix as $U^{(\nu)}$, then we want that

$$\lim_{\substack{\mu, \nu \to \infty \\ \mu \le \nu}} (U^{(\nu)} - U'^{(\mu)}) = 0. \tag{62}$$

(We will not give here an exhaustive analysis of this "limit"-relation.)

We have seen that the matrices entering the theory obey two types of restriction. Those representing observables must satisfy (57) while those representing mixtures satisfy (61). We shall refer to those of the first type as being of class L_∞ and those of the second type as of class L_1. We can introduce a restriction intermediate between (57) and (61) by requiring that the mean squared proper value shall be bounded. This condition can be stated in the form that for all ν,

$$\frac{1}{2^\nu} \sum_{i,j} |a_{ij}^{(\nu)}|^2 \quad \text{shall be bounded,} \tag{63}$$

and matrices satisfying it may be referred to as of class L_2. We have

$$\text{Class } L_\infty \subset \text{Class } L_2 \subset \text{Class } L_1, \tag{64}$$

but the advantage of (63) is that it is self-dual. Under it the range of the statistical matrices would be the same as that of the matrices of the observables.

We must now discuss the formal properties of our matrices. The following terminology will be useful; we say that two consecutive matrices in a sequence $A^{(1)}, A^{(2)}, A^{(3)}, \ldots$, are *identical* if they are connected by expansion, for expansion is merely the adding of a system concerning which we know nothing new. If we take some matrix of order 2^ν and then form those of higher order by expansion, those of lower order being, of course, related to it by contraction, then we shall remain in class L_∞ and will have a system which refers to knowledge of the first ν systems. We shall call it a finite system of order ν. The totality of such systems will form a basis and any element can be approximated by such matrices. If we take two matrices A, B of class L_∞, then

$$\lim \frac{1}{2^\nu} \sum_{i,j} a_{ij} \bar{b}_{ij} \quad \text{exists} \tag{65}$$

and since this quantity has the algebraic properties of an inner product, the matrices we are considering form something analogous to a *Hilbert* space.

We ask, what elements of order $(\nu + 1)$ are orthogonal to all elements of order ν? We may consider the element of order ν as

	$\leftarrow \quad 2^\nu \quad \rightarrow$				
a	b	c	–	–	–
–	–	–	–	–	–
–	–	–	–	–	–

which can be expanded into the identical element

						$\leftarrow \qquad 2^\nu \text{ matrices} \quad \rightarrow$		
a	0	b	0	c	0	–	–	–
0	a	0	b	0	c			
–		–		–		–	–	–
–		–		–				
–		–		–		–	–	–

Thus the general matrix of order $(\nu + 1)$

						$\leftarrow$		2^ν matrices		$\rightarrow$
a_{11}	a_{12}	b_{11}	b_{12}	c_{11}	c_{12}			$-$	$-$	$-$
a_{21}	a_{22}	b_{21}	b_{22}	c_{21}	c_{22}					
$-$		$-$		$-$				$-$	$-$	$-$
$-$		$-$		$-$						
$-$		$-$		$-$				$-$	$-$	$-$

will be orthogonal to this if the traces of the individual matrices $\begin{vmatrix} a_{11} & a_{12} \\ a_{21} & a_{22} \end{vmatrix}$, etc.
are zero. This means that if we take a matrix of order $(\nu + 1)$ that is orthogonal to the matrices of order ν and make a contraction, the result will be zero. Thus those matrices that are identical with the matrices of order ν are essentially unaffected by contraction while those that are orthogonal to them are destroyed by this process.

We may make use of the *Pauli* spin matrices in forming a basis. For $\nu = 0$ we need only the unit matrix 1. For $\nu = 1$ we take, in addition to the expansion of the unit matrix, the three spin matrices $\sigma_1^{(1)}, \sigma_2^{(1)}, \sigma_3^{(1)}$. For $\nu = 2$ we may divide all the possible matrices into two classes, those that are identical with the previous ones and those that are orthogonal to them. The matrices of the first type are just the expansions of the previous matrices. To obtain the matrices of the second type we note that since they are orthogonal to the original matrices they must be direct products of these matrices with two-rowed square matrices whose traces are zero, and a basis for the latter is given by the *Pauli* spin matrices. Thus for $\nu = 2$ we have, in addition to the expansions of the previous matrices, $1, \sigma_1^{(1)}, \sigma_2^{(1)}, \sigma_3^{(1)}$, also their direct products with the *Pauli* matrices which we write at this stage as $\sigma_1^{(2)}, \sigma_2^{(2)}, \sigma_3^{(2)}$.

Thus at each stage we add a set of spin matrices $\sigma_1^{(\nu)}, \sigma_2^{(\nu)}, \sigma_3^{(\nu)}$, form direct products with all the matrices of the previous stage, and take these new matrices together with the expansions of the new σ's and of all the previous matrices to form the new basis.

Consider some $\sigma^{(\nu)}$. It will be a 2-rowed square matrix having zero trace and being orthogonal to all previous σ's. It will therefore be orthogonal to any linear combination of them and since any matrix will be a linear combination of all the σ's, it will tend to zero in the weak sense that

$$\lim_{\nu \to \infty} \frac{1}{2^\nu} Trace(\mathcal{A}, \sigma^{(\nu)}) = 0 \qquad (66)$$

for all $\mathcal{A}$. This means roughly that most of the elements in the matrix will be zero in the limit, and such a result is also evident from the fact that the matrices form a *Hilbert* space and (66) expresses the orthogonality of $\sigma^{(\nu)}$ to all previous elements.

We may here make a comparison with the second quantized theory of *Fermi-Dirac* assemblies, as described by *Jordan and Wigner*, with the theory of holes and with *Jordan*'s neutrino theory of light. We need consider only two new σ's at each stage since the third is always the product of the other two. The σ's previously considered are such that they anticommute if the upper indices are the same, but commute if these indices are different. However, when we are concerned with only two new σ's at each stage we can take all the σ's so formed as anti-commuting by taking the new σ's at each stage as, say, $\sigma_2^{(\nu)}, \sigma_3^{(\nu)}$ and multiplying on the left by $\sigma^{(1)}, \sigma^{(2)}, \sigma^{(3)}, \ldots \sigma^{(\nu-1)}$. We may call the σ's so obtained again $\sigma_2^{(\nu)}, \sigma_3^{(\nu)}$ and now all these σ's will anticommute. We may now define new matrices α, α^+ in the *Jordan-Wigner* manner by the equations

$$\sigma_2^{(\nu)} + i\sigma_3^{(\nu)} = \alpha_\nu, \qquad \sigma_2^{(\nu)} - i\sigma_3^{(\nu)} = \alpha_\nu^+ \qquad (67)$$

Since $\sigma_2^{(\nu)} \to 0$ and $\sigma_3^{(\nu)} \to 0$ it follows that $\alpha_\nu \to 0$ and $\alpha_\nu^+ \to 0$. Again we have

$$
\begin{aligned}
\sigma_1^{(\nu)} &= i\sigma_2^{(\nu)}\sigma_3^{(\nu)} = \frac{1}{2}i[\sigma_2^{(\nu)}, \sigma_3^{(\nu)}] = \\
&= [\sigma_2^{(\nu)} - i\sigma_3^{(\nu)}, \sigma_2^{(\nu)} + i\sigma_3^{(\nu)}] = [\alpha_\nu^+, \alpha_\nu].
\end{aligned}
$$

Hence as $\nu \to \infty$ we find

$$\alpha_\nu^+ \alpha_\nu - \alpha_\nu \alpha_\nu^+ \to 0.$$

But since the α's obey the commutation rules

$$[\alpha_\mu^+, \alpha_\nu]_+ = \delta_{\mu\nu},$$

it follows that

$$\alpha_\nu^+ \alpha_\nu + \alpha_\nu \alpha_\nu^+ = 1$$

and therefore that

$$\alpha_\nu^+ \alpha_\nu \to \frac{1}{2}, \qquad \alpha_\nu \alpha_\nu^+ \to \frac{1}{2}.$$

Thus our α's will satisfy the following limiting conditions:

$$\left.\begin{array}{ll} \alpha_\nu \to 0, & \alpha_\nu^+ \to 0; \\ \alpha_\nu^+ \alpha_\nu \to \frac{1}{2}, & \alpha_\nu \alpha_\nu^+ \to \frac{1}{2} \end{array}\right\} \qquad (68)$$

In the original *Jordan-Wigner* theory we have, on the contrary, the different limiting conditions that

$$\left.\begin{array}{ll} \alpha_\nu \to 0, & \alpha_\nu^+ \to 0; \\ \alpha_\nu^+ \alpha_\nu \to 0, & \alpha_\nu \alpha_\nu^+ \to 1, \end{array}\right\} \qquad (69)$$

while in the theory of holes we have (69) if ν runs over "positive" states, and

$$\left.\begin{array}{ll} \alpha_\nu \to 0, & \alpha_\nu^+ \to 0; \\ \alpha_\nu^+ \alpha_\nu \to 1, & \alpha_\nu \alpha_\nu^+ \to 0 \end{array}\right\} \qquad (70)$$

if ν runs over "negative" states. The situation is the same in *Jordan*'s neutrino theory. Thus in the theory of holes most of the positive high-energy levels are assumed to be filled, while the negative ones are assumed to be empty; but in the present theory there is an even chance that they are occupied or not.

When a statement is made within the framework of the usual quantum theory, we can transform it into a statement belonging to the present theory by simply replacing the α's of the old theory by the α's of the new.

We must now consider the role of the *Schrödinger* equation. In the ordinary quantum theory it is

$$\frac{\partial}{\partial t}\varphi = iA\varphi \qquad (A = \frac{2\pi}{h}H),\tag{71}$$

where A is a Hermitean operator, being a multiple of the energy operator H. If H is independent of the time we can solve this equation explicitly:

$$\varphi_t = e^{itA}\varphi_0.\tag{72}$$

For the statistical matrix corresponding to this state we have

$$U_t = e^{itA}U_0e^{-itA}.\tag{73}$$

For an observable we obtain the result

$$A_t = e^{itA}A_0e^{-itA}\tag{74}$$

These give an isomorphism of the system on itself of the type called *inner*. (An inner automorphism is of the type

$$K \to K' = V^{-1}KV,$$

where V is a fixed unitary operator.)

In the present theory the *Schrödinger* equation cannot have this form since no *total* energy operator will exist. We must see therefore to what extent the form of the *Schrödinger* equation can be modified.

With the lapse of time a system, if left undisturbed, must undergo an automorphism. The observable $A = A_0$ will thereby become A_t. The meaning of A_t is merely that instead of making the observation A_0 now, we wait a time t seconds and then perform it. Let us denote this isomorphism by

$$A_t = I_tA_0.\tag{75}$$

Then since there is no difference between the moment 0 and the moment s, we have

$$A_{s+t} = I_tA_s$$

and hence

$$I_{s+t} = I_s I_t, \tag{76}$$

i.e. the I_t form a one-parameter group.

Now if the I_t are inner isomorphisms, that is if

$$A_t = I_t A_0 = V_t A_0 V_t^{-1} \qquad (V_t \text{ unitary}), \tag{77}$$

then (76) implies that

$$V_{s+t} = c_{st} V_t V_s \qquad (c_{st} \text{ a constant}).$$

From this one can derive, purely mathematically, the result

$$V_t = d_s e^{itA},$$

where d_s is a constant and A a fixed operator. Thus

$$A_t = I_t A_0 = e^{itA} A_0 e^{-itA}. \tag{78}$$

This gives for the observables

$$\frac{\partial}{\partial t} A_t = i(A A_t - A_t A); \tag{79}$$

and, from the invariance of $Tr(A_t U_t)$ we obtain for the statistical matrices

$$\frac{\partial}{\partial t} U_t = i(A U_t - U_t A). \tag{80}$$

Although the deduction does not necessitate the identification of A with $\frac{2\pi}{h} H$, where H is the total energy operator, the well-known physical (correspondence) arguments make it very hard to have A be anything else.

So we see that the customary form of the *Schrödinger* equation is essentially dependent on all isomorphisms I_t being inner isomorphisms.

In the usual theory, where all bounded operators correspond to observables, this is inescapable since it can be shown mathematically that the system of *all* bounded operators possesses inner isomorphisms only.

However, the system of matrix sequences such as we have described can be shown to possess also non-inner isomorphisms. We can thus avoid the present form of the *Schrödinger* equation and therefore we no longer need the existence of a total energy operator. This is necessary for formal reasons also, for it can be shown that in our present system no four-operators can form a four-vector for any unitary representation of the *Lorentz* group – not even if we introduce unbounded operators. Thus, if the resulting theory is to be special-relativistically invariant, the total energy and the total momenta *cannot* be observables.

Consider the systems $\sum_1, \sum_2, \sum_3, ---$ with the corresponding proper energies $\epsilon_1, \epsilon_2, \epsilon_3, ---$ divergent. If we try to handle this as a *Fermi-Dirac* assembly, we must form an energy

$$\sum_\nu \epsilon_\nu \alpha_\nu^+ \alpha_\nu.$$

The only reason that this does not diverge is that in the usual theory we have $\alpha_\nu^+ \alpha_\nu \to 0$ sufficiently strongly to overcome the divergence due to the ϵ's. In the present theory, since $\alpha_\nu^+ \alpha_\nu \to \frac{1}{2}$, we would get a divergent result. But though the Hamiltonian is divergent and the *Schrödinger* equation would thus have no meaning, we can still obtain an automorphism which will leave both the commutation rules and the limiting values invariant. We merely replace α_ν by $e^{i\epsilon_\nu t}\alpha_\nu$. The point is that this is *not* an inner automorphism of the system and yet it is actually the common type of automorphism in the present theory; the inner isomorphisms are not the usual ones, as our simple example has shown, and thus we cannot look for a differential equation in the present theory analogous to the *Schrödinger* equation.

We cannot treat *Bose-Einstein* assemblies directly by the method we have just described since the corresponding distinguishable classical oscillators each have an infinite number of dimensions. It is necessary to use the *De Broglie-Jordan* trick of replacing each *Bose-Einstein* particle by several *Fermi-Dirac* particles. If we write β_ν, β_ν^+ for the matrices describing the *Bose-Einstein* particles, we have

$$\left. \begin{array}{l} \beta_\mu \beta_\nu^+ - \beta_\nu^+ \beta_\mu = \delta_{\mu\nu}, \\ \beta_\mu \beta_\nu - \beta_\nu \beta_\mu = 0, \end{array} \right\} \tag{81}$$

while for *Fermi-Dirac* particles we have

$$\left. \begin{array}{l} \alpha_\mu \alpha_\nu^+ + \alpha_\nu^+ \alpha_\mu = \delta_{\mu\nu}, \\ \alpha_\mu \alpha_\nu + \alpha_\nu \alpha_\mu = 0. \end{array} \right\} \tag{82}$$

Jordan succeeded in obtaining (81) from (82) by means of an algebraic trick. He wrote a relation of the type

$$\beta_\mu = \sum_{\lambda,\nu} A_{\lambda\nu}^\mu \alpha_\lambda^+ \alpha_\nu \tag{83}$$

and from this it is seen that the sums of β's are given by the sums of their corresponding A's, while though the relation of the products of β's to those of the A's is not simple, the commutators of the β's correspond to the commutators of their A's. Thus the equations (81) become

$$\left. \begin{array}{l} A^\mu A^{\nu+} - A^{\nu+} A^\mu = \delta_{\mu\nu}, \\ A^\mu A^\nu - A^\nu A^\mu = 0, \end{array} \right\}$$

where A^μ is the matrix $(A^\mu_{\lambda\nu})$. These matrix equations possess solutions. Actually, *Jordan* took

$$
A^k \sim
\left[
\begin{array}{cccccc|cccccccccccc}
\bullet & & 0 & 0 & 0 & 0 & 0 & 0 & 0 & 0 & 0 & 0 & 0 & 0 & 0 & 0 & 0 & 0 \\
 & \bullet & & 0 & 0 & 0 & 0 & 0 & 0 & 0 & 0 & 0 & 0 & 0 & 0 & 0 & 0 & 0 \\
0 & & \bullet & & 0 & 0 & 0 & 0 & 0 & 0 & 0 & 0 & 0 & 0 & 0 & 0 & 0 & 0 \\
0 & 0 & & \bullet & & 0 & 0 & 0 & 0 & 0 & 0 & 0 & 0 & 0 & 0 & 0 & 0 & 0 \\
0 & 0 & 0 & & \bullet & & 0 & 0 & 0 & 0 & 0 & 0 & 0 & 0 & 0 & 0 & 0 & 0 \\
0 & 0 & 0 & 0 & & \bullet & 0 & 0 & 0 & 0 & 0 & 0 & 0 & 0 & 0 & 0 & 0 & 0 \\ \hline
0 & 1 & & & & \uparrow & \bullet & m & & 0 & 0 & 0 & 0 & 0 & 0 & 0 & 0 & 0 \\
0 & 0 & 1 & & & k & & \bullet & a & & 0 & 0 & 0 & 0 & 0 & 0 & 0 & 0 \\
0 & 0 & 0 & 1 & & \downarrow & 0 & & \bullet & i & & 0 & 0 & 0 & 0 & 0 & 0 & 0 \\
0 & 0 & 0 & 0 & 1 & & 0 & 0 & & \bullet & n & & 0 & 0 & 0 & 0 & 0 & 0 \\
0 & 0 & 0 & 0 & 0 & 1 & 0 & 0 & 0 & & \bullet & d & & 0 & 0 & 0 & 0 & 0 \\
0 & 0 & 0 & 0 & 0 & 0 & 1 & 0 & 0 & 0 & & \bullet & i & & 0 & 0 & 0 & 0 \\
0 & 0 & 0 & 0 & 0 & 0 & 0 & 1 & 0 & 0 & 0 & & \bullet & a & & 0 & 0 & 0 \\
0 & 0 & 0 & 0 & 0 & 0 & 0 & 0 & 1 & 0 & 0 & 0 & & \bullet & g & & 0 & 0 \\
0 & 0 & 0 & 0 & 0 & 0 & 0 & 0 & 0 & 1 & 0 & 0 & 0 & & \bullet & o & & 0 \\
0 & 0 & 0 & 0 & 0 & 0 & 0 & 0 & 0 & 0 & 1 & 0 & 0 & 0 & & \bullet & n & \\
0 & 0 & 0 & 0 & 0 & 0 & 0 & 0 & 0 & 0 & 0 & 1 & 0 & 0 & 0 & & \bullet & a \\
0 & 0 & 0 & 0 & 0 & 0 & 0 & 0 & 0 & 0 & 0 & 0 & 1 & 0 & 0 & 0 & & \bullet & l
\end{array}
\right]
$$

these matrices commute so that it seems impossible that we should get anything but zero on the right of (81). However, the actual result is $\alpha^+_\nu \alpha_\nu]^{\nu=+\infty}_{\nu=-\infty}$ and since $\alpha^+_\nu \alpha_\nu \to 1$ for $\nu \to +\infty$ and $\to 0$ for $\nu \to -\infty$, we get the $\delta_{\mu\nu}$ we need.[1] It really amounts to summing the infinite series

$$\ldots + 1 - 1 + 1 - 1 + 1 - \ldots$$

and in accordance with the principles of the subtraction physics we may say that since the two ends of the series are each (+1) the sum is (+1)!

Finally, a remark concerning *Lorentz* invariance. The systems we have described can be considered as generated by abstract elements α_ν which satisfy (a) the *Jordan-Wigner* commutation rules, and (b) the limiting relations (68). Under the *Lorentz* group the α's are transformed by

$$\alpha_\nu \to \alpha'_\nu = \sum_\mu \sigma_{\nu\mu} \alpha_\mu, \tag{84}$$

where σ is unitary. We can show that this converges and that it gives an isomorphism of the system on itself, and this is enough to ensure the *Lorentz* invariance since the matrices $(\sigma_{\nu\mu})$ may be chosen so as to form a representation of the *Lorentz* group. The isomorphism (84) is not an inner isomorphism so that we shall not have infinitesimal generators for these isomorphisms.

EDITORS' NOTES

In the mimeographed lecture notes, von Neumann's contribution appears on pp. 147-172. For the present publication the equations have been renumbered.

1. Here we omit the sentence "(This is explained also on p. 109 of these notes.)" which refers to a technical point in the contribution of Maurice Pryce. We thank Karl von Meyenn for making the full notes available to us.

ADOLF GRÜNBAUM

A NEW CRITIQUE OF THEOLOGICAL INTERPRETATIONS OF PHYSICAL COSMOLOGY

This paper is a *sequel* to my "Theological Misinterpretations of Current Physical Cosmology" (*Foundations of Physics*, vol. 26, 1996; revised in *Philo*, vol.1, No.1, 1998). It draws on portions of my earlier (Grünbaum, 2000) but goes beyond it and modifies the overlapping portions.*

In that earlier (1998) paper, I argued that the Big Bang models of (classical) general relativity theory as well as the original 1948 versions of the steady state cosmology are each *logically incompatible* with the time-honored theological doctrine that *perpetual* divine creation ("*creatio continuans*") is *required* in each of these two theorized worlds to keep them in existence. Furthermore, I challenged the perennial theological doctrine that there must be a divine creative cause (as distinct from a transformative one) for the very existence of the world, *a ratio essendi*. This doctrine is the theists's reply to the question "Why is there something, rather than just nothing?"

I begin my present paper by arguing against the response by the contemporary Oxford theist Richard Swinburne and by Leibniz to my *counter*-question: "But why should there be just nothing, rather than something?"

I. INTRODUCTION

My first aim in this paper is to pinpoint the defects of the time-honored arguments for *perpetual* divine creation given by a succession of theists including Aquinas, Descartes, Leibniz, Locke, as well as by the present-day theists Richard Swinburne and Philip L. Quinn. One of these defects will also turn out to vitiate a pillar of the medieval Arabic Kalam argument for a creator (Craig, 1979).

II. THE NONEXISTENCE OF THE ACTUAL WORLD AS ITS PURPORTED "NATURAL" STATE

A. Swinburne and Leibniz on the Normalcy of Nothingness

In Richard Swinburne's extensive writings in defense of (Christian) theism, notably in (Swinburne 1991;1996), he presents two versions of his argument for

M. Rédei and M. Stöltzner (eds.),
John von Neumann and the Foundations of Quantum Physics, 269–288.
© *Adolf Grünbaum. Printed in the Netherlands.*

his fundamental thesis that the most natural state of affairs of the existing world and even of God is *not* to exist *at all*! As he put it (1996, p. 48):

It is extraordinary that there should exist anything at all. Surely the most natural state of affairs is simply nothing: no universe, no God, nothing. But there is something.

It will be expeditious to deal first with the more recent 1996 version of his case, and then with his earlier (1979;1991) substantial articulation of Leibniz's argument from *a priori* simplicity.

Surprisingly, Swinburne deems the existence of something or other to be "extraordinary," i.e., literally out of the ordinary. To the contrary, surely, the most pervasively ordinary feature of our experience is that we are immersed in an ambiance of existence. Swinburne's initial assertion here is, at least *prima facie*, a case of special pleading in the service of a prior philosophical agenda. Having made that outlandish claim, Swinburne builds on it, averring that "surely the most natural state of affairs is simply nothing." Hence he regards the question "Why is there anything at all, rather than just nothing?" as paramount.

As we know, the Book of Genesis in the Old Testament starts with the assertion that, in the beginning, God created heaven and earth from scratch. And, as John Leslie (1978, p. 185) pointed out, "when modern Western philosophers have a tendency to ask it [i.e., the existential question above], possibly this is only because they are heirs to centuries of Judeao-Christian thought."

This conjecture derives added poignancy from Leslie's observation that "To the general run of Greek thinkers the mere existence of a thing [or of the world] was nothing remarkable. Only their changing patterns provoked [causal] inquisitiveness." And Leslie mentions Aristotle's views as countenancing the acceptance of "reasonless existence." It is a sobering fact that, before Judaism and Christianity molded the philosophical intuitions of our culture, those of the Greeks were basically different.

Yet in post-Hellenic culture, there is a long history of sometimes emotion-laden, strong puzzlement, even on the part of atheists such as Heidegger, about the mere existence of our world (Edwards, 1967). Thus, Wittgenstein (1993, p. 41) acknowledged the powerful *psychological* reality of wondering at the very existence of the world. But logically, he rejected the question altogether as "nonsense," because he "cannot imagine its [the world's] not existing" (pp. 41-42), by which he may perhaps have meant not only our world, but more generally, as Rescher (1984, p. 5) points out, some world or other: Wittgenstein could be convicted of a highly impoverished imagination, if he could not imagine the non-existence of just our particular world.

Before turning to the logical aspects of the cosmic existential question, let me mention a psychological conjecture as to why not only theists, but also some atheists, find that question so imperative. For example, Heidegger (1953, p. 1) deemed "Why is there anything at all, rather than just nothing" the most fundamental question of metaphysics, yet he offered no indication as to an answer to it, and he saw its source in our facing nothingness in our existential anxiety.

I gloss this psychological hypothesis as surmising that our deeply instilled fear of death has prompted us to wonder why we exist so *precariously*. And we may then have extrapolated this precariousness, more or less unconsciously, to the existence of the universe as a whole.

Psychological motivations aside, let me tentatively recast Swinburne's afore-cited rather vague statement "The most natural state of affairs is simply nothing" to read instead as "The most natural state of our existing world is *not* to exist at all." This reformulation may avoid the hornet's nest inherent in the question as to the sheer intelligibility of nothingness qua purportedly *normal state* of our world.[1] Yet my reformulation is still conceptually troublesome by being incoherent: How can *non*-existence coherently be a genuine state – natural or otherwise – of the *actually existing* world?

In any case, I shall be in the position to offer my own reasons for endorsing Henri Bergson's injunction as follows: We should never assume that the "natural thing" would be the existence of nothing. He rested this proscription on grounds radically different from mine, when he declared[2]: "The presupposition that *de jure* there should be nothing, so that we must explain why *de facto* there is some-thing, is pure illusion." But Bergson's reasons for charging illusoriness are conceptual and *a priori*, whereas mine will turn out to be empirical.

As we know, a long theistic tradition has it that this *de jure* presupposition is correct *and* that there must therefore be an explanatory cause *external* to the world; furthermore, it is argued that this cause is an omnipotent, omni-benevo-lent, and omniscient personal God. Yet, as Gerald Massey has pointed out to me, the stated *de jure* presupposition must be qualified in the case of Thomas Aquinas: Unlike Swinburne, Aquinas held that, *except for God, de jure* there should be nothing.

But what are good grounds for postulating just what is the natural, spontane-ous, normal state of the world in the absence of an intervening external cause? In opposition to an *a priori* conceptual dictum of naturalness, I have argued in previous writings from the history of science that changing evidence makes the verdict inevitably *empirical* rather than *a priori* (Grünbaum, 1996; 1998). Here, a summary will have to suffice.

I welcome Swinburne's use of the phrase "natural state of affairs" (1996, p. 48) to the extent that it dovetails with the parlance I used, when I elaborated on the notion of "natural state" by speaking of it as the "spontaneous, externally un-disturbed, or normal" state. To within the qualification regarding the very exist-ence of God as such, as stated above, Swinburne's claim that "the most natural state of affairs is simply nothing" had been enunciated by Aquinas, Descartes, Leibniz, and a host of other theists. Hereafter, I shall designate their thesis as asserting "the spontaneity of nothingness," or "SoN" for brevity.

In my parlance, the terms "natural," "spontaneous," "normal," and "exter-nally unperturbed" serve to characterize the historically dictated *theory-relative* behavior of physical and biological systems, when they are *not* subject to any ex-*ternal* influences or forces. In earlier writings (Grünbaum, 1954; 1989; 1990;

1996; 1998), I called attention to the *theory-relativity* of such naturalness by several examples from physics and biology.

Thus, in (Grünbaum, 1996 and 1998, Sections 3 and 4), I pointed out that the altogether "natural" behavior of suitable subsystems in the now defunct original Bondi & Gold Steady-State World is as follows: Without any interference by a physical influence external to the subsystem, let alone by an external matter-*creating* agency or God, matter pops into existence spontaneously in violation of Lavoisier's matter-conservation. This spontaneous popping into existence follows deductively from the conjunction of the theory's postulated matter-*density*-conservation with the Hubble law of the expansion of the universe. For just that reason, I have insisted on the use of the *agency-free* term "matter-accretion" to describe this process, and have warned against the use of the *agency-loaded* term "matter-*creation*."

In the same vein, I emphasized that according to Galileo and to Newton's first law of motion, it is technically *"natural"* that a *force-free* particle moves uniformly and rectilinearly, whereas Aristotle's physics asserted that a force is required as the external cause of a sublunar body's non vertical uniform rectilinear motion. In short, Aristotle clashed with Galileo and Newton as to the "natural," spontaneous, dynamically unperturbed behavior of a body, which Aristotle deemed to be one of rest at its proper place. Thus, Galileo and Newton *eliminated* a *supposed external dynamical cause* on *empirical* grounds, explaining that uniform motion can occur spontaneously without such a cause.

But, if so, then the Aristotelean demand for a causal explanation of *any* motion whatever by reference to an external perturbing force is predicated on a *false underlying assumption*. Clearly, the Aristoteleans then begged the question by tenaciously continuing to ask: "What net external force, pray tell, keeps a uniformly moving body going?" Thus, scientific and philosophical questions can be anything but innocent by loading the dice with a *petitio principii!*

I omit a biological example concerning the "spontaneous generation of life" from inorganic chemicals (Grünbaum, 1973, pp. 573-574).

As illustrated by the ill-conceived question put to Galileo by his Aristotelean critics, it is altogether misguided to demand an external cause of the *deviations* of a system from the pattern that an empirically discredited theory tenaciously affirms to be the "natural" one (Grünbaum, 1973, pp. 406-407).

The proponents of the spontaneity of nothingness (SoN) have not offered any *empirical* evidence whatever for it! Yet the lesson from the history of science seems to be that just such evidence is required for tenability. However, some of the advocates of SoN have offered an *a priori* argument from *conceptual* simplicity in its defense. I now turn to such a defense.

B. Leibniz's and Swinburne's A Priori Simplicity Argument for SoN

The imposition of *a priori* notions of naturalness is sometimes of-a-piece with imposing tenaciously held criteria of what mode of scientific explanation is required for *understanding* the world. The demise of Laplacean determinism in physics, and its replacement by irreducibly stochastic, statistical models of micro-physical systems, is another case of empirical discreditation of a tenacious demand for the satisfaction of a previously held ideal of explanation: It emerges *a posteriori* that the universe just does not accommodate rigid prescriptions for explanatory understanding that are rendered otiose by a larger body of evidence.

Relatedly, the theory of radioactive decay in nuclear physics, for example, runs counter to Leibniz's demand for a "sufficient reason" for all logically contingent states of affairs. Thus, since the existence of our actual world is logically contingent, he insisted that there *must* be a sufficient reason for its existence.

Leibniz makes this demand in his 1714 essay, "The Principles of Nature and of Grace Based on Reason" from which I now cite just a few sentences:

Now we must advance to *metaphysics,* making use of the great principle, little employed in general, which teaches that *nothing happens without a sufficient reason;* ... This principle [of sufficient reason] laid down, the first question which should rightly be asked will be, *why is there something rather than nothing?*

But note that so far, Leibniz does not answer *my counter-question* "But why, oh why, *should* there be just nothing, rather than something?"

To *justify his* question, he now resorts to an *a priori* conceptual argument for SoN from *simplicity*:

For *nothing is simpler and easier than something* [emphasis added]. Further, suppose that things must exist, we must be able to give a reason *why they must exist so* and not otherwise.

8. Now this sufficient reason for the existence of the universe cannot be found *in the series of contingent things* ...

Two important points need to be noted here: (i) In English, the term "nothing" in "nothing is simpler and easier than something" is intended to stand for "nothing-*ness*", as is clear from the noun "Das Nichts" in the German original in the Felix Meiner edition of 1956, and (ii) Leibniz is *not* content to legitimate his question "Why is there something rather than nothing?" by mere recourse to the principle of sufficient reason; instead, he relies *explicitly* on the *a priori* conceptual simplicity of nothingness and *implicitly* on SoN!

As we just saw, Leibniz wrote in 1714 (Wiener, 1951, p. 525): "... why is there something rather than nothing? For nothing is *simpler* and *easier* than something" (italics added). But why, one must ask, is this *conceptual* claim, if granted, mandatory for what is *ontologically* the spontaneous, externally undisturbed state of the *actual* world? Alas, Leibniz does not tell us here. Yet I argued above that, according to our best scientific knowledge, our notion of *warranted* spontaneity is relative to changing *empirically-based* scientific theories. In short, conceptual simplicity does not necessarily bespeak ontological spontaneity.

The epistemic moral of this sketchy history is two-fold: (i) The character of just what behavior of the actual world and of its subsystems is "natural" is an empirical *a posteriori* matter, rather than an issue that can be settled *a priori*, yet (ii) SoN has no *empirical* credentials at all, as acknowledged, in effect, by the purely conceptual arguments for it which have been offered by its more "recent" defenders.

But the question could be and has been asked why this supposed form of "scientism" should be mandatory. Friedrich von Hayek and his acolytes have characterized scientism as a doctrine of explanatory scientific imperialism with utopian pretensions. Much more precisely, Richard Gale and Alexander Pruss (forthcoming) defined scientism as implying that everything that *is* explained is explained by either science or some kind of explanation having strong affinities to actual scientific explanation. Thus, in their construal, scientism is *not* taken to assert that everything is explained by science *tout court*, but only that everything that is actually explained, is explained by science.

It is easy enough, as theists like Leibniz, Swinburne, and Philip L. Quinn (1993) have done, to disavow scientism as just defined, although Swinburne insists (1991) that his version of theism is *methodologically* of-a-piece with various modes of scientific inference, such as the use of Bayes's theorem to credibilify scientific hypotheses. And he then marshals that theorem fallaciously to aver that God probably exists (Grünbaum, 2000, sections 5.2 and 5.3).

But such a theistic disavowal of scientism places the heavy probative burden of justifying the theistic explanatory alternative on the theistic rejectionists. The most prominent alternative they have proffered is modeled on volitional agency-explanations of actions, as distinct from ordinary event-causation.

I shall argue, however, that a divine volitional explanation of the actual topmost or most fundamental laws of nature, of their constants, and of the pertinent boundary or initial conditions founders: It will turn out to be abortive, because it is epistemically ill-founded. And it founders as well ethically, because, as Hume has emphasized, no omni-benevolent and omnipotent God would ever create a world with so overwhelmingly much gratuitous and uncompensated natural evil such as cancer, evil that is not due to human decisions and actions. In particular, evil comprises both moral and natural evil.

This egregious difficulty is attested by wide agreement, even among theists, that no extant theodicy has succeeded in neutralizing it. True enough, Swinburne (1991, ch. 11 and p. 284; 1996, ch. 6) offers his own theodicy, but Quentin Smith (1991, pp. 165-168; 1992; 1997, pp. 137-157) has discredited his effort.

Having gotten no help from Leibniz toward a cogent defense of SoN, I turn to Swinburne's quite general argument from simplicity for which he claims multiple sanction from science (1996).

Let me present just *some* of my own objections to Swinburne's appeal to simplicity.

As we know from Occam's razor, his injunction is to abstain from postulating entities beyond necessity. Mindful of this proscription, Swinburne (1991, p. 84) characterizes the simplicity and complexity of hypotheses in terms of the *number* of entities, the sorts of entities, and the kinds of relations among entities that they postulate. But clearly, in scientific theorizing, the *regulative ideal of Occam's razor is subject to the crucial proviso of heeding the total available evidence, including its complexity.*

Thus, in the history of actual science, increasingly greater theoretical faithfulness to the facts required the frequent *violation* of Swinburne's criterion of simplicity with respect to the number of postulated entities. *And the decisive moral will be that such simplicity as can be achieved while explaining the phenomena is an empirical matter, and not subject to Swinburne's ontologically legislative a priori conceptual simplicity.*

Examples from actual science that violate Swinburne's mandating of conceptual simplicity abound. Let me enumerate just a couple of them. Additional telling examples are given in (Grünbaum, 2000).

(i) By Swinburne's normative criterion of numerical simplicity, the pre-Socratic Thales's *monistic* universal hydro-chemistry of the world's substances is about a hundredfold simpler than the empirically discovered periodic table of the elements. And there are even two isotopes of Thales's water, one heavier than the other. Furthermore, in organic chemistry, isomerism is a complication. Moreover, the frequency of monochromatic light is simple, but white light is composed of a whole range of spectral frequencies.

(ii) Among laws of nature, van der Waals's laws for gases are more complicated than the Boyle-Charles law for ideal gases. Again, in the Newtonian two-body system of the earth and the sun, Kepler's relatively simple laws of planetary motion are replaced by more complicated ones that take account of the sun's own acceleration. And, in the case of the ten-body motions of the sun and its nine planets, only complicated infinite series can represent the equations of motion.

Moreover, Einstein's field equations are awesomely complicated, non-linear partial differential equations, and as such are enormously *more complicated* than the ordinary second order differential equation in Newton's law of universal gravitation. Remarkably, Swinburne himself mentions this greater complexity of Einstein's gravitational field equations, but his comment on it does not cohere with his demand for *a priori* simplicity as an avenue to the truth. He says (1991, p. 79):

Newton's laws ... are (probably) explained by Einstein's field equations of General Relativity [presumably he means as special approximations under specified restrictive conditions]. In passing from Newton's laws to Einstein's there is I believe a considerable loss of [*a priori*] simplicity ... But there is some considerable gain in explanatory power.

When considering simplicity, it is crucial to specify the particular respect(s) in which one theory is held to be simpler than another. After all, theory B might be simpler than theory A in *one* respect, while being more complicated in another. Thus, an *explanatory unification* of previously disparate sorts of theoretical elements, when achieved by a new theory, could be taken as a mark of its greater simplicity vis-à-vis its predecessor, even though the two theories might exhibit an *inverse* simplicity-ordering in some other respect. In this vein, the general theory of relativity ("GTR") effected a unification of the description of the physical geometry with a theory of gravitation, which had been quite distinct in its Newtonian predecessor. The special theory of relativity had already introduced the notion of an invariant four-dimensional space-time metric *via* an invariant speed of light, but without a theory of gravitation. Thereupon, the GTR absorbed gravitation in the metric tensor g_{ik} of its space-time geometry.

But observe that this simplifying unification does *not* owe its legitimacy at all to some greater *a priori* simplicity à la Swinburne; instead it derives its warrant from such empirical promptings as the equivalence of gravitational and inertial mass, which Eötvös had demonstrated experimentally.

Note that the *sacrifice* of *a priori* simplicity for the sake of greater explanatory power is dictated by *empirical* constraints, and that empirical facts can play a vital epistemic role in achieving simplifying theoretical economies. Thus empirical facts override Swinburne's *a priori* simplicity qua the governing heuristic criterion of theory-formation. And it is unavailing for Swinburne to reason ill-foundedly that greater *a priori* simplicity makes for greater *a priori* prior probability, and that this supposed greater prior probability, in turn, is ontologically *mandatory as to truth.* Yet in a recent monograph *Simplicity as Evidence of Truth,* Swinburne (1997) claims that the inductive *methodological* injunction of Occam's razor is *ontologically* legislative as to truth!

Evidently, Swinburne presents us with a tendentiously *a priori* misdepiction of the use of simplicity criteria in actual science, although he claims continuity with actual scientific theory-construction for his conceptual standard of simplicity. Just as the lesson spelled by scientific theoretical progress undermined his *a priori* conceptual avowal of SoN as a basis for external divine creation, so also his pseudo-Occamite argument for normative *a priori* simplicity fails. It emerges that Swinburne's attempt to underwrite SoN by recourse to simplicity is abortive.

C. The Role of SoN in the Medieval Arabic Kalam Argument

To conclude my contention that the appeal to SoN wrought philosophical mischief in several of the major theistic cosmological arguments, let me just mention the medieval Arabic *Kalam* argument for a creator, as articulated by William Craig (1979).

As I show in (Grünbaum, 2000), contrary to the assertion of the contemporary theist William L. Craig, the so-called Kalam version of the cosmological argument, which he defends, is likewise predicated, though only quite tacitly and insidiously, on the baseless SoN.

III. Critique of the "Explanation" of the Most Fundamental Laws of Nature by Divine Creative Volition

Philip Quinn wrote (1993, pp. 607-608, italics added):

... The conservation law for matter-energy is logically contingent. So if it is true, the question of why it holds rather than not doing so arises. If it is a fundamental law and only scientific explanation is allowed, the fact that matter-energy is conserved is an inexplicable brute fact. For all we know, the conservation law for matter-energy may turn out to be a derived law and so deducible from some deeper principle of symmetry or invariance. But if this is the case, the same question can be asked about this deeper principle because it too will be logically contingent. If it is fundamental and only scientific explanation is allowed, then the fact that it holds is scientifically inexplicable. Either the regress of explanation terminates in a most fundamental law or it does not. If there is a deepest law, it will be logically contingent, and so the fact that it holds rather than not doing so will be a brute fact. If the regress does not terminate, then for every law in the infinite hierarchy there is a deeper law from which it can be deduced. In this case, however, the whole hierarchy will be logically contingent, and so the question of why it holds rather than some other hierarchy will arise. So if only scientific explanation is allowed, the fact that this particular infinite hierarchy of contingent laws holds will be a brute inexplicable fact. Therefore, on the assumption that scientific laws are logically contingent and are explained by being deduced from other laws, there are bound to be inexplicable brute facts if only scientific explanation is allowed.

There are, then, genuine explanatory problems too big, so to speak, for science to solve. If the theistic doctrine of creation and conservation is true, these problems have solutions in terms of agent-causation. *The reason why there is a certain amount of matter-energy and not some other amount or none at all is that God so wills it, and the explanation of why matter-energy is conserved is that God conserves it.*

Quinn and Swinburne quantify the "amount of matter-energy". But even in elementary physics, qua intergral of the equation of motion, the energy clearly depends on the constant of integration, i.e., on the zero of energy and on the choice of units. Does God's supposed creative decree contain such mundane specifications?

In the same vein as Quinn, Swinburne (1996, pp. 21-22) characterizes *ultimate* explanation as "intentional" or "*personal*".

Swinburne asserts theistic *pan*-explainability, declaring (ibid.): "using those same [scientific] criteria, we find that the view that there is a God explains *everything* we observe, not just some narrow range of data. It explains the fact that there is a universe at all [via SoN], that scientific laws operate within it, ..." (cf. also his 1991, ch. 4 on "Complete Explanation.")

Note, however, that Swinburne and others who offer divine volition explanations would offer precisely such an explanation, if the laws of our world were radically different, or even if, in a putative world-ensemble of universes, each of them had its own laws, vastly different from the respective laws in the others.

Their schema of theistic volition explanations "postulates that all explanation is reducible to personal explanation" on the model of the agent-causation Aristotle's practical syllogism for intentional action. For brevity, I shall hereafter denote the practical syllogism by the acronym "PS."

Swinburne maintains that the hypothesis of divine creation just "moves beyond" scientific explanations via *the very same epistemological criteria*. As against that contention, let me now set forth some explanatory discrepancies between them:

Neither Swinburne nor Quinn *spelled out* the provision of a deductive theistic volitional explanation, which they claim for the hypothesis of divine creation. I now offer a reconstruction of essentially the deductive explanatory reasoning that, I believe, they had in mind. And I am glad to report that Quinn (private communication) authenticated my reconstruction, at least in regard to himself.

Premise 1.   God freely *willed* that the state of affairs described in the *explanandum* below ought to materialize.

Premise 2.   Being omnipotent, he was able to cause the existence of the facts in the *explanandum* without the mediation of other causal processes.

Conclusion: Our world exists, and its contents exhibit its specific most fundamental laws.

At least two basic considerations jeopardize the viability of this proffered volitional theological explanation schema:

(i) Epistemologically, it will succeed *only if* the theist can produce cogent evidence, *independent of the explanandum,* for the very content of the volition that the proffered explanation imputes to the Deity; failing that, the deductive argument here is not viable epistemically; but where has the theist produced such independent evidence? Moreover, Premise 2 unwarrantedly assumes the availability of a successful cosmological argument for the existence of the God of theism.

(ii) Relatedly, the explanation is inherently and conspicuously *ex post facto*, because the content of the volition imputed to God is determined only retrospectively, depending entirely on what the specifics of the most fundamental laws have turned out to be.

(iii) Indeed, there is a vicious *epistemic* circularity in the logical structure of the premises vis-à-vis the *explanandum*, which renders it *explanatorily trivial:* In order to know what God willed and then did, Quinn and Swinburne need to insert the *explanandum identically* into *each* of the premises, and that *trivializes* the deductive explanation!

In ordinary action-explanations on the model of Aristotle's Practical Syllogism, we often, indeed typically, do have independent evidence – or at least access to independent evidence – as to the *content* of the agent's *motives*. That is to say, we have *evidence for the imputed motives other than* the action taken by the agent. And, absent such independent evidence, we reject the proffered action-explanation as epistemologically viciously circular. In *this* sense, I claim that the attribution of the existence of the world and of its particular laws to God's willing-that-it-exist is unacceptably *ex post facto* or epistemically viciously circular.

William James has beautifully encapsulated the *ex post facto* character of the relevant sort of theological explanation, in which God is Hegel's Absolute. Speaking of the facts of the world, James (1975, p. 40) declared:

Be they what they may, the Absolute will father them.

But no such *ex post facto* deficiency is found in typical explanations in physics or biology such as (i) the Newtonian gravitational explanation of the orbit of the moon, (ii) the deductive-nomological explanations of optical phenomena furnished by Maxwell's equations, which govern the electromagnetic field, or in a statistical context, (iii) the genetic explanations of hereditary phenotypic human family resemblances.

As illustrated by these examples, the complaint against vicious epistemic circularity cannot be disarmed by pointing out that deductive validity as such guarantees the logical containment of the *explanandum* in the premises of the *explanans*: As I have just shown, the presence or absence of vicious epistemic circularity depends crucially on the particular logical structure of that containment. The advocates of this theological explanation claim that it conforms essentially to Aristotle's PS, subject to the modification that God's creative action is direct, rather than mediated by other causal processes, as in the case of human actions (Swinburne, 1991, p. 294).

But according to familiar scientific evidential criteria which Swinburne tirelessly professes to employ – as in his appeal to Bayes's theorem, – by being epistemically viciously circular and *ex post facto*, his and Quinn's deductive argument is *epistemologically frivolous*. Relatedly, the same derogatory verdict on such epistemic frivolity would be rendered by the rules of evidence imposed in courts of law.

Nonetheless, Swinburne (1991, p. 109), speaking of the explanandum *e*, is explicitly satisfied with such an *ex post facto* mode of explanation: "… clearly whatever *e* is, God being omnipotent, has the power to bring about *e*. He will do so, if he chooses to do so." Yet since we obviously have no independent evidential access to God's choices, Swinburne has to *infer* completely *ex post facto* from whether or not *e* is actually the case whether God's choices included *e*.

Let me emphasize that, as I lodge it, the complaint that *ex post facto* explanations afford no independent empirical check on their premises is *not* focused on their being *non-predictive.* Non-predictiveness is *not* tantamount to untestability. For example, (neo)-Darwinian evolutionary theory is essentially unpredictive, but it retrodicts numerous previously unknown past facts. But an explanation that is neither retrodictive nor predictive and whose premises have no corroboration by evidence independent of the given *explanandum,* is paradigmatically *ex post facto.*

Alas, Swinburne is entirely unfazed in the face of the *ex post facto* character and vicious epistemic circularity of his deductive theistic explanation above. Having harped on God's omnipotence earlier, Swinburne develops it further (1991, p. 295, emphases added):

God, being omnipotent, cannot rely on causal processes outside his control to bring about effects, so his range of easy control must coincide with his range of direct control and *include all states of affairs which it is logically possible for him to bring about.*

Precisely because God is omnipotent, however, he could clearly have chosen any one of the logically possible sets of fundamental laws to achieve his presumed aims, rather than the actual laws. Yet, if so, then exactly that latitude shows that, if the stated vicious epistemic circularity is to be avoided, the theological explanatory scenario fails to satisfy Leibniz's demand for a "full reason why", if there is a world at all, "it should be such as it is". To assert peremptorily in the face of that latitude that God inscrutably willed the actual world as a matter of brute fact evades Leibniz's question, and just baptizes that evasion by an *ipse dixit.*

Furthermore, Swinburne himself concedes that the theistic explanation explains too much: "It is compatible with too much. There are too many different possible worlds which a God might bring about" (1991, p. 289). Thus, God's supposed will to create the actual world is presumably an unexplained brute fact.

How then does Swinburne reason that his theological deductive brute fact explanation improves upon a scientific system in which explanation is envisioned as departing from the most fundamental laws of nature, which are themselves taken to hold as a matter of brute fact? In the face of the epistemic flaws I have set forth, Swinburne's and Quinn's theological superstructure appears to be an explanatorily barren step backward.

The best I can do on their behalf to make the supposed divine creative process intelligible is to construe their direct divine causation as taking the following form: God is in the injunctive mental state "let there be the existing world," with its laws including the Biblical "let there be light." And this mental state instantaneously causes the world to exist.

But in all of our ordinary and scientific reasoning, it would be regarded as magical thinking, in the mode of primitive savages, to suppose that any *mere thought* could bring about *the actual physical existence* of the thought-object out of nothing. Indeed, in Freud's theory, such a belief in "the omnipotence of thoughts" is a hallmark of obsessional neurosis (*Standard Edition* of *The Complete Psychological Works of Sigmund Freud*, vol. 13, p. 85).

Hence I can only welcome the assertion of the Jesuit theologian Michael Buckley (1990, p. 314) that, as for divine volitional creation, "We really do not know how God 'pulls it off.'" But then Buckley continues in an apologetic mode: "Catholicism has found no great scandal in this admitted ignorance." While I accept this account of the attitude of the exponents of Catholic doctrine, I must regard the admitted explanatory gaping lacuna in its very ambitious context as a kind of explanatory scandal.

IV. THE "ANTHROPIC PRINCIPLE"

A. The Scientific Status of the Anthropic Principle

The so-called "Weak Anthropic Principle" (WAP) has been construed in a number of naturalistic, non-theological ways, whose nub is the following: *Given the currently postulated laws of nature, very sensitive physical conditions, going back to the earliest stages of a big bang universe, are causally necessary for the cosmic evolution and existence of carbon-based humanoid life.* And these very delicate initial or boundary conditions are *a priori* exceedingly improbable, where the *a priori* probabilities are presumably defined on the set of all logically possible values of the pertinent physical conditions, which include the physical constants in the above laws of nature.

Various authors speak of these sensitive initial conditions as being "fine-tuned" for life. But John Leslie, an exponent of the Design-interpretation of WAP, issued the disclaimer ".... talk of 'fine tuning' does not presuppose that a divine Fine Tuner, or Neoplatoism's more abstract God, must be responsible" (1990, p. 3).

Yet the Roman Catholic theist Ernan McMullin rightly cautions (1993, p. 602): "*Fine tuning* has something of the ambiguity of the term *creation*; if it be understood as an action, then the existence of a "fine tuner" seems to follow. Perhaps a more neutral term would be better." Indeed, for just that reason, I myself suggest the use of the term *"bio-critical values"* in lieu of the locution "fine tunings."

In an admirably thorough and cogent article, John Earman (1987) gave the following appraisal of WAP, but *without* any evaluation (p. 314) of a teleological construal of it in a theistic argument from Design (pp. 314-315):

Conclusion
Insofar as the various anthropic principles are directed at the evidentiary evaluation of cosmological theories they are usually interpretable in terms of wholly sensible ideas, but the ideas embody nothing new, being corollaries of any adequate account of confirmation. And insofar as anthropic principles are directed at promoting Man or Consciousness to a

starring role in the functioning of the universe, they fail; for either the promotion turns out to be an empty tease or else it rests on woolly and ill-founded speculations.

B. Critique of the Theistic Design Interpretation of WAP

In his *Dialogues on Natural Religion*, David Hume famously warned against sliding from an ordered universe to one featuring theistic Design.

I shall now examine critically the theistic Design interpretation of WAP, with which Earman did not deal.

Recall (from Section III) that divine omnipotence has the defect of failing to explain the explanandum by explaining too much. So also, I shall now argue, the theists Swinburne and Leslie undermine their teleological explanation of WAP by invoking God's Design to explain the *a priori* very improbable bio-critical values ("fine-tuning").

It is crucial to note that, as a point of departure of their theistic argument, both Swinburne and McMullin take the laws of nature as given and fixed, no less than philosophical naturalists do. Thus, Swinburne wrote (1991, p. 306): "... *given* the actual laws of nature or laws at all similar thereto, boundary conditions will have to lie within a narrow range of the present conditions if intelligent life is to evolve ..." Elsewhere, Swinburne (1990, p. 160) had made the same claim. Moreover, Swinburne averred (1991, p. 312): "... the peculiar values of the constants of laws and variables of initial conditions are substantial evidence for the existence of God, *which alone can give a plausible explanation of why they are as they are*" (italics added). And divine action supposedly does so by teleologically transforming the *a priori* very low probability of the bio-critical conditions into probable ones.

Clearly, Swinburne operates with the assumption that the laws of nature are *given* as fixed. And he does so twice: (i) in his initial inductive inference of divine teleology from the bio-critical values, whose critical role is predicated on a given set of laws, and (ii) in the theistic teleological explanation of why these values 'are as they are'. And my challenge will be that this explanation does not cohere with divine omnipotence, which negates the fixity of the laws.

Relatedly, though far more cautiously than Swinburne, McMullin wrote (1993, p. 603):

According to the Biblical Account of creation, God chose a world in which human beings would play an important role, and would thus have been committed to whatever else was necessary in order for this sort of universe to come about. If fine-tuning was needed this would present no problem to the Creator.

Although McMullin is careful to speak of fine-tuning conditionally, notice that fine-tuning would be needed *only* in the context of the *givenness* of the actual laws of nature. The logical structure of Swinburne's explanation of the existence of the bio-critical values can be schematized essentially as follows: the premisses are (i) the laws of nature are given, (ii) God wants to create human life, and (iii) under the constraints of the given laws, 'fine tuning' is necessary for the existence of human life. And the conclusion is that God selected the bio-critical values in his creation, and therefore they materialized.

But just this explanation founders on the shoals of divine omnipotence, just as did the theistic volitional explanation of the ultimate laws of nature. As noted earlier, Swinburne's account of omnipotence included the following creative nomological latitude (1991, p. 295): "God, being omnipotent, ... his range of easy control must ... include all states of affairs [including laws] which it is *logically possible* for him to bring about" (italics added). Yet this conclusion undermines the theistic teleological interpretation of WAP, a philosophical reading that was predicated on the *contrary* assumption that God confronts fixed, rather than disposable laws of nature. Instead, God can achieve any desired outcome by any laws of his choosing. This result demonstrates the logical incoherence of the theistic anthropic argument for Design.

Moreover, divine omnipotence makes the causal necessity of the bio-critical values *irrelevant* to the divine teleological scenario: In the context of suitably different natural laws, relating their corresponding variables to the existence of humanoid life, the *a priori* very low probability of the standard bio-critical values is no longer an issue at all, since the critical role of these values is relative to a specified fixed set of laws and is not played by them *per se*.

Furthermore, as Earman remarked (1987, p. 309), "the selection function [of initial/boundary conditions]" is served just as well by the existence of stars and planetary systems supporting a carbon-based chemistry but no life forms."[3] Thus, the so-called "anthropic" coincidences, which are purely physical, are *coextensive* with those that are critical for the formation and primordial existence of stars and galaxies. Not being distinctively anthropic, we must beware that the so-called "Weak Anthropic Principle" (WAP) not be allowed to give the misleading impression that the "fine-tunings" are *anthropically* unique, qua being necessary for *our* existence.

V. CRITIQUE OF SWINBURNE'S BAYESIAN ARGUMENT FOR THE EXISTENCE OF GOD

A. The Incoherence of Swinburne's Apologia

Quinn has given a concise summary of Swinburne's Bayesian argument for the existence of God. In Quinn's words (1993, p. 622):

He argues that the hypothesis of theism is more probable given the existence over time of a complex physical universe than it is on tautological evidence alone, and he further contends that this argument is part of a cumulative case for theism whose ultimate conclusion is that "on our total evidence theism is more probable than not" (1979, 291).

But Swinburne hedged his Bayesian plaidoyer, declaring (1991, ch. 13, p. 244): "Certainly one would not expect too evident and public a manifestation [of the existence of God], ... If God's existence, justice and intentions became items of evident common knowledge, then man's freedom to choose [belief *or* disbelief] would in effect be vastly curtailed."

Richard Gale (1994, p. 39) clarified this claim of Swinburne's as follows: "While Swinburne's overall aim is to establish that the [Bayesian] probability that God exists is greater than one-half, he does not want the probability to be too high, for he fears that this would necessitate belief in God on the part of whoever accepts the argument, thereby negating the accepter's freedom to choose not to believe."

Yet Swinburne's argument here is *flatly incoherent*: He appeals to the need for free choice of belief to justify God's not giving us evidence for a high probability of his existence, *on the grounds that a high probability would compel us after all to believe in him*! But why, oh why, would a high probability *necessitate* our belief at all, *if* we have free choice to believe or not in the first place? In the process of saving our freedom to believe, Swinburne inconsistently assumes the causal determination of our beliefs (cf. Grünbaum, 1972, Section II, B and C). His argument fares no better, if the pertinent freedom to choose is *between good and evil*.

Gale (op. cit., p. 40) cites Swinburne as having asserted that "S believes that *p* if and only if he believes that *p* is more probable than any alternative," where the alternative is usually not-*p*. But Gale rightly disputes that claim: He gives examples from Tertullian to Kierkegaard, and from his own life, to the following effect (ibid.): "It certainly is possible for someone to believe a proposition while believing that it is improbable, even highly improbable."

In sum, Swinburne's apologia for God's evidential coyness is deeply incoherent.

B. Swinburne's Bayesian Argument for the Existence of God

Swinburne takes it for granted that Bayes's theorem, which is derived from the formal (Kolmogorovian) probability axioms, is applicable to the probability of *hypotheses*, and can thus serve as a paradigm for probabilifying scientific and theological hypotheses. This kind of use of Bayes's theorem has been challenged and even rejected, if only because of the well-known problems besetting the determination of *non*-subjective values of the so-called "prior probability" of the hypothesis at issue. Yet Swinburne (1990, p. 155) asserts exaggeratedly that "Bayes's Theorem ... is a crucial principle at work for assessing hypotheses in science, history and all other areas of inquiry."

But if we do use Bayes's theorem to probabilify hypotheses, we must be mindful of a crucial distinction made by Hempel, yet unfortunately not heeded by Swinburne. W.C. Salmon (forthcoming) has lucidly articulated it as follows:

... Bayes's theorem belongs to the context of confirmation, not to the context of explanation. ... This is a crucial point. Many years ago, Hempel made a clear distinction between two kinds of why-questions, namely, *explanation-seeking* why questions and *confirmation-seeking* why-questions. Explanations-seeking why-questions solicit answers to questions about why something occurred, or why something is the case. Confirmation-seeking why-questions solicit answers to questions about why *we believe* that something occurred or something is the case. The characterization of nondemonstrative inference as inference to the best explanation serves to muddy the waters – not to clarify them – by fostering confusion between these two types of why-questions. Precisely this confusion is involved in the use of the "cosmological anthropic principle" as an explanatory principle. [Here Salmon cites his (1998)].

Swinburne likewise muddies the waters by his failure to heed the Hempel-Salmon distinction. He does so when he tries to use that theorem both to probabilify (i.e., to increase the confirmation of) the existence of God, on the one hand, and, on the other, to show that theism offers the best explanation of the known facts, assuming that God exists.

It is vital to distinguish the *absolute* confirmation of h from its *incremental* confirmation (Salmon, 1975, p. 6). The former obtains iff the probability of h is fairly close to 1; the latter, also called "relevant confirmation", is defined by the condition that the posterior probability of h be *greater* than its prior probability.

Upon dividing both sides of Bayes's theorem by the prior probability $p(h/k)$, it is evident that h is incrementally confirmed by e, iff $p(e/h \cdot k) > p(e/k)$, i.e., iff the likelihood exceeds the *expectedness*. In this context, h is the hypothesis that "God exists" (Swinburne, 1991, p. 16).

In his *magnum opus* on *The Existence of God*, Swinburne had argued for the following conclusion (1991, p. 291): "On our total evidence theism is more probable than not." Yet to reach this conclusion, he proceeds as follows: "... to start without any factual background knowledge (and to feed all factual knowledge gradually into the evidence of observation), and so to judge the prior probability of theism solely by *a priori* considerations, namely, in effect, simplicity" (1991, p. 294 and p. 63n).

I must refer the reader to (Grünbaum, 2000, sections 5.2 and 5.3), where I present major arguments that lead to the following conclusion: Swinburne failed multiply to establish that the posterior probability of the existence of God exceeds one-half. *A fortiori,* he has failed to show, in turn that the hypothesis of the existence of God can serve at all as a warranted premiss to provide explanations. There I also advanced considerations that undermine Swinburne's theodicy. And I now add that the weakness of that theodicy is shown further by the content of the Roman Catholic Exorcist Rite. It brings in Satan as a counterweight to God in order to reconcile divine omni-benevolence with such natural evils as death, declaring that Satan "hast brought death into the world" ("Exorcism", *Encyclopedia Britannica*, 1929).

VI. CONCLUSION

None of the cross-section of diverse theists, past and present, whose arguments I have considered have presented even partly cogent evidence for the existence of their God.

NOTES

*   This substantially modified version of (Grünbaum, 2000) appears here by permission of Oxford University Press, the publisher of the *British Journal for the Philosophy of Science* in which the original had appeared in vol. 51, pp. 1-43.

1.   In 1931, Rudolf Carnap (Schleichert, 1975) explained in a major paper that the noun "Nothingness" is a product of logical victimization by the grammar of our language.
2.   Quoted in Leslie, 1978, p. 181 from Bergson's *The Two Sources of Morality and Religion*, Part 2.
3.   For numerous relevant details, see Leslie (1990, ch. 1, Section 1.4).

REFERENCES

Buckley, M. (1990), "Religion and Science: Paul Davies and John Paul II", *Theological Studies* **51**: 310-324.
Carnap, R. (1931), "Überwindung der Metaphysik durch Logische Analyse der Sprache", in H. Schleichert, (ed.), *Logischer Empirismus – der Wiener Kreis. Ausgewählte Texte mit einer Einleitung*. Munich: W. Fink, 1975, pp. 149-171.
"Conservation of Mass-Energy" (1965), in J. Newman, (ed.), *International Encyclopedia of Science*. Vol. 1. Edinburgh: Thomas Nelson, pp. 276-277.
Craig, W. (1979), *The Kalam Cosmological Argument*. New York: Harper & Row.
Craig, W. (1991), "Pseudo-Dilemma?" Letter-to-the-Editor, *Nature* **354** (December 5): 347.
Craig, W. (1992), "The Origin and Creation of the Universe: A Reply to Adolf Grünbaum", *British Journal for the Philosophy of Science* **43**: 233-240.
Craig, W. (1994a), "Creation and Big Bang Cosmology", *Philosophia Naturalis* **31**(2): 217-224.
Craig, W. (1994b), "A Response to Grünbaum on Creation and Big Bang Cosmology", *Philosophia Naturalis* **31**(2): 247.
Davidson, D. (1982), "Paradoxes of Irrationality", in R. Wollheim and J. Hopkins, (eds.), *Philosophical Essays on Freud*. New York: Cambridge University Press.
Davies, P. (1999), *The Fifth Miracle: The Search for the Origin and Meaning of Life*. New York: Simon and Schuster.
Earman, J. (1987), "The SAP Also Rises: A Critical Examination of the Anthropic Principle", *American Philosophical Quarterly* **24**(4): 307-317.
Earman, J. (1992), "The Problem of Old Evidence." Chap. 5 in *Bayes or Bust?* Cambridge, MA: MIT Press.
Earman, J. (1993), "The Cosmic Censorship Hypothesis." Chap. 3 in J. Earman et al., (ed.), *Philosophical Problems of the Internal and External Worlds: Essays on the Philosophy of Adolf Grünbaum*. Pittsburgh and Konstanz: University of Pittsburgh Press and University of Konstanz Press, pp. 45-82.
Earman, J. (1995), *Bangs, Crunches, Whimpers, and Shrieks: Singularities and Acausalities in Relativistic Spacetimes*. New York: Oxford University Press.

Edwards, P. (1967), "Why", in P. Edwards, (ed.), *The Encyclopedia of Philosophy*. Vol. 8. New York: The Macmillan Company & The Free Press, pp. 296-302.

"Exorcism" (1929), in *Encyclopedia Brittanica*. vol. 8, Fourteenth Edition, pp. 972-973.

Freud, S. (1953-1974), *The Standard Edition of the Complete Psychological Works of Sigmund Freud*, translated by J. Strachey et al. London: Hogarth Press.

Gale, R. (1991), *On the Nature and Existence of God*. New York: Cambridge University Press.

Gale, R. (1994), "Swinburne's Argument from Religious Experience." Chap. 3 in A. Padgett, (ed.), *Reason and the Christian Religion, Essays in Honor of Richard Swinburne*. Oxford: Oxford University Press.

Gale, R. (1999), "Santayana's Bifurcationist Theory of Time", in *Bulletin of the Santayana Society*, 16, Fall.

Gale, R. and Pruss, A. (forthcoming), "A New Cosmological Argument", *Religious Studies*.

Glymour, C. (1980), *Theory and Evidence*. Princeton, NJ: Princeton University Press.

Grünbaum, A. (1954), "Science and Ideology", *Scientific Monthly* 79(1) (July): 13-19.

Grünbaum, A. (1972), "Free Will and Laws of Human Behavior." Part VIII, Chap. 62 in H. Feigl, W. Sellars, and K. Lehrer, (eds.), *New Readings in Philosophical Analysis*. New York: Appleton-Century-Crofts, pp. 605-627.

Grünbaum, A. (1973), *Philosophical Problems of Space and Time*. Dordrecht: Reidel Publishing Co.

Grünbaum, A. (1990), "The Pseudo-Problem of Creation in Physical Cosmology", in J. Leslie, (ed.), *Physical Cosmology and Philosophy*. Philosophical Issues Series. New York: Macmillan Publishing Co., pp. 92-112. This is a reprint of the article in *Philosophy of Science* (1989).

Grünbaum, A. (1994), "Some Comments on William Craig's 'Creation and Big Bang Cosmology'", *Philosophia Naturalis* 31(2): 225-236.

Grünbaum, A. (1995), "The Poverty of Theistic Morality", in K. Gavroglu, J. Stachel, and M.W. Wartofsky, (eds.), *Science, Mind and Art: Essays on Science and the Humanistic Understanding in Art, Epistemology, Religion and Ethics, Vol. III, in Honor of Robert S. Cohen*. Boston Studies in the Philosophy of Science, vol. 165. Dordrecht, The Netherlands: Kluwer Academic Publishers, pp. 203-242.

Grünbaum, A. (1996), "Theological Misinterpretations of Current Physical Cosmology", *Foundations of Physics* 26(4) (April): 523-543.

Grünbaum, A. (1998), "Theological Misinterpretations of Current Physical Cosmology", *Philo* 1(1) (Spring/Summer): 15-34. This article is a revision of Grünbaum (1996).

Grünbaum, A. (2000), "A New Critique of Theological Interpretations of Physical Cosmology", *The British Journal for the Philosophy of Science* 51: 1-43.

Heidegger, M. (1953), *Einführung in die Metaphysik*. Tubingen: Niemeyer.

James, W. (1975), *Pragmatism*. Cambridge, MA: Harvard Press.

Leibniz, G. W. (1714), "The Principles of Nature and of Grace Based on Reason", in P. P. Wiener, (ed.), *Leibniz Selections*, 1951. New York: Charles Scribner's Son. The German original *Vernunftprinzipien der Natur und der Gnade* is available in a 1956 publication of Verlag Felix Meiner, Hamburg.

Leslie, J. (1978), "Efforts to Explain All Existence", *Mind* 87(346): 181-194.

Leslie, J. (1990), *Universes*. New York: Routledge.

Leslie, J. (1996), *The End of the World*. London; New York: Routledge.

Leslie, J. (1998), "Cosmology and Theology." *Stanford Encyclopedia of Philosophy*. <http://plato.stanford.edu/entries/cosmology-theology/>

Mayo, D. (1996), "The Old Evidence Problem." Section 10.2 in *Error and the Growth of Experimental Knowledge*. Chicago: University of Chicago Press.

McMullin, E. (1993), "Cosmology & Religion." Chap. 31 in N. Hetherington, (ed.), *Cosmology*. New York: Garland Publishing.

Parsons, K. M. (1989), *God and The Burden of Proof*. Buffalo, NY: Prometheus Books.

Quinn, P. L. (1993), "Creation, Conservation, and the Big Bang." Chap. 23 in J. Earman et al., (ed.), *Philosophical Problems of the Internal and External Worlds: Essays on the Philosophy of Adolf Grünbaum*. Pittsburgh and Konstanz: University of Pittsburgh Press and Universitätsverlag Konstanz, pp. 589-612.

Rescher, N. (1984), *The Riddle of Existence : An Essay in Idealistic Metaphysics*. Lanham, MD: University Press of America.

Salmon, W. C. (1975), "Confirmation and Relevance", in G. Maxwell and R. M. Anderson, (eds.), *Induction, Probability, and Confirmation*. Minnesota Studies in the Philosophy of Science, vol. VI. Minneapolis: University of Minnesota Press, pp. 3-36.

Salmon, W. C. (1978), "Religion and Science: A New Look at Hume's DIALOGUES", *Philosophical Studies* 33: 143-176.

Salmon, W. C. (1979), "Experimental Atheism", *Philosophical Studies 35*: 101-104.

Salmon, W. C. (1998), *Causality and Explanation*. New York: Oxford University Press.

Salmon, W. C. (forthcoming), "Inference to the Best Explanation".

Schellenberg, J. L. (1993), *Divine Hiddenness and Human Reason*. Ithaca: Cornell University Press.

Schick, T., Jr. (1998), "The 'Big Bang' Argument for the Existence of God", *Philo* 1: 95-104.

Smith, Q. (1991), "An Atheological Argument from Evil Natural Laws", *International Journal for the Philosophy of Religion* 29: 159-174.

Smith, Q. (1992), "Anthropic Coincidences, Evil and the Disconfirmation of Theism", *Religious Studies* 28: 347-350.

Smith, Q. (1994), "Anthropic Explanations in Cosmology", *Australasian Journal of Philosophy* 72(3) (September): 371-382.

Smith, Q. (1995), "Internal and External Causal Explanations of the Universe", *Philosophical Studies* 79: 283-310.

Smith, Q. (1997), *Ethical and Religious Thought in Analytic Philosophy of Language*. New Haven: Yale University Press.

Smith, Q. (1998), "Swinburne's Explanation of the Universe", *Religious Studies* 34: 91-102.

Swinburne, R. (1979), *The Existence of God*. New York: Oxford University Press.

Swinburne, R. (1989), "Violation of a Law of Nature." Chap. 8 in R. Swinburne, (ed.), *Miracles*. New York: Macmillan Publishing Co.

Swinburne, R. (1990), "Argument from the Fine Tuning of the Universe." Chap. 12 in J. Leslie, (ed.), *Physical Cosmology and Philosophy*. New York: Macmillan Publishing Co., pp. 154-173.

Swinburne, R. (1991), *The Existence of God*, Third Edition. New York: Oxford University Press.

Swinburne, R. (1996), *Is There a God?* New York: Oxford University Press.

Swinburne, R. (1997), *Simplicity as Evidence of Truth*. Milwaukee: Marquette University Press.

Wilson, P. (1993), "The Anthropic Principle", in N. Hetherington, (ed.), *Cosmology*. New York: Garland Publishing.

Wittgenstein, L. (1993), "Lecture on Ethics", in J. Klagge and A. Nordmann, (eds.), *Philosophical Occasions 1912-1951*. Cambridge, England: Hackett Publishing Co., pp. 37-44.

Worrall, J. (1997), "Is the Idea of Scientific Explanation Unduly Anthropocentric: The Lessons of the Anthropic Principle", in *Discussion Paper Series from the London School of Economics*. DP25/96.

Center for Philosophy of Science
2510 Cathedral of Learning
University of Pittsburgh
Pittsburgh, PA. 15260
U.S.A.
E-mail: grunbaum+@pitt.edu

WESLEY C. SALMON

SCIENTIFIC UNDERSTANDING IN THE TWENTIETH CENTURY

As we come to the end of the century and the millennium, there is an irresistible temptation, especially for those of us who are old, to look back and try to assess the progress, if any, that has accrued. When I refer to scientific understanding, my aim is not to examine the understanding of science – i.e., how adequately people understand physics, psychology, biology, geology, etc. – but rather, to consider the kind of understanding that science furnishes to us. I shall focus on the understanding of the natural phenomena that scientific knowledge yields. In this endeavor, I want to emphasize the intellectual dimension of science, not the technology that has sprung from it, although I shall comment on the technological aspect toward the end.

Looking back to the beginning of the *millennium*, we find scientific understanding in a sad state. Even the great scientific accomplishments of the ancient Greeks had been lost. It is best not to dwell on that deplorable situation.

Looking back to the beginning of the *century,* we see an altogether different picture. By the end of the nineteenth century, various branches of science were highly developed. Yet, strangely, from our present perspective, many philosophically-minded scientists and scientifically-minded philosophers denied that science provides any *understanding* of the world. Among those who held this view were Pierre Duhem, Ernst Mach, and Karl Pearson. Pearson put the point concisely: "Nobody now believes that science *explains* anything; we all look on it as a shorthand description, as an economy of thought" ([1911] 1957, p. xi, emphasis in original). The sole function of science, on this view, is to systematize our knowledge; it enables us to describe, organize, and predict – in the way in which, for example, astronomy makes possible the prediction of eclipses.

Along with the rejection of scientific explanation early in the century, many knowledgeable physical scientists, such as Ernst Mach and Wilhelm Ostwald, denied the reality of such microentities as molecules, atoms, and sub-atomic particles. Here is Pearson again: "[M]ay there not be some danger that the physicist of today may treat his electron, as he treated his old unchangeable atom, as a reality of experience, and forget that it is only a construct of his imagination" ([1911] 1957, p. xii). Ostwald (1906) is even more concise: "Atoms are only hypothetical things" (quoted by Nye, 1972, p. 151). According to various authors of that period, genuine understanding, if it exists, can be found only in such extra-scientific fields as metaphysics or theology.

Obviously, the situation at the end of the century is altogether different. In scientific journals and the popular press, we find instance after instance of scien-

M. Rédei and M. Stöltzner (eds.),
John von Neumann and the Foundations of Quantum Physics, 289–304.
© 2001 *Kluwer Academic Publishers. Printed in the Netherlands.*

tific explanation. As I write, today's issue of the *New York Times* (12 October 1999) carries a front page story on the Nobel Prize in medicine awarded to Günter Blobel "for helping [to] explain diseases like cystic fibrosis." The redshift of light from distant galaxies is explained by their recession from us. Sickle-cell anemia is explained by a defective gene. Characteristics of living organisms are explained in terms of evolution. Scientists and philosophers now seem to agree pretty largely that science can and does provide explanations. A prominent example is Nobel-laureate Steven Weinberg's book, *Dreams of a Final Theory* ([1992] 1994). The final theory for which he yearns, which we do not possess at present, is a fundamental physical theory that will, in principle, *explain* everything. He believes that such a theory is attainable because of *explanatory arrows* that seem to be pointing toward a final unifying theory. The whole argument of the book depends on the presumption that science can explain. He sums up by declaring, "Whether or not the final laws of nature are discovered in our lifetime, it is a great thing for us to carry on the tradition of holding nature up to examination, asking again and again *why* it is the way it is" ([1992] 1994, p. 275). Earlier in the book, he expresses his view on the possibility of scientific explanation in a colorful metaphor: "To tell a physicist that the laws of nature are not explanations of natural phenomena is like telling a tiger stalking prey that all flesh is grass" (ibid., p. 28). I would add a variation on his theme: "To tell a physicist that 'his electron is only a construct of his imagination' is also like telling a tiger stalking prey that all flesh is grass."

Some people doubt that there is any such thing as progress in philosophy; for such individuals I would point to the transition regarding scientific explanation, from the beginning to the end of the century, as a genuine and important example. Obviously, the twentieth century has seen stupendous scientific progress; I am suggesting that there has been philosophical progress as well. The purpose of the present paper is to examine this philosophical transition. I will argue that, although scientific progress has played an important part, progress in philosophy of science has also contributed essentially.

I. CONCEPTS OF UNDERSTANDING

I have used the term "understanding" in the title and opening paragraphs of this paper. To many people, this word seems to have no definite meaning – it is vague, ambiguous, and subjective. Perhaps we should avoid it altogether in discussions of science. I do not share this view. Of course it is ambiguous, but this is no reason to throw up our hands in dismay and ban the term. Rather, for the philosopher, it poses a challenge to sort out the various meanings in a reasonably clear and definite way. It is not meaningless or hopelessly vague. Figure 1 offers a sort of road map to help us find the way through the tangle of meanings. It has three major columns. The first two, which will be discussed in this section, pertain mainly to nonscientific understanding. To say this is not to denigrate these types; many important kinds of understanding lie outside of the practice of science. The third main column, which will be discussed in the next section, includes the sorts of understanding that science can yield. It should be emphasized that this chart is not meant to be exhaustive of all different senses of the term; it is intended, rather, to be suggestive regarding ways of attacking the problem.

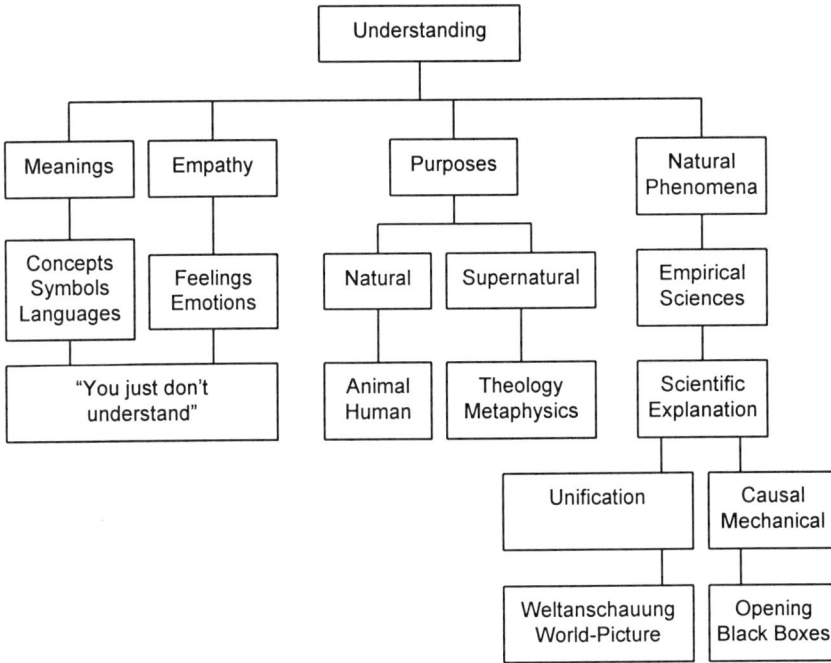

Figure 1: Types of human understanding

In 1991, the noted linguist Deborah Tannen published a book entitled *You Just Don't Understand*. It furnishes an excellent point of departure for material contained in the first major column. The widespread popularity of this book is indicated by the fact that it appeared in the *New York Times* list of best-selling books for about three years. Her thesis is that men and women do not understand each other because they speak different languages; the result is a failure to grasp the different meanings that are expressed. One could add, I think, that a similar failure accounts for the 'generation gap'; parents cannot understand their children, and children cannot understand their parents. Both types of failure of understanding seem to be fairly persistent features of western culture. A similar phenomenon crops up in other contexts as well; for example, bosses do not understand their employees, and teachers do not understand their students.

Problems of understanding meanings are not confined to linguistic contexts. To understand renaissance paintings one must know the iconography of the period. To understand philosophy, one must know the meanings of various concepts. The meanings of cultural and political symbols are required to understand other civilizations and nations. A dreadful example of failure to understand occurred in an American high school, where Jewish students were forbidden to display the Star of David because, the school authorities said, it could be mistaken by other students as the symbol of a teenage gang. Clearly, some education about meaning was urgently needed.

A closely related sort of understanding – one that is often intermingled with linguistic understanding – is empathy. If you say to your friend, who has just suffered a painful conclusion to a love affair, "I understand," it means that you are aware of the feelings he or she is experiencing, and, quite possibly, that you too have undergone a similar unpleasant breakup. If you have expressed anger at your psychotherapist, and he or she says, "I understand," this indicates that the therapist knows how you feel, and why you are experiencing that sort of emotion on this occasion.

Moving to the middle column of figure 1, we find another familiar form of understanding – one that involves an awareness of the motives or purposes that lie behind someone's action. If, for example, you ask why I purchased a certain book, I might explain that I am about to depart on a long airplane flight, and that I want to have diverting reading to help pass the time. I might go on to say that I have read other novels by the same author and have found them enjoyable. We all engage in a great deal of purposive activity; we understand such behavior, our own or someone else's, when we know what these purposes are. When we say to someone, "I don't understand you," or to yourself, "I don't know why I said that," we are expressing a lack of understanding that can only be cured by further knowledge of motives or purposes. Unconscious motives or purposes are not to be ruled out. Understanding of this sort is not confined to human behavior. When a dog brings its leash to its mistress or master, we understand that it wants to go for a walk. When a male bird puts on a courting display before a group of females, we know that the goal is procreation.

Understanding in terms of goals is relevant to scientific explanation in two crucial ways. In the first place, explanation in terms of conscious purposes is so familiar to humans that it may be tempting to suppose that we cannot understand any phenomenon unless we can cite a purpose. This may be connected to the idea that we cannot *really* understand natural phenomena simply on the basis of laws of nature; instead, it is necessary to have a lawgiver and to know its purposes. It has long been held – and still is in many quarters – that the will of God is the only *genuine* explanation. Perhaps this is the fundamental reason for the reluctance of scientists and philosophers of science in the early part of the twentieth century to accept the notion that there is any such thing as *scientific* explanation. They did not want to make science animistic or anthropomorphic. They wanted to avoid the appeal to entelechies or vitalistic principles in biology.

They wanted to eliminate supernatural explanations in science – the sort of thing that goes under the label "creation science" in America.

In the second place, the question arises whether functional explanations can be admitted as scientific. Certainly humans create devices, such as thermostats, that are explicitly designed to fulfill a function, namely, temperature control. There is no mystery here. However, we believe that the ears of elephants also function as temperature-control mechanisms, but they were not designed by human engineers. A great deal of philosophical attention, both positive and negative, has been devoted to functional explanation during the twentieth century. I strongly believe, largely on the basis of work by Larry Wright (1976), that functional explanations have a legitimate place in the biological sciences, and that they are fully compatible with mechanical explanations of the same phenomena. In addition, human institutions and practices are said to have functions. In the behavioral sciences we find explanations in terms of *latent functions* – practices that fulfill functions that the practitioners are not explicitly aware of. For instance, a rain dance may have no effect on the weather, but it may produce social solidarity in a group that is suffering stress because of serious drought. Here, too, I believe, legitimate scientific explanations may be functional. Thus, there is a straightforward sense in which "natural purposes" in the middle column may well deserve a place in the category of understanding furnished by the empirical sciences.

II. TWO TYPES OF SCIENTIFIC EXPLANATION

Moving on to the third main column of figure 1, we find ourselves in the domain of the kind of understanding the sciences can furnish. Scientific explanations provide scientific understanding. I shall maintain that there are at least two major categories of explanation. Those in the first category seek to fit natural phenomena into a coherent world picture; these involve scientific explanations that unify diverse phenomena by providing a unifying schema. Those in the second category explain phenomena by exposing the mechanisms that produce them. These are known as causal/mechanical explanations.

a.

To introduce explanation of the *unification* variety, we note that every culture with which we are familiar has a cosmology – a world picture. In the Judeo-Christian tradition it can be found in the book of Genesis. Tony Hillerman, in his best-selling novels, provides a sympathetic account of the Navajo world picture. Ancient Greece had its well-known myths. In ancient Greece, however, something unique occurred around 600 B.C., namely, the birth of scientific cosmology based upon systematic empirical evidence. As far as we know, the first attempt was made by Thales of Miletus, who held that everything is composed of water. This was certainly a modest beginning, but by the times of Democritus, Plato, and Aristotle, the cosmological systems were highly sophisticated. The works of Democritus have not survived, but we have a detailed atomistic account in the poem of Lucretius, *De Rerum Natura.* Aristotle's cosmology was adapted and adopted by the medieval Catholic Church; its poetic expression can be found in Dante's *Divine Comedy.*

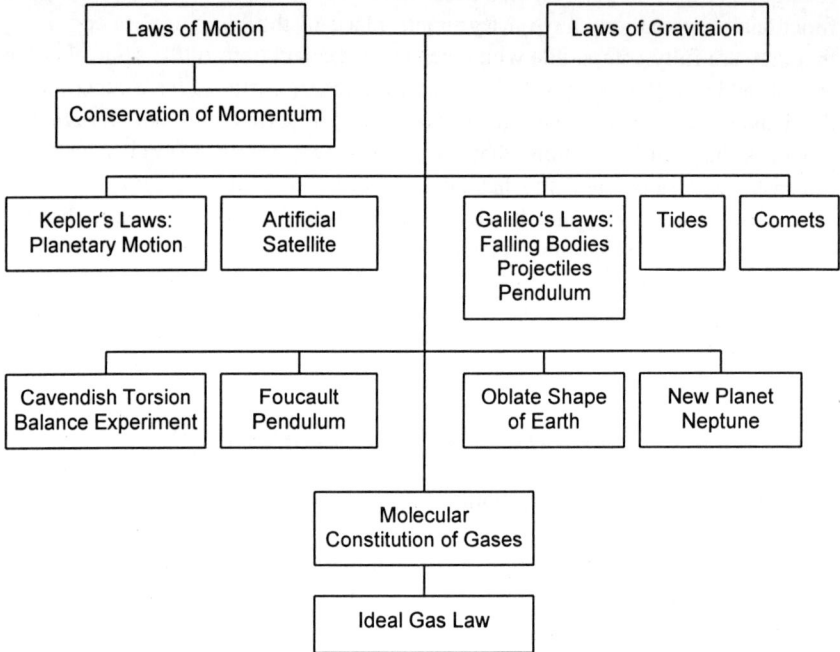

Figure 2: Newtonian Synthesis

The Copernican Revolution, aided by Kepler, Galileo, and Newton, produced an extraordinarily accurate and comprehensive cosmological theory. (Of course, it differed drastically from our late-twentieth-century "big bang" models.) In this connection, I want to emphasize the unification of the world picture that resulted; figure 2 maps out some major parts of the Newtonian Synthesis. We notice first that, on the top row we have three simple laws of motion and the law of universal gravitation. The three laws of motion,

1. The law of inertia,

2. $F = ma,$

3. The law of equal and opposite action and reaction,

describe the behavior of bodies upon which forces of unspecified kinds impinge (force equals zero in the first law). From these three laws the law of conservation of momentum follows as a deductive consequence. Although we no longer believe that Newton's laws are correct, we still retain the law of conservation of momentum.

The law of universal gravitation states that the force of attraction between two bodies is proportional to the product of their masses and inversely proportional to the square of the distance between them. When it is added to the list of laws, we see that the resulting theory explains a diverse collection of phenomena. Those on the next row of the schema were phenomena that were known in Newton's time. Kepler's laws of planetary motion govern the behavior of the solar system. Galileo's laws concerning falling bodies, the motions of projectiles, and the synchrony of the pendulum govern the behavior of small bodies close to the surface of the earth. It will be noted that I have listed artificial satellites among phenomena familiar to Newton. He understood the fundamental principle quite clearly; he even provided a diagram showing how they work. Go to the top of a high mountain. Throw a rock; notice that it falls to earth near the base of the mountain. Take another rock and throw it harder; notice that it falls to earth farther from the base. Continue the process. Eventually, a rock thrown hard enough will, in its fall, entirely circle the earth; it might even come around and hit you in the back of the head. True: Newton had no modern rocket fuel or complex electronics, but he knew the physical principle. Even today, as far as I know, Newtonian mechanics is used to calculate orbits of satellites and paths of space vehicles; general relativity is not required for this purpose.

In addition to these phenomena, which had been studied scientifically prior to Newton, I'm implicitly giving Galileo credit for the artificial satellite on account of his work on projectiles and falling bodies. The tides and comets are a different story. Although Galileo claimed to have a theory of the tides, which he believed to constitute positive proof of the motion of the earth, it was entirely mistaken. The regularities were known to mariners, but no explanation had been given. Comets were even worse; they seemed to be altogether irregular, and they were widely dreaded as signs of heavenly displeasure, even among well-educated people. Edmund Halley, for whom the famous comet is named, gave Newton credit for eliminating this source of superstitious fear, but as subsequent events have shown, superstition is not that easily overcome.

In the next row of the figure, we have phenomena that were not known in Newton's lifetime. In all of the cases so far mentioned in which gravitation plays a part, we have either an interaction between two bodies of astronomical dimensions (e.g., the planets and the sun; the earth and the moon) or between one large object and one small object (Newton's apple and the earth; Galileo's pendulum). In the eighteenth century, Henry Cavendish performed an extremely delicate experiment in which he demonstrated the gravitational attractions among objects small enough to be manipulated in his laboratory. This experiment measured the universal gravitational constant G (not to be confused with Galileo's constant g).

As a result, the masses of the earth, the sun, the moon, and all of the planets in the solar system could be ascertained.

The Foucault pendulum, a product of the nineteenth century, finally provided direct evidence for the rotation of the earth on its axis, and this rotation explains why the earth has an oblate shape, bulging at the equator. In addition, Newtonian celestial mechanics, together with observations of Uranus, led to the discovery of Neptune, a planet that had never previously been observed. Finally, if we assume that gases are composed of small particles that behave in accordance with Newton's laws, the ideal gas law results.

The point of the discussion of the Newtonian synthesis is to illustrate, in an extraordinarily vivid way, the explanatory power of a system that reduces an enormous variety of natural phenomena to a very small number of rather simple laws. The passion humankind possesses for cosmological systems seems to be a longing to see how we fit into the general scheme of things, and how the objects we encounter fit into the same general pattern. Dante's *Divine Comedy* fulfills the same function, but it does so in an Aristotelian-Christian cosmology. As has often been noted, the transition – instigated by the Copernican revolution – from the Aristotelian-Christian to the Newtonian cosmology was dramatic and traumatic.

b.

To get the basic idea behind explanations of the *causal/mechanical* type, consider an old-fashioned watch that operates by means of springs, cog wheels, escapements, etc. You can gain an understanding of this device by taking it apart, observing the relationships among the parts, and successfully putting it back together so that it functions properly. More generally, we find that nature confronts us with many 'black boxes'; to understand various phenomena, we need, so to speak, to open up these black boxes to examine their contents and inner workings. For example, earlier in the twentieth century, many psychologists adhered to behaviorism, the doctrine that psychology is concerned only with more or less directly observable stimuli and responses – inputs and outputs – without speculating about the inner workings of the human (or other animal) organism. Contemporary neuroscience adopts an opposite stance; it seeks to discover the inner mechanisms and how they operate. Behaviorism advocated leaving the black box closed without trying to peer inside; neuroscience advocates opening the black box in order to study its inner workings.

One major problem concerning nature's black boxes is that their internal workings often involve entities that are not directly observable – not because they are inside of the black box, but because, in many cases, they are too small. There are microscopic entities like cells and submicroscopic entities like molecules, atoms, and subatomic particles. Recall that, at the outset, I mentioned two views that were widely held at the beginning of the twentieth century, namely, the impossibility of scientific explanation and the unreality of atoms and molecules. Remember the quotations from Pearson and Ostwald. Where causal/mechanical explanations are at issue, it is easily seen that these two notions go hand in hand.

Scientific realism, sometimes known as theoretical realism, has been the subject of much discussion in philosophy of science during the twentieth-century. This is the doctrine that such things as atoms and molecules exist, and that we can know something about them. In spite of this continuing controversy, I think that the issue was settled in approximately the first decade of the century – by 1911, to be precise. This was an extremely important scientific development that has been insufficiently appreciated by philosophers of science. The result was achieved by a combination of theoretical work by Albert Einstein and Maryan Smoluchowski with experimental work by Jean Perrin. In its earliest stages, Brownian motion was the phenomenon under investigation.

It is well known that the botanist Robert Brown, early in the nineteenth century, observed the apparently random dance of particles of pollen suspended in liquid. At first he thought he was seeing some sort of vital phenomenon, but he observed the same kind of motion in small particles of inorganic substances suspended in fluid. It is said that he even obtained a small fragment of the Sphinx, which he ground into tiny particles. They, too, exhibited the same behavior. If anything is dead, the Sphinx is dead. Throughout the remainder of the nineteenth century, various attempts were made to explain Brownian motion, but all of them failed. In one of his famous 1905 papers, Einstein offered the theory that microscopic particles would execute the kind of motion Brown had described. Because he did not have a sufficiently exact account of Brownian motion at that time, he claimed to be unsure whether this theory would apply to Brownian particles. The motion was attributed to random collisions of molecules of the suspending fluid with the suspended particles. He saw the possibility of an empirical experiment that could decide between phenomenological thermodynamics and the molecular-kinetic theory.

At about the same time, quite independently of Einstein and Smoluchowski, Perrin began his experimental studies of Brownian motion. Perrin realized that by observing the vertical distribution of suspended particles, the rate at which they diffuse through the suspending fluid, and their rotational motion he could calculate Avogadro's number N, the number of molecules in a mole (gram molecular weight) of any substance in three distinct ways. Einstein had wondered whether the crucial experiment he had described would be too difficult to carry out in the laboratory. Perrin, in effect, performed the experiment, showing that it could be done. In 1908 – exactly a century after John Dalton's atomic theory came out – he published the results of these three types of experiments relating to Brownian motion that he had actually carried out, which had yielded consistent values for N. In addition, Perrin cited Max Planck's work on blackbody radiation and Ernest Rutherford's work on alpha radiation, each of which yielded a completely independent ascertainment of *N*. These results agreed with his.

Perrin continued his work on Avogadro's number, and by 1911, he had at his disposal 13 separate and distinct ways of determining *N,* all of which yielded results in agreement with one another. Such agreement would be miraculous if molecules were not real. In his 1913 book, *Les Atomes,* he listed all of these methods in a table, commenting that these results give the molecular hypothesis "a probability bordering on certainty." The community of knowledgeable physical scientists soon came to agree almost unanimously. The significance of this development should not be underrated. Avogadro's number is *the link* between the macrocosm and the microcosm; with its aid we can calculate microquantities from macroquantities and vice versa. It thus turns out that we *can* open nature's black boxes and learn about their internal mechanisms. A superb historical account of Perrin's work can be found in Mary Jo Nye's *Molecular Reality* (1972).

III. THE POSSIBILITY OF SCIENTIFIC EXPLANATION

Between the triumph of the atomic theory of matter early in the century and the middle of the twentieth century it would not have been incoherent to claim that we can *describe* the nature and behavior of atoms, molecules, and subatomic particles, and that we can make successful predictions on the basis of such knowledge, but to deny that we have achieved anything that deserves the honorific title of explanation or understanding. To appreciate the transition from this position to our current *fin de siècle* confidence in the possibility of scientific explanation and understanding, we must turn to the work of philosophers.

According to an old doctrine, going back at least to Aristotle, we seek to know, not only *what,* but *why.* A major obstacle to that goal was the fact that there was no precise and well-developed account of the nature of scientific explanation. In my view, the dividing line between the prehistory and the history of the philosophical theory of scientific explanation came in 1948, with the publication of a landmark paper by Carl G. Hempel and Paul Oppenheim. To be sure, similar ideas had been advanced in the nineteenth century by John Stuart Mill and earlier in the twentieth century by Karl Popper, but these presentations lacked the precision and rigor of the Hempel-Oppenheim paper, which proved to be the fountainhead from which almost all subsequent work has flowed, directly or indirectly. I have traced many of these developments in *Four Decades of Scientific Explanation* (1990). Strangely, to my mind, this paper was virtually ignored for about a full decade, but in the late 1950s and early 1960s a plethora of critical papers appeared. Some authors, most notably Michael Scriven, sought to refute the Hempel-Oppenheim approach, but many others were put forward mainly in the way of 'friendly amendments', by authors who were basically sympathetic, but who found what they saw as correctable flaws. The most significant aspect of the story from the standpoint of the present discussion is that, to the best of my knowledge, *none of the critics argued that scientific explanation is impossible in principle.*

Prior to spelling out their precise logical explication of one deductive model of explanation, Hempel-Oppenheim offered four preliminary criteria of adequacy for any such characterization of scientific explanation:

1. An explanation must be a valid deductive argument; the conclusion of the argument describes the fact to be explained, and the premises present the facts offered to explain the conclusion.

2. At least one of the premises must state a general law of nature.

3. The explanans (the premises of the argument) must have empirical content.

4. The premises must be true.

Various strong challenges were brought against these criteria (see Salmon, 1990, §2.1–2.3).

Hempel and Oppenheim explicitly stated that they were not attempting to provide a model for all legitimate scientific explanations. They were dealing with deductive explanations of particular facts. They acknowledged that there are acceptable scientific explanations that are inductive or statistical, and they recognized that many legitimate explanations are explanations of general regularities rather than particular facts. In his 1965 book, *Aspects of Scientific Explanation and Other Essays in the Philosophy of Science,* Hempel offered four distinct models:

1. Deductive-nomological (D-N) explanations of particular facts.

2. Deductive-nomological (D-N) explanations of general regularities.

3. Inductive-statistical (I-S) explanations of particular facts.

4. Deductive-statistical (D-S) explanations of statistical regularities.

At this point, Hempel declared that all and only explanations that conform to one of these models are legitimate scientific explanations. This view was widely accepted, and it became the 'received view' of scientific explanation for at least a couple of decades.

Without going into all of the criticisms that have been leveled at this doctrine, I think it is fair to say that the 'received view' is no longer received. Many of the details are spelled out in my (1990). Few, if any, philosophers who are at present writing seriously on scientific explanation accept Hempel's account. Nevertheless, the fact that such an initially attractive view had been put forth in explicit detail convinced philosophers that scientific explanation could be sensibly discussed. It might be said that Hempel provided a clear target for other philosophers to shoot at. They could say, in effect, that, even if Hempel's account was not quite right, it does give us some idea of what scientific explanation might be like. No longer was it appropriate to moan that "understanding" and "explanation" are hopelessly vague and subjective terms. Although Hempel made many significant contributions to twentieth-century philosophy of science, this was, in my opinion, the most important. It enables us to say that, no matter how we char-

acterize scientific explanation in detail, we can firmly maintain that *science does
provide genuine understanding of natural phenomena.*

IV. RELATIONS BETWEEN THE TWO CONCEPTS OF EXPLANATION

In the 1960s and 1970s, there was a strong opposition between those who
supported Hempel's 'received view' and those who advocated a causal/mechani-
cal approach to scientific explanation. Hempel's models encountered two major
difficulties, among others. First, as I have already mentioned, we seem to find
genuine functional explanations in the biological and behavioral sciences. These
have no place in the 'received view'. Second, Hempel's system does not incorporate any causal constraints. For this reason, it encounters many counterexam-
ples it cannot easily handle. One of the most famous involves a flagpole and its
shadow. Given a sunny day and a flagpole set on level ground, and given the
height of the flagpole and the elevation of the sun in the sky, it is easy to explain
why it casts a shadow of a certain length. However, given the same flagpole
under the same circumstances, it is easy to deduce the height of the flagpole from
the length of its shadow. Although this latter argument fulfills all of Hempel's
conditions, it goes against strongly entrenched intuitions to say that the length of
shadow explains the height of the flagpole. To put the matter simply, the presence of the flagpole causes the shadow, but the shadow does not cause the flagpole to have its particular height. It turns out that Hempel's characterization is
too narrow and too broad. It excludes functional explanations that seem to play a
legitimate role in many branches of science, and it includes the explanation of
the height of the flagpole by the length of the shadow.

At the same time, those who opposed the 'received view' were unable to provide an acceptable account of causality. As is well known, in the eighteenth century David Hume enunciated a profound critique of the concept of causality. It
may be that the difficulty in coming to terms with Hume's problem induced
Hempel and many other logical empiricists to steer clear of that concept. However, precisely the same consideration places an obligation on the advocate of a
causal theory of explanation to provide an analysis of causality that adequately
deals with Hume's problem. It seems to me that those who offered causal accounts of explanation failed to meet this requirement.

By the end of the twentieth century important developments have occurred
on both sides of this opposition. The natural successor to Hempel's program is
the *unification* conception, which, roughly speaking, takes scientific understanding to lie in the reduction of the number of independent assumptions or argument
types that are required to infer the facts we intend to explain. This approach,
which may well be implicit in Hempel's program, was spelled out explicitly by
Michael Friedman (1974). His detailed formulation proved to be flawed, but
Philip Kitcher has offered a variant that may be acceptable (1993). On the causal
side of the issue, I have devoted a good deal of effort to developing an account of

causality that meets Hume's problem squarely, and that can be used in a theory of scientific explanation (Salmon, 1998). It must be frankly acknowledged, I think, that, at the century's end, both programs face unsolved problems. We hope they are not insoluble.

This story has another twist. Although there was fundamental conflict in earlier periods between the 'received view' and the causal theory, it appears that the successors to these two approaches may be quite compatible with one another. The easiest way to make this point is with a concrete illustration. I have often used the following true story for this purpose.

A number of years ago, a congenial colleague in physics – a friendly physicist – was sitting on a jet airplane awaiting take-off. Across the aisle sat a young boy who was holding a string to which was attached a helium-filled balloon. To pique the child's curiosity, the friendly physicist asked him what he thought would happen to the balloon when the plane accelerated for take-off. After a bit of thought the boy said he believed that it would move toward the back of the cabin. My friend opined that it would move toward the front of the cabin. A number of adults in the vicinity, listening to the conversation, insisted that the friendly physicist was wrong; indeed, a cabin attendant wagered a miniature bottle of scotch. As it turned out, when the plane accelerated, the balloon did move forward, and my friend enjoyed his free drink.

How can the behavior of the balloon be explained? There are at least two ways. First, it could be noted that as the plane begins to move, the rear wall of the cabin exerts a force on the air molecules in its vicinity, and that from this a pressure gradient from the rear to the front occurs. Since helium is less dense than the air, the unbalanced force will push the balloon forward. This is a causal/mechanical explanation; it appeals to mechanical forces exerted on physical objects.

A second explanation appeals to the equivalence principle of general relativity. According to this theory, an acceleration is locally equivalent to a gravitational field. Thus, when the airplane accelerates, the objects inside react as if a massive gravitating body were placed behind the plane. Since the helium-filled balloon tends to rise in the earth's atmosphere under the influence of the earth's gravitation, the balloon will tend to behave similarly to the equivalent of a new gravitational field – i.e., it will 'rise' toward the front of the cabin. Since this explanation comprehends this little incident under one of nature's most general principles, it qualifies as a unification-type explanation.

Which of these two explanations is right? They both are. It would be a mistake to think that any given fact has only one correct explanation. Obviously, there are pragmatic criteria that determine which is best in a given situation; it would be quite unsuitable to appeal to general relativity to explain to a ten-year-old boy why his balloon behaved in this surprising way. Nevertheless, from a scientific standpoint, each provides a kind of understanding. Indeed, it is possible that the two modes of explanation will converge. As we pursue the investigation of causal mechanisms, we often find that the same kind of underlying mechanism explains widely different kinds of phenomena. Such a discovery tends to unify our understanding of these phenomena. If physicists could find the *final theory* to which Weinberg refers, it would unify our understanding of the world in terms of a single underlying mechanism. But that is still a dream, and one that may be far from becoming a reality.

V. CONCLUSION

I have claimed that, with the combined efforts of scientists and philosophers, the twentieth century has seen two major developments, namely, the recognition that science can achieve knowledge of unobservable entities and their properties, and the recognition that science can – by furnishing explanations – yield understanding of natural phenomena. Without the first of these achievements, I strongly doubt that the second could have occurred, because many of our best explanations rely essentially on the behavior of such things as electrons, atoms, and molecules.

Let me compare scientific understanding with the kind of understanding that rests on other grounds. Several years ago, an anthropologist friend gave me a fascinating semi-popular book, *Seven Clues to the Origin of Life* by A. G. Cairns-Smith, a noted geneticist. Posing the problem of the origin of life on earth as a Sherlock Holmes type of mystery, he pursues the clues and offers the hypothesis that living organisms originated ultimately from clay. Although it is not obvious to our unaided senses, clay has a crystalline structure at the microscopic level. He lays out the physical and chemical details of the mechanisms by which the evolution of life from clay could have come about. He does not claim that this hypothesis is solidly confirmed by the evidence; I do not know whether it is widely accepted today. But he is offering a genuinely scientific possible explanation that is open to empirical test. If it did turn out to be true, it would furnish concrete understanding of the origin of life on earth.

The Book of Genesis, in contrast, tells us, "And the Lord God formed man of the dust of the ground and breathed into his nostrils the breath of life; and man became a living soul" (2:7). It is not my intention to disparage *literary expressions* of this sort. If people find spiritual inspiration or psychological consolation therein, so be it. The value, however, is quite different from that of *scientific understanding.* For one thing, the scientific explanation is open to confirmation or disconfirmation on the basis of objective evidence. This point applies, not only to Cairns-Smith's hypothesis about the origin of life, but also to the reality of molecules and the explanation of Brownian motion. It applies equally to the explanations furnished by Newtonian physics and to the fruits of science at the end of the twentieth century.

As we embark on the twenty-first century, we know that we face many global problems: global warming, depletion of the ozone layer, population growth, famine, inadequate supplies of clean water, atmospheric pollution, etc. Take global warming. First, we need to find out whether this phenomenon is actually occurring. This involves the collection of vast quantities of data about present and past climatic conditions and weather patterns. If it turns out – as seems to be the case – that global temperatures are rising, it is imperative to ascertain to what degree this climatic trend is the result of human activities, and to what degree it is a result of general secular tendencies that have nothing to do with human agency. In other words, we need to explain the warming trend. If we succeed, then we may find actions that we can take to modify the trend or mitigate its effects. If we did not believe that science could provide explanations, it seems to me, we would deprive ourselves of the practical tools that might enable us to cope effectively with these major problems.

It has often been said that among the main aims of science are prediction and control. Such claims have been made by those who believe in the possibility of scientific explanation and those who do not. It seems to me, however, that we are in a much better position to exercise control when we can explain how nature works. If we know that global warming is occurring, we can predict the melting of the polar ice caps and the rise in sea level. We can then build dikes and move away from present coastlines. If we establish that emissions of greenhouse gases as a result of human activities explain the warming, we can work to reduce such emissions by using alternative energy sources. To put the point in the simplest possible terms: we are in a much better position to react intelligently to phenomena that we understand than those we merely recognize.

The theme of this paper is that recognition of the possibility of scientific explanation and of the attempts of clarify its nature have constituted a major step in philosophy in the twentieth century. This is an intellectual achievement of great proportions. In addition, I believe that this intellectual progress carries major practical rewards. It seems to me that philosophers and scientists can provide important benefits to humanity by spreading the word throughout the world that major problems can be addressed and understood scientifically. For this reason, I strongly applaud the Programme on Capacity Building in Science of the International Council for Science (ICSU – formerly known as the International Council of Scientific Unions). This effort is designed to disseminate scientific information and education globally. Twenty-five International Scientific Unions, representing a broad range of disciplines, and 95 nations, through their National Academies of Science, Research Councils, or similar agencies, adhere to this organization. Among its members is the International Union of History and Philosophy of Science. ICSU is the largest nongovernmental and noncommercial interdisciplinary scientific organization in the world, with access to impressive scientific resources – literally, hundreds of thousands of scientists worldwide. ICSU is not a policy-making body, but it seeks to make relevant and reliable scientific information available to those who will be making policy in the future. Scientific understanding of the problems, of their causes, and of the possibilities for their remediation can help concretely in coping with the work that needs to be done in the twenty-first century. As scientists promulgate the scientific understanding that is required to deal with the global problems we inevitably face, philosophers and historians of science can contribute to this effort by disseminating an understanding of the nature of science.

REFERENCES

Cairns-Smith, A. G., 1985, *Seven Clues to the Origin of Life*. Cambridge: Cambridge University Press.

Friedman, Michael, 1974, "Explanation and Scientific Understanding," *Journal of Philosophy* **71**, pp. 5-19.

Hempel, Carl G., and Paul Oppenheim, 1948, "Studies in the Logic of Explanation," *Philosophy of Science* **15**, pp. 135-175.

Hempel, Carl G., 1965, *Aspects of Scientific Explanation and Other Essays in the Philosophy of Science*. New York: Free Press.

Kitcher, Philip, 1993, *The Advancement of Science*. New York: Oxford University Press.

Lucretius, 1951, *On the Nature of the Universe*. Trans. R. E. Latham. Baltimore: Penguin Books.

Nye, Mary Jo, 1972, *Molecular Reality*. London: Macdonald.

Pearson, Karl, [1911] 1957, *The Grammar of Science*. New York: Meridian Books.

Perrin, Jean, 1913, *Les Atomes*. Paris: Alcan. Trans. D. L. Hammick, 1916, *Atoms*. London: Constable & Co.

Salmon, Wesley C., 1990, *Four Decades of Scientific Explanation*. Minneapolis: University of Minnesota Press.

Salmon, Wesley C, 1998, *Causality and Explanation*. New York: Oxford University Press.

Tannen, Deborah, 1991, *You Just Don't Understand*. New York: Ballantine.

Weinberg, Steven, [1992] 1994, *Dreams of a Final Theory*. New York: Vintage Books.

Wright, Larry, 1976, *Teleological Explanations*. Berkeley: University of California Press.

University of Pittsburgh
1001 Cathedral of Learning
Pittsburgh, PA 15260-6125
U.S.A.

HENRIQUE JALES RIBEIRO

FROM RUSSELL'S LOGICAL ATOMISM TO CARNAP'S *AUFBAU:* REINTERPRETING THE CLASSIC AND MODERN THEORIES ON THE SUBJECT*

The theme of this paper was inspired by studies related to the subject of my doctoral dissertation,[1] and, more specifically, by the work of A. Richardson and M. Friedman on the same subject presented in their two recently published books.[2] The material in these books which addresses the connection between Russell and Carnap's *Der logische Aufbau der Welt* reveals the same basic perspective in both authors and, in fact, represents the first in depth enquiry of this connection, despite certain fairly essential limitations which I hope to reveal in this paper. The line of investigation I intend to take in the following may therefore be outlined as such; to examine, albeit briefly, the extent to which Richardson's and Friedman's perspective can offer us a correct historical and philosophical approach to the influence of Russell's philosophy on the *Aufbau*, and, by confirming the existence of the limitations alluded to, to determine whether this perspective may be adequately reformulated independently of their existence, and to determine how, in general terms, it may in fact be reformulated. Thus, although my analyses and commentaries do not fall within a strictly historiographical framework, as is the case in the work developed by Richardson and Friedman, it is nevertheless possible to achieve certain objectives characteristic of this framework which, eventually, may become the subject of a future historiography of the philosophy of Russell and his influence on the *Aufbau*.[3] In order to achieve these objectives my main aim is neither negative nor destructive but essentially philosophical; I see myself (somewhat immodestly perhaps) together with the authors in question as partners in the investigation and resolution of problems arising from the presentation, discussion and testing of competitive theories, an example of what occurs in the scientific enquiry. In these circumstances, and from this point of view, I see myself as a philosopher who points to certain difficulties and problems in the theory put forward by these authors to explain the *Aufbau*, and, with particular reference to Russell's philosophy, concludes by suggesting an alternative theory.

M. Rédei and M. Stöltzner (eds.),
John von Neumann and the Foundations of Quantum Physics, 305–318.
© 2001 *Kluwer Academic Publishers. Printed in the Netherlands.*

I. SOME PRESUPPOSITIONS OF RICHARDSON'S AND FRIEDMANS'S WORK ON CARNAP'S *AUFBAU*, AND ITS LIMITS

In terms of the work of Richardson and Friedman under discussion, and, more specifically, taking the traditional interpretation of the relationship of Carnap's *Aufbau* to Russell's philosophy as their point of departure, the authors propose three fundamental theses which are considered, in general terms, to establish the rehabilitation of the book and its place in the history of analytical philosophy, which had been questioned by this previous interpretation. They are as follows:

A) Contrary to the traditional interpretation shared by certain contemporary analytical philosophers such as Quine or Putnam, the *Aufbau* is not the result of the influence of Russell's philosophy of logical atomism, or, in particular, of Russell's external word program elaborated in the book *Our Knowledge of the External World* (1914),[4] given that this program is based upon clearly empiricist and foundationalist presuppositions which are not present in Carnap's book;[5]

B) Although Russell's philosophy had supplied Carnap with the logical-mathematical methods necessary for the development of his own thought in the *Aufbau* (namely, the theory of types and the logical-mathematical apparatus which this involves), the essential philosophical context of this book, that is to say, the context in which these methods are expressed or, generally speaking, independently adapted from Russell's philosophy itself, is the German neo-Kantianism of Cassirer and others at the beginning of the century;[6]

C) One of the fundamental reasons for assuming A) and B) is that Russell's logical atomism is basically reductionist and empiricist whereas Carnap's *Aufbau*, on the contrary, arises from an idealist and holistic perspective which is absent in Russell but significantly present in an explicit form in neo-Kantianism (for example, in the Marburg school of neo-Kantianism). This thus explains the importance given in the book to the problem of the objectivity of the physical and natural sciences or to the thesis of the essentially structural nature of our knowledge, and which, in fact, once it has been reformulated by Carnap, becomes an entirely new vision of the relationship between logic and epistemology in the context of current investigations of the foundations of logic and mathematics from the beginning of the 1930s onwards.[7]

Any one of these theses has, according to the authors, fairly profound philosophical implications for an understanding of the *Aufbau*, which, obviously, cannot be discussed here in its entirety. The critique of the interpretations of the *Aufbau* held by Quine, Putnam and others led these authors to accept the theory that Russell was a quite ingenuous empiricist or else philosophically unaware of the presuppositions of his own empiricism, whilst, at the same time, under the pretext of the neo-Kantian influence, they reject the identification of the *Aufbau* with empiricism. However, setting aside for now the assumption that Russell was an empiricist, the aforementioned interpretations are not anywhere analyzed and discussed in terms of their own nature and philosophical significance. The only aspect referred to and criticised in this respect is, obviously, the complete devaluation of the importance of the *Aufbau* for contemporary analytical philosophy, which runs throughout all these interpretations. While the connection between the *Aufbau* and an ingenuous and completely outdated empiricism is destroyed, at the same time the essential conditions for contesting the theses of Quine and others and, in broader terms, for reviving the originality and the impact of the book on the history of philosophy, are created.

In the same way, in B) an entire set of fundamental implications for an understanding of the *Aufbau* and its aforementioned originality and impact are elaborated. However, I would suggest that this does not happen without a certain degree of philosophical perplexity on our part with regard to the generally implicit distinction between the *logical-mathematical context* of the *Aufbau*, within which, according to the authors, we must include Russell's mathematical philosophy, and, in particular, the theory of types, and the *philosophical context* itself within which, again according to the authors, we must include neo-Kantianism.[8] The perplexity arises from the fact that Russell's external world program, as I will attempt to demonstrate later, is a program which is far from being easily reduced to logical-mathematical terms, and establishes, in fact, a much vaster philosophical context within which, under the aegis of "logic as the essence of philosophy" and, in part, under the influence of Wittgenstein's criticisms of Russell around 1913, a possible integration may be found for the Russellian theories of acquaintance, logical constructions and descriptions, that is to say, for a whole set of philosophical theories which had a profound and indisputable influence (regardless of what it may be) on the *Aufbau*. Another way of affirming the same argument, particularly with regard to the theory of types, would be to say that the theory of types was conceived indirectly by Russell, as early as in the "Introduction" of the first edition of the *Principia Mathematica*, but, in more general terms, after 1918 and the lectures on *The Philosophy of Logical Atomism*, as an onto-epistemological theory of types. In other words, as a theory which dealt *not only with symbols but also with things*, although the former were the main concern of logic, and thus as a theory which fundamentally demands, alongside the logical justification, an epistemological one in its own right with regard to the complementary nature of the three theories mentioned above.[9] Thus, the strategy adopted by Richardson and Friedman in order to justify the divorce between these two different types of context, consists in devaluing the distinctly philosophical significance of the theory of types, and, surprisingly enough, attributing originally to neo-Kantianism what may be considered to be, for example, the Carnapian versions of the doctrine of "logic as the essence of philosophy" and of the doctrine of the neutrality of logic regarding metaphysics, which, as we know, are characteristically and unequivocally Russellian.[10]

We may now comprehend the essence of thesis C), namely that, according to the authors, only one philosophical orientation of an idealist and holistic nature such as neo-Kantianism may constitute the basis for the development of a reflec-

tion which, even though it is located at the junction between logic and epistemology, as is Russell's logical atomism, will nevertheless show a tendency towards
a pure epistemology which is completely independent of the basic presuppositions of the theory of knowledge in general, and which, when fully understood, is
not supposed to have, as its objective, any ontological implications. In this reformulation of Russell's doctrine of "logic as the essence of philosophy", which
breaks as much with the traditional empiricism of Locke, Berkeley, Hume and
Mach as with Kantianism itself in general philosophical terms, Richardson and
Friedman in fact anticipate the logical and semantic problem of Carnap's *Die
logische Syntax der Sprache*, and even that of Carnap's "Empiricism, Semantics,
and Ontology".[11] According to them, this necessitates a reassessment of the present relevance of the *Aufbau* (namely, in light of the challenge to the traditional
interpretation of Quine and others to whom I have already referred).[12]

In the context just outlined, the fairly obvious objection is that Richardson's
and Friedman's interpretations do not in fact respect the repeated affirmations of
Carnap in the *Aufbau*, as they do not respect his latter claims in his *Autobiography* with regard to this book and to Russell's influence, according to which it is
the philosophy of the latter and, in particular, the external world program presented in *Our Knowledge of the External World* which constitute the fundamental context for his own logical construction of the world.[13] The basic pretext they
offer, i.e. that, as far as the *Autobiography* is concerned, Carnap was thinking of
the way in which the *Aufbau* was originally understood by the Vienna Circle,
that is to say, in terms of the problem of the status of the so-called "basic statements" or "protocol-sentences", is not very convincing since it leaves the evident
connection between the *Autobiography* and the *Aufbau* itself unexplained, and,
moreover, because it is hardly likely that Carnap could have thought so clearly
about discussions that took place at the beginning of the thirties, and, at the same
time, to our surprise, have been equivocal about the essential context of the
development of his own philosophy in general in the twenties. The crucial difficulty, however, is that this explanation ignores the reiterated affirmations concerning the material of the *Aufbau* itself in relation to different aspects of Russell's program.[14] It is possible to assume that some different reconstruction of the
Aufbau as profound as Richardson's and Friedman's may show that Carnap had
been at least partly equivocal about these affirmations, but it is hard to believe,
as these authors claim, that he had been *completely* equivocal. Perhaps the answer to this problem, as I will go on to suggest, does not lie in an explicit opposition between the influences of Russell and those of neo-Kantianism, as we have
been presented with here, but, rather, in the idea that these influences essentially
complement each other in the prevailing context of the program of *Our Knowledge of the External World*, as Carnap suggests in the *Aufbau*.

The conclusion that may immediately be drawn from these observations, with
respect to the methodological presuppositions of Richardson's and Friedman's
interpretations is, by default, that these finally appear, not as a limited historiographical study, but as a "rational reconstruction" of the *Aufbau*.[15] It arises from

a set of philosophical presuppositions that are not necessarily Carnapian, and therefore do not necessarily have a historical base. From this point of view, thesis C) (the presupposition that the *Aufbau* triggered off a completely original reflection on the nature of the a priori problematic of philosophy that is at the base of our systems of representation in general, from language to the physical-natural sciences and mathematical logic, thereby constituting, in Friedman's words, a "scientifically respectable replacement for traditional epistemology")[16] seems to be essentially motivated not so much by Carnap's work itself, but, as Friedman says in another work, by the need for a return, in new terms, to a type of "reasonable" speculative reflection, in the face of the naturalist and relativist decay of contemporary philosophical thinking.[17]

II. A SURVEY OF SOME WORK TO BE DONE CONCERNING RUSSELL'S INFLUENCE ON CARNAP'S *AUFBAU*

To show the limits and difficulties of each of the theses described above, it would be necessary to examine certain presuppositions of Richardson and Friedman which are not in fact really addressed anywhere, and which again refer to Russell's philosophy. In particular, it would be necessary to show the erroneous, if not actually largely false, nature of the identification of Russell's philosophy with empiricism, which is reiterated unquestioningly at every step by these authors in their respective works. In addition, it would be necessary to argue with some force either against the interpretation itself that is presented to us in relation to the limits of the traditional theory of Quine and others on the *Aufbau*, or against that which refers explicitly to the philosophy of Russell and its historical and philosophical connection with this book. At this point, circumstances seem to suggest the need for a reformulation of certain fundamental aspects of the theory on the *Aufbau* advanced by these authors. Indeed, if it were shown that the influence of Russell on Carnap is generally poorly understood, the fundamental theses (theses A and B) of Richardson and Friedman would fall apart, and that theory would have to be reformulated, even though it could always continue to incorporate the important analyses done by both of them on neo-Kantianism.[18]

To achieve this goal it would obviously be essential to thoroughly examine the schematic and simplistic, if not downright crude, framework of Russell's so-called "external world program", which is given by Richardson and Friedman. In particular, while always bearing the *Aufbau* in mind, it would be essential to proceed with the study of the following aspects, which have mostly been left untouched in the studies on Russell:

I) the relationship between the multiple relation theory of judgment, presented in the Introduction to the *Principia Mathematica* (1910), the theory of types developed in the same work, and Russell's logical atomism in *Our Knowledge of the External World*;[19]

II) the connection between the theory of types, thus understood, and the theories of acquaintance, descriptions and logical constructions in the sphere of what Russell, in *The Philosophy of Logical Atomism*, called "logically perfect language";[20]

III) the relationship between the philosophies of Russell and Wittgenstein in the framework of the doctrine of logical atomism, in order to be able to assess, not the impact itself of the *Tractatus Logico-Philosophicus* on the *Aufbau* (we know today that that impact had no relevance to Carnap's book), but the philosophical pertinence, in general, of Wittgenstein's criticism on Russell;[21]

IV) the final development (understood by Carnap, at the time he wrote the *Aufbau*) of the theory of logical constructions in Russell's *The Analysis of Mind* (1921), particularly the concept of neutral monism described in that book.[22]

Understandably, it is impossible to deal with any of the themes listed above, even very cursorily. I shall confine myself, therefore, firstly to making some observations on the main reasons why, nowadays, specialised research on the philosophy of Russell, and I myself in particular, regard it as false to identify Russell with empiricism. Finally, I shall try to suggest the fundamental philosophical setting for Russell's ideas that led Carnap, in the *Aufbau* and after it, to claim as his, during the 1920s, the philosophical project of *Our Knowledge of the External World*.

III. THE MYTH OF RUSSELL'S "EMPIRICISM":
REREADING THE HISTORY OF ANALYTICAL PHILOSOPHY

With respect to the identification of Russell with empiricism, suggested by Quine and others from the 1950s on and reiterated without real justification in the works of Richardson and Friedman, it is essential to point out that this identification has emerged in the history of analytical philosophy as a meta-historical and meta-philosophical thesis that seeks to place the problem of the legitimacy of empiricism not in Russell's philosophy itself, but in the subsequent analytical contexts and the respective conceptions of analysis, and to exclude, in practice, that philosophy from the strictly analytical field. This identification is based on the assumption that Russell's own conception of analysis, in contrast with later conceptions of those contexts, would be contaminated by his markedly psychological and epistemological perspectives, and therefore could not constitute the more or less ideal historical norm for the development of analytical philosophy. This conception, in particular, would hark back to the tradition of the so-called "British empiricism" of Locke, Berkeley, and Hume, a tradition that is essentially foreign to the true spirit of philosophical analysis.[23]

This was precisely the perspective argued by English analytical philosophy (M. Black, A. J. Austin, P. Strawson, etc.) from the 1950s on, following a theory on the topic presented by A. J. Ayer.[24] Indeed it was to this that Quine and others (Putnam and Goodman, for instance) appealed indirectly, having been clearly influenced in this respect by the English philosophical context of the era. It should be noted that neither Quine and Putnam nor the English philosophers themselves, ever gave an adequate historical and philosophical explanation of the assimilation of Russell and empiricism, which indiscriminately embraced all the vast philosophical work of that author.[25] In fact, they took it more or less for granted, on the assumption that the Russellian concept of analysis was (in the words of J.O. Urmson, the official historian of the English analytical movement) a "classic" or "metaphysical" concept of analysis to be rejected outright.[26] This empiricist imputation of Russell's philosophy, therefore, becomes even more strange or paradoxical once it is ascertained that Russell himself never defended this type of connection either explicitly or implicitly, and had even appeared to expressly reject it on various occasions.[27]

In the last twenty years specialised research on Russell has clearly suggested the falsity of such an imputation. Recently what has emerged as the context for Russell's thought at the end of the previous century has not proved to be the British empiricism of Locke, Berkeley and Hume, as the English analytical philosophers erroneously believed, but the English neo-Hegelian idealism of Bradley, McTaggart and Bosanquet.[28] In addition, the realism and pluralism of Russell's philosophy of logic and mathematics between *The Principles of Mathematics* and the *Principia Mathematica* was never intended to have any epistemological significance, in the sense in which empiricism is generally understood.[29] Equally, as far as the foundations of mathematics in particular is concerned, logicism was essentially conceived in hypothetico-deductive terms, that is to say, in very different terms from traditional foundationalism in philosophy.[30] Finally, as for the so-called "Russell's external world program" and logical atomism in general, in other words, as for Russell's philosophy after *Our Knowledge of the External World*, it seems clear today that, for him, the concept of acquaintance was more one possible methodological principle for general philosophical investigation in the face of the holistic pretensions of pragmatism (W. James, J. Dewey, and others) at the beginning of the century rather than an epistemological principle in itself, like phenomenalism.[31] Russell himself, bearing in mind Mach and his theories, clearly rejected this assimilation at a certain point.[32] Moreover, if it is true that the epistemological significance of acquaintance is largely residual in Russell's work, it is nonetheless also true that it is transitory, being limited to a period in this work of no more than five or six years. After 1918, principally through the concept of "vagueness", there emerged in Russell's philosophy what may be called a partial semantic holism, within the framework of which the concept of acquaintance shed its previous foundationalist connotations.[33]

IV. THE ONTO-EPISTEMOLOGICAL INTERPRETATION OF RUSSELL'S THEORY
OF TYPES AND CARNAP'S *AUFBAU*: THE KEY OF RUSSELL'S INFLUENCE

Finally, we reach on the interpretation of Russell's external world program presented by Richardson and Friedman, and which I considered as very simplistic
and crude. Far from being limited to the well-known book *Our Knowledge of the
External World* and being developed and finalized, even in partial form (as these
authors assert), therefore far from being initially a 'program' in the strict
meaning of the word, it appears especially as a project or sketch of intentions
(more or less elaborated in some aspects) about which Russell did not have at the
time the necessary philosophical insight as regards their real nature or meaning.
This happens because, shortly beforehand, between 1912 and 1913, Wittgenstein's criticisms of Russell's philosophy in the manuscript *Theory of Knowledge*,
and in the *Principia Mathematica*, had led Russell to attempt a general reformulation of his philosophy. The book *Our Knowledge of the External World* is a
first example.[34] Part of Russell's difficulties in that work consists in the problem
of knowing how to relate and harmonize in a philosophically intelligible way the
skeleton of a logic that goes from atomic to general propositions with the philosophy of logic of the *Principia Mathematica*, and, in particular, with the theory of
types. Given the presuppositions that Russell shared in the introduction to the
Principia (particularly, the theory of judgement as multiple relation) and in the
Theory of Knowledge, this is an epistemological problem in a wide sense: Although he had at the time the logical and mathematical means that were poten-
tially adequate to deal with it, he did not in fact have the epistemological means,
strictly speaking. Another difficulty that is no less important, to which I have
already referred, is the problem of knowing how it is possible to integrate the
doctrines of acquaintance, descriptions and logical constructions (which supposedly have a close relationship from the point of view of logic and epistemology), once a consistent philosophical relation has been achieved between the
logical atomism of the theory of propositions and the theory of types of the
Principia. A possible solution, presented by Russell in the lectures on *The Philosophy of Logical Atomism* through the concept of "logically perfect language",
involves returning to some of the suggestions of the "Introduction" of the
Principia, and interpreting the theory of types onto-epistemologically – that is to
say, conceiving it as an ontological theory of types, as types that are concerned
not only with symbols but also with things, and also attributing in some way the
hierarchy of types, thus considered, to the proper subject of knowledge, which is
thus supposedly constituted by it, and in some way constructing it.[35] In the
"Introduction", the philosophical framing that Russell gives to the theory of
judgement as multiple relation points towards this solution, in that it considers
that hierarchy as, in a way, a hierarchy of judgement (on the first level we have
judgements about particulars, on the second, judgements about classes of

particulars etc.).[36] But, naturally, we understand the scale of the task required for such an interpretation of the theory of types, a task that, in fact, is no different, *grosso modo*, from that which Carnap completed in the *Aufbau*.[37] The fact is that in 1914 Russell did not have any solution to these basic problems, nor did he really try to suggest one, inebriated as he probably was with the partial successes achieved by the application to space of the doctrine of logical constructions, or, in a more general way, by the originality of his conception of "logic as the essence of philosophy" when interpreted in epistemological terms.[38]

If this is true for Russell, we could imagine it happening with a reader who was not sufficiently versed in philosophy. But with Carnap? If it is true that Russell's external world *program* is above all a *project* – that is to say, it is the project of integrating into a coherent philosophical structure (the theory of types onto-epistemologically conceived) the different doctrines constituting the basis of the development of that program, and, more generally, the basis of "logic as the essence of philosophy" –, what attitude did Carnap have, or could he have had, in relation to such a project? As he himself asserts repeatedly in *Aufbau,* and reiterates later in his *Autobiography,* his attitude in fact consisted in taking up again Russell's project, and obviously reframing it philosophically, which almost makes not recognizable Russell's intuition about a theory of onto-epistemologically interpreted types, and, specially, his doctrines of acquaintance, descriptions and logical constructions. The "constitutional system" presented in the *Aufbau* is, in truth, the development and reformulation of Russell's project in *Our Knowledge of the External World* in the new conditions of philosophical and scientific thought particular to Carnap's context in the middle of the 1920s, including, of course, German neo-Kantianism.

NOTES

* Paper presented at the Institut 'Wiener Kreis', 15th November 1999, with the support of *Fundação Calouste Gulbenkian* (Av. de Berna, 45-A Lisboa Portugal).

1. H. Ribeiro, *Bertrand Russell and the Origins of Analytical Philosophy. The Impact of L. Wittgenstein's 'Tractatus Logico-Philosophicus' upon Russell's Philosophy.* Coimbra: Universidade de Coimbra 1999.

2. A. Richardson, *Carnap's Construction of the World. The 'Aufbau' and the Emergence of Logical Empiricism.* Cambridge: Cambridge University Press 1998; M. Friedman, *Reconsidering Logical Positivism.* Cambridge: Cambridge University Press 1999.

3. It is necessary to say, right at the outset of this paper that apparently, Russell didn't read the *Aufbau* until very late, probably the forties and the writing of his book *Human Knowledge: Its Scope and Limits.* In this book he makes for the first time some (important) references to Carnap's *Aufbau,* but the problem of the influence of his own philosophy on this book is not discussed. (See B. Russell, *Human Knowledge: Its Scope and Limits,* London: George Allen and Unwin 1966, p. 90ff.) As a reading of *The Collected Papers of Bertrand Russell* could demonstrate, not a single reference to the *Aufbau* exists in all the books and papers of Russell before 1948. Furthermore, no copy of the *Aufbau* exists in the *Bertrand Russell Archives* (McMaster University, Hamilton-Ontario, Canada). Russell may have read the *Aufbau* when he was in the U.S.A. (1939-1943), but we cannot be sure of that.

4. See W. V. Quine, "The Two Dogmas of Empiricism: Empiricism without Dogmas", in: *From a Logical Point of View. Logico-Philosophical Essays.* Cambridge-Massachusetts: Harvard University Press 1994, p. 24ff.; and H. Putnam, *Mind, Language and Reality*, Philosophical Papers, vol. 2, Cambridge: Cambridge University Press 1984, pp. 441-451. N. Goodman, without mentioning Russell, apparently also subscribed to such an interpretation. See "The Significance of *Der logische Aufbau der Welt*", in: P. A. Schilpp (Ed.), *The Philosophy of Rudolf Carnap.* La Salle-Illinois: Open Court 1963, pp. 545-558.

5. M. Friedman sums up in the following words the traditional interpretation: "The epistemological point of the *Aufbau* is to develop a traditional phenomenalist or reductionist solution to this problem: the external world does not lie behind, or correspond to, the immediate sense data; rather, it is nothing but a complex logical construction out of such immediate data. Our claims about the external world are, in the end, complex claims about the immediate sense data and hence thus justifiable in principle. What then distinguishes the *Aufbau* within the empiricist tradition is simply the greater detail and rigor with which it attempts to carry out this phenomenalist program." In: *Reconsidering Logical Positivism*, p. 117.

6. *Ibid.*, p.124ff.

7. A. Richardson says, for example: "No concerns of an epistemological nature about logic are in evidence in his book. Indeed, the role that structure plays in the account of objectivity indicates that logic must be in place before any epistemological question can be raised." And: "For Carnap, unlike Russell, the rejection of metaphysics is not governed by the acceptance of an ontology of objects of acquaintance and a method that shows how to do without anything else. Carnap seeks to reject all questions of ontology; epistemology has nothing to say about such questions." In: *Carnap's Construction of the World*, pp. 25-26.

8. M. Friedman says in this regard: "The *Aufbau* is not best understood as starting from fundamentally empiricist philosophical motivations and then attempting to put these into effect – on the basis of the new mathematical logic of *Principia Mathematica* – in a more precise and rigorous way than had been previously possible. The epistemological motivations of the *Aufbau* begin rather with the concerns and problems of the neo-Kantian tradition ... In the *Aufbau*, however, the new mathematical logic of *Principia Mathematica* provides Carnap with all the philosophical concepts and distinctions he needs. Carnap thereby achieves a standpoint that is both non-psychological and truly metaphysically neutral, and, at the same time, he transforms the neo-Kantian tradition into something essentially new: 'logico-analytic' philosophy." In: *Reconsidering Logical Positivism*, p. 141.

9. Thus conceived, this is the main project of Russell's lectures on "The Philosophy of Logical Atomism", and, in particular, of his conception of a "logically perfect language", as I will try briefly to show later (section 4). The onto-epistemological nature of the theory of types is a consequence of Russell's theory of meaning. As he asserts: "I think that the notion of meaning is always more or less psychological, and that it is not possible to get a pure logical theory of meaning, nor therefore of symbolism ... At any rate I am pretty clear that the theory of symbolism and the use of symbolism is not a thing that can be explained in pure logic without taking account of the various cognitive relations that you may have to things." In: John Slater (Ed.), *B. Russell. The Philosophy of Logical Atomism and Other Essays: 1914-1919*, The Collected Papers of Bertrand Russell, vol. 8, London: George Allen and Unwin 1986, p. 167.

10. See A. Richardson, *Carnap's Construction of the World*, specially, chap. V, "The Fundamentals of neo-Kantian Epistemology", pp. 116-138. Nevertheless, Richardson is forced to admit some essential difficulties in his own interpretation. At a certain moment, he acknowledges *"the inchoate account of logic found in the neo-Kantian literature"*, and the fact that "Despite calling their project 'the logic of objective knowledge' and investigating 'the logical conditions of measurement', *there is very little by way of delimiting the principles of logic.*" Further, he acknowledges that the framework of Cassirer's philosophy ignores "Frege's conceptual notation" or "Russell's type theory", and that, finally, "the neo-Kantian account of the primacy of logic is hollow; ultimately, logic itself requires an ontological foundation". In: *ibid.*, pp. 136-137 (all the italics are mine).

11. See M. Friedman, *Reconsidering Logical Positivism*, p. 124.

12. See A. Richardson, *Carnap's Construction of the World*, chap. 9, "After Objectivity: Logical Empiricism as Philosophy of Science", pp. 217-229.

13. I recall here some relevant passages of Carnap's *Autobiography*. Concerning the influence of Russell's books on the *Aufbau*, and, in particular, of *Our Knowledge of the External Word*, he says (without mentioning any distinction whatsoever between a logical and mathematical context, and a philosophical one): "Some passages of the book made an especially vivid impression on me because they formulate clearly and explicitly a view of the aim and method of philosophy which I had implicitly held for some time ... I felt as if this appeal [Russell's appeal to the study of logic as the central study in philosophy presented in the very last past of the book] had been directed to me personally. To work in this spirit would be my task from now on! And indeed henceforth the application of the new logical instrument for the purposes of analyzing scientific concepts and of clarifying philosophical problem has been the essential aim of my philosophical activity. I now began an intensive study of Russell's books on the theory of knowledge and the methodology of science. I owe very much to his work, not only with respect to philosophical method, but also in the solution of special problems. I also continued to occupy myself with symbolic logic ... In 1924 I wrote the first version of the later book, *Abriss der Logistik* [1929]. It was based on *Principia* [Russell's *Principia Mathematica* 1910-1913]. Its main purpose was to give not only a system of symbolic logic, but also to show its application for the analysis of concepts and the construction of deductive systems." In: *The Philosophy of Rudolf Carnap*, p. 13f.

14. Ignoring, for the moment, Russell's theory of types, some of these affirmations concern, for example, Russell's "structuralism" in logic and mathematics. Nevertheless, in Richardson's and Friedman's works the relevance of the concept of "structure" in the *Aufbau*, which is connected with the problem of the objectivity of the empirical sciences, is *only explained by the influence of neo-Kantianism*. In this regard, the fact is that in the *Aufbau* Russell's influence is always quoted in the first place several times: "From the relations, we must go on to the structure of relations if we want to reach totally formalized entities. Relations themselves, in their qualitative peculiarity, are not intersubjectively communicable. *It was not until Russell ... that the importance of structure for the achievement of objectivity was pointed out.*" (In: R. Carnap, *The Logical Structure of the World. Pseudo-problems in Philosophy*, London: Routledge and Kegan Paul 1967, p. 29, italics mine.) On the other hand, concerning the importance of the concept of objectivity, which, in contrast, has been stressed especially by neo-Kantianism, Carnap explicitly saw his philosophy in the *Aufbau* as a *development* or *enlargement* of Russell's philosophy of mathematics (not as a rupture with it, as these authors held): "Whitehead and Russell, by deriving the mathematical disciplines from logistics, have given a strict demonstration that mathematics (viz., not only arithmetic and analysis, but also geometry) is concerned with nothing but structure statements. However, the empirical sciences seem to be of an entirely different sort ... " (*Ibid.*, p. 23) Again, as I anticipated previously, this relationship between the concepts of *structure* and *objectivity* points out to the existence of an essential complementarity between the influences of Russell's and neo-Kantian's philosophies on the *Aufbau*.

15. I use here the concept in the sense used by R. Rorty, "The Historiography of Philosophy: Four Genres", in: R. Rorty/J. B. Schnewind/Q. Skinner (Eds.), *Philosophy in History*. Cambridge: Cambridge University Press 1984, pp. 49-75.

16. M. Friedman, *Reconsidering Logical Positivism*, p. 5.

17. See M. Friedman, "Philosophy and the Exact Sciences. Logical Positivism as a Case Study", in: J. Earman (Ed.), *Inference, Explanation and Other Frustrations. Essays in the Philosophy of Science*. Berkeley/Los Angeles/Oxford: University of California Press 1992, pp. 84-98. Unfortunately, this very important paper was not published in Friedman's book. When compared with his other papers, it seems to develop a different line of thought, because logical positivism is accused of being at the origins of contemporary philosophical relativism. On the other hand, also in contrast with his later book, Friedman clearly insists in that paper on the failure of logical positivism, and, in general, on its negative aspects.

18. This is not, however, A. Richardson's view: he thinks that he can hold precisely the same theory even if Russell was not an empiricist, as I claim. In some public discussion on this subject, he wrote to me: "1.) I am interested in Carnap's project in the *Aufbau* and in avoiding an easy assimilation of it to Russell's project in *OKEW* [*Our Knowledge of the External World*]. 2.) The connection to *OKEW* (and not all the other writings you mention) [see the points I-IV listed below in my text] is made both by Carnap in his autobiography and by Quine in his many

writings. 3.) Thus, for the purposes of the book, I am interested only in Russell's *OKEW* project and that only as a source of influence on Carnap or as a way to understand what Carnap's is doing. 4.) A careful reading of my book would see that I acknowledge that Russell's work is more complicated than Quine's reading of it allows and that, thus, Quine's account of Carnap's motivations may be doubly wrong – i.e. *wrong about Carnap in part of being wrong about Russell* [the italics are mine]. See the footnotes on p. 21, for example [where Richardson, having as a base N. Griffin's and P. Hylton's books on Russell, mentions Russell's idealistic perspectives]. In short: It is wonderful that you care so deeply about Russell's project and want to get it right. I, on the other hand, care deeply about rejecting a certain story about Carnap's project. This story starts from a certain understanding of Russell's project. In order to be clear in my exposition, I wanted to motivate *that* reading of Russell and then explain why *that* project is not what Carnap is up to. *Whatever more complete version of Russell you eventually come up with, I feel quite confident that is not Carnap's project either* [the last italics are mine]." Quotation of A. Richardson's message to "Russell-l" (received 7 March, 1999), the electronic list of the Bertrand Russell Society.

19. Each of these points (the multiple relation theory of judgment, the theory of types, and the theory of logical atomism) is completely ignored. Perhaps the excuse (unacceptable, as we shall see) would be to say that such theories are not *explicitly* presented in *Our Knowledge of the External Word*. But what about the *Principia Mathematica*? And what about the explicit connection, acknowledged by Carnap in his *Autobiography*, between the *Principia* and *Our Knowledge*?

20. I will develop this point in the section 4 of this paper: "The onto-epistemological interpretation of Russell's theory of types and Carnap's *Aufbau*: the key of Russell's influence". Note that, as I have already said, neither Richardson nor Friedman study anywhere that kind of connection in their respective works on the *Aufbau*; in particular, the theory of types is clearly ignored, be it in its logico-mathematical aspects or in its philosophical ones.

21. Apparently, Carnap did not study the *Tractatus* until the completion of the manuscript of the *Aufbau*. But, of course, insofar as Wittgenstein's contribute for Russell's logical atomism was important, and in so far as his criticism of Russell in the *Tractatus* were also essential, he cannot be ignored by any serious investigation. Note that A. Richardson, for example, explicitly ignores the study of the relationship between Carnap's *Aufbau* and Wittgenstein's *Tractatus*.

22. Russell's neutral monism, as Mach's, is completely ignored by Richardson and Friedman. This concept seems to anticipate, in its own way, Carnap's idea of logic as the neutral framework of different (metaphysical) languages on the problem of the existence of the external world (such as realism and idealism). When applied to just the very same problem throughout the theory of logical constructions, neutral monism, too, arrives at the idea that logic (that is, the theory in question) is a neutral framework within which the traditional metaphysical conflicts can be solved. We must emphasize that neutral monism, in Russell's philosophy, is not simply a philosophical theory amongst others, but is, in fact, *the final step of his external world program*. After an initial rejection of neutral monism in *Theory of Knowledge* (1913), Russell adheres to it in "On Propositions" (1919), and specially in *An Analysis of Mind* (1921).

23. Concerning English analytical philosophy in the 1950s, see J. O. Urmson, *Philosophical Analysis. Its Development Between the Two World Wars*. Oxford: Clarendon Press 1956; and J. O. Urmson, "Histoire the L'Analyse", in *La Philosophie Analytique*. Paris: Minuit 1962, pp. 11-22. Concerning analytical philosophy afterwards, M. Dummett's views are surely an essential reference. See M. Dummett, *Frege: Philosophy of Language*. Worcester-London: Duckworth 1981, pp. 664-684. This conception is clearly suggested by some references of A. Richardson. At a certain point, he says about Carnap's project: "The new logic is, thus, not a toll to use in pursuit of a reductive epistemological-cum-ontological project bequeathed to us by the British empiricists, but rather a way of reformulating the whole question of what is at stake in philosophy. Carnap's antimetaphysics is surely the consequence of a much more fundamental understanding of 'logic as the essence of philosophy' than is Russell's empiricism of 1914." In: *Carnap's Construction of the World*, pp. 26-27.

24. The theory was first presented by Ayer in the paper "The Analytic Movement in Contemporary British Philosophy", in Actes du *Congrès International the Philosophie Scientifique* (Paris, 1935). It was developed in several works after that, as, for instance, in Ayer's *Language, Truth*

and Logic. London: Victor Gollancz, 1936. Recall Ayer's first words in this book, where we can find too a surprising identification between Wittgenstein and empiricism: "The views which are put forward in this treatise derive from the doctrines of Bertrand Russell and Wittgenstein, which are themselves the outcome of the empiricism of Berkeley and D. Hume." *Ibid.*, p. 11.

25. As I showed in my dissertation, neither Quine and Putnam, nor Goodman, in spite of the originality of their respective philosophies, have a reading of the history of analytical philosophy autonomous and independent of the English analytical philosophy of the fifties. This explain why they have accepted, in general, the views of the English philosophers on the subject, without even trying to discuss them.

26. See J. O. Urmson, "Histoire de l'analyse", p. 17ff.

27. See, for example, B. Russell's words in the paper "A Microcosm of British Philosophy" (1919): "Traditional British Philosophy, as represented by Locke, Berkeley, Hume, Mill and Spencer, never became technical. It could be read by gentleman of leisure, and was read by artisans. It started from common sense, criticizing its inconsistencies with more or less severity in ordinary language and usually in a excellent literary style. It arrived in the end at scepticism – at least that was its logical outcome, explicit in Hume, but concealed from the others in proportion to their muddle-headness. Dr. Moore is an admirable representative of this method, by no means sceptical in temperament, but often driven into sceptical conclusions by his perfect intellectual integrity." In: *B. Russell. Essays on Language, Mind and Matter: 1919-1926*, p. 385.

28. See P. Hylton, *Russell, Idealism, and the Emergence of Analytic Philosophy.* Oxford: Oxford University Press 1990; and N. Griffin, *Russell's Idealist Apprenticeship.* Oxford: Oxford University Press 1991.

29. See P. Hylton, "Logic in Russell's Logicism", in D. Bell and N. Cooper (Eds.), *The Analytic Tradition: Meaning, Thought and Knowledge.* Cambridge-Massachusetts: Basil Blackwell, 1990, pp. 137-172.

30. See B. Russell, "The Regressive Method of Discovering the Premises of Mathematics" (1907), read before the Cambridge Mathematical Club, 9 March 1907. In B. Russell, *Essays in Analysis.* London: George Allen & Unwin 1973. And A. D. Irvine, "Epistemic Logicism and Russell's Regressive Method", in A. D. Irvine (Ed.), *Bertrand Russell. Critical Assessments.* London and New York: Routledge 1999, pp. 172-195.

31. It is in this sense that the concept of "acquaintance" is introduced in the manuscript *Theory of Knowledge.* See Part I, chap. II, "Neutral Monism", in particular, p. 22ff. In the Part I, "Preliminary Description of Experience", Russell presents even, at a certain moment, a refutation of empiricism, or as he calls it, "the older empiricism philosophy": "it is certain that the world contains some things not in my experience, and highly probable that it contains a vast number of such things." In: E. R. Eames (Ed.), *B. Russell. Theory of Knowledge: The 1913 Manuscript.* (The Collected Papers of Bertrand Russell, vol. 7), London and N. York: Routledge 1993, p. 11.

32. See B. Russell, "The Philosophical Analysis of Matter" (1925), in *Bertrand Russell. Essays on Language, Mind and Matter: 1919-1926*, pp. 275-284. Russell says, for example: "There is a philosophy called 'phenomenalism' which is attractive, but to my mind not practically feasible. This would base physics upon phenomena alone. I think those who advocate this philosophy have hardly realized its implications." Ibid., p. 281.

33. In my dissertation I studied the emergence and development of the concept of "vagueness" in Russell's philosophy, and its historical and philosophical implications, from the manuscript *Theory of Knowledge* to his later works. My point is that, contrary to a well-known interpretation (according to which such concept must be opposed to the "exactness" and "precision" of an ideal language to be constructed), the vagueness of ordinary language, as the vagueness of our "systems of representation" In general, for Russell, is essentially the result of an inevitable mediation of the data by language (regardless of what the data and the language may in fact be, for example, in the context of the hypothetico-deductive systems of the empirical sciences). By "vagueness", what Russell intended to say (mainly after 1918) was really what Quine much more later will define as the "indetermination of translation". Russell's partial semantic holism led him in "On Propositions" (1919) to neutral monism and to the theory that meaning, in general, has its basis in usage.

34. In the "Preface" of this book Russell explicitly acknowledges the importance of Wittgenstein's influence when, after a reference to Whitehead, he says: "In pure logic, which, however, will be

very briefly discussed in these lectures, I have had *the benefit of vitally important discoveries, not yet published, by my friend, Mr. Ludwig Wittgenstein.*" In: B. Russell, *Our Knowledge of the External World.* London: George Allen & Unwin 1949, p. 9 (italics mine).

35. Russell presents his "logically perfect language" in the lectures on "The Philosophy of Logical Atomism" from this very wide perspective, that is to say, as something which should embrace all the human experience of each subject of knowledge. It should embrace because we must understand it, theoretically, as constructed by each subject from the acquaintance basis, using, in some way not explained by Russell, logical constructions and descriptions. The enormous complexity of such language means, for him, that it is impossible for philosophy to construct (or re-construct) it. We could perhaps construct a logically perfect language if this language were constituted only by syntax (but this is not the case): "In a logically perfect language, there will be one word and no more for every simple object, and everything that is not simple will be expressed by a combination of words, by a combination derived, of course, from the words for the things that enter in, one word for each simple component. A language of that sort will be completely analytic, and will show at a glance the logical structure of the facts asserted or denied. The language which is set forth in *Principia Mathematica* is intended to be a language of that sort. It is a language which has only syntax and no vocabulary whatsoever. ... It aims at being that sort of language that, *if you add vocabulary*, would be a logically perfect language. Actual languages are not logically perfect in this sense, and they cannot possibly be, if they are to serve the purposes of daily life. A logically perfect language, if it could be constructed, would not only be intolerably prolix, but, as regards its vocabulary, would be very large private to one speaker. That is to say, all the names that it would use would be private to that speaker and could not enter into the language of another speaker. ... Altogether, you would find that it would be a very inconvenient language indeed. That is one reason why logic is so very backward as a science, because the needs of logic are so extraordinarily different from the needs of daily life. ... I shall, however, assume that we have constructed a logically perfect language, and we are going on State occasions to use it." In: *B. Russell. The Philosophy of Logical Atomism and Other Essays: 1914-1919*, p. 176.

36. See B. Russell, *Principia Mathematica*. Cambridge: Cambridge University Press 1910, p. 45ff.

37. Russell's theory of types can be extended to what Carnap in the *Aufbau* calls the "heteropsychological levels", and can include, therefore, social and cultural objects. Russell himself, sometimes, was tempted to interpret his theory in this sense, but, in fact, he never did that. This development of Russell's theory of types, by Carnap, can be compared to the development of that theory, by Tarski, in terms of a "hierarchy of languages". In both cases, Russell had the *intuition* of these possible interpretations of his theory.

38. In this regard, significantly, part of the texts of *Our Knowledge of the External World* to be quoted here have already been quoted by Carnap himself in his *Autobiography*. I would only add, briefly, a passage (partly omitted by him) of Russell's last remarks in that book: "... It is in this way that the study of logic becomes the central study in philosophy: it gives the method of research in philosophy, just as mathematics provides the method in physics. And as physics, which, from Plato to the Renaissance, was unprogressive, dim, and superstitious as philosophy, became a science through Galileo's fresh observation of facts and subsequent mathematical manipulation, so philosophy, in our days, is becoming scientific through the simultaneous acquisition of new facts and logical methods." In: B. Russell, *Our Knowledge of the External World*, pp. 243-244.

Instituto de Estudos Filosóficos
Faculdade de Letras
Universidade de Coimbra
P-3049 Coimbra
Portugal

MASSIMO FERRARI

RECENT WORKS ON OTTO NEURATH

A. SOULEZ, F. SCHMITZ, J. SEBESTIK (Eds.), *Otto Neurath, un philosophe entre guerre et science*, Paris, L'Harmattan, 1997. [cited as ON]

J. BERNARD, F. STADLER (Eds.), *Neurath: Semiotische Projekte & Diskurse*, "Semiotische Berichte", XXI, 1997, Heft 1. [cited as N]

M. OUELBANI (Ed.), *Otto Neurath et la Philosophie Autrichienne*, Tunis, Cérès Editions, 1998. [cited as ONPA]

E. NEMETH, R. HEINRICH (Eds.), *Otto Neurath: Rationalität, Planung, Vielfalt*, Wien-Berlin, Oldenbourg-Akademie Verlag, 1999. [cited as RPV]

I.

The ignorance about Otto Neurath's thought and action within the Vienna Circle as well as within the "scientific philosophy" definitely belongs to the past. In the last two decades the rediscovery of Neurath's legacy represents one of the most significant aspects of the historical and theoretical studies on logical empiricism. As Friedrich Stadler has recently pointed out, "the rediscovery of Neurath [is] not merely a phenomenon of academic nostalgia, but itself constitutes research into the conditions and possibilities of changing a paradigm in the philosophy of science"[1]. Particulary in the last ten years this re-evaluation has been the subject of numerous books and contributions, which offer a radically new view of the life, the philosophy of science and knowledge, the political and economical thought of Otto Neurath, and, last but not least, of his enthusiastic activity for the Encyclopedia-project. It is sufficient for the readers of this *Yearbook* to recall the enlightening reconstruction of the emergence of Neurath's naturalism within the Vienna Circle offered in 1992 by Thomas E. Uebel and the extensive, critical description of Neurath's work and life which was published in 1996 by Nancy Cartwright, Jordi Cat, Lola Fleck and Thomas E. Uebel.

The books and contributions to be dealt with in this review seem to represent a new phase of the Neurath-research. Whereas the "revolutionary" phase of Neurath's rediscovery is drawing to an end, lately the critical work on Neurath

M. Rédei and M. Stöltzner (eds.),
John von Neumann and the Foundations of Quantum Physics, 319–327.

has begun to outline – so to speak – his "normal" image. "Back to Neurath" (Rudolf Haller) was the provocative motto of the *Neurath renaisssance*; but today "normal science" is moving towards the delineation of a relatively established "paradigm", in contrast with previous appraisals of Neurath's thought. Nowadays, no scholar can any longer approach the history of logical empiricism and of analytical philosophy since Quine without a detailed account of Neurath's contribution to themes such as holism, historical development of science, fallibilism, anti-foundationalist reconstruction of knowledge process, and so on. Moreover Neurath belongs definitively to the "gallery" of 20th century philosophers of science, so that he represents an object of interest for research programs flourishing not only in the German and English-speaking world. As we can see from two of the publications under review, in the last years French studies added to the Italian studies on Neurath, such as that of Danilo Zolo about "reflexive epistemology" as the heart of Neurath's philosophical legacy.[2] Antonia Soulez emphazises in her "Qui était Otto Neurath?" (ON, 12) that the name of Neurath is still unknown to most French philosophers, so that the 20th century heir of D'Alembert's and Diderot's encyclopedia urgently needs to be rediscovered. In other words, the interest on Neurath's life and thought seems to constitute an international trend within the more recently historical and theoretical approach to the development of philosophy of science and logical empiricism. Certainly, Neurath is no more the "great unknown" of the 20th century philosophy.[3]

II.

Obviously it is almost impossible to give a detailed account of all the contributions included in the collections of essays under review. It is better perhaps to thematically group this rich material which during the last two or three years has piled up on the scholar's desk stimulating him or her to survey the new research in this field and to draw up the agenda for the next, desirable work on Neurath. To begin with, Neurath's intellectual biography seems to be a research field of great interest, also as regards many facets of his early activity as social scientist and his contribution to the "first Vienna's Circle". On the other hand, Neurath's intellectual life is particularly exciting: both from a philosophical point of view in terms of considering the different economical, historical, sociological, political and epistemological interests which characterize from the very beginning the unusual intellectual trajectory of this polymath, of whom Rainer Hegselmann offers a stimulating portrait in his "Otto Neurath. War economist, social engineer, empiricist philosopher, and visual educationalist" (ONPA, 97-124). In many cases the "genesis" of a thinker proves to be crucial for the understanding of his *entire* work: to be sure, this is particularly true in the case of Neurath, so that it is undoubedtly correct when in her "Qui était Otto Neurath?" Antonia Soulez stresses that "the essential traits of Neurath's thought can be found *in nuce* in his earlier writings" (ON, 15). Therefore it is a good idea to provide a

French translation of Neurath's fundamental, youthful essay on Descartes and the auxiliary motive, which contains the first formulation of "pseudorationalism" as central epistemological theme and which plays – as Elisabeth Nemeth emphasizes – a leading role in the development of Neurath's thought (cfr. ON, 19-33 and E. Nemeth, "Les voyageurs égarés de Descartes": ONPA, 139). But it will be appropriate to remember that in addition to this translation the French reader has now at his disposal other translations too: on the one hand, the biographical sketch written by Paul Neurath (together with Neurath's bibliography and chronology compiled by Denis Lelarge), and, on the other hand, the pamphlet *Jüdische Planwirtschaft in Palästina*, which was published by Neurath in 1921 under the pseudonym of Karl Wilhelm offering a good specimen of his attempt of "technical-social construction" immediately after the end of World War I (cfr. ON, 167-196, 199-215, 217-226).

Both these biographical and scientific aspects require a consideration of Neurath's work not only with regard to his epistemological views and to his navigation "on the early boat", but also with regard to his economical, political and social thought as well, especially during the war period and in the 1920s. First of all, the point is to establish the real importance of Neurath's acquaintance with the social sciences (that is the *non*-"hard" sciences, today often called special sciences) in order to understand the epistemological status of natural sciences (that is the "hard" sciences). According to Michel Rosier, social sciences strike to Neurath *as richer* than natural sciences, first of all because they enable one to establish a close link between "political activity and scientific activity" – a link which represents the core of Neurath's empiricism ("Empirisme logique, économie de guerre et utopies sociales": ON, 129). For her part, Elisabeth Nemeth ("Auf dem Weg zur Wiedergewinnung eines wissenschaftlichen Gegenstandes": RPV, 148) maintains that to introduce the historical perspective into the field of natural sciences doesn't mean weakening them, since Neurath, on the contrary, seems to be convinced that the social sciences have to be built taking natural sciences as their model (and "physicalism" would be exactly the proof of this methodological adaptation). In spite of this it still remains clear that Neurath's "diversity" (as well as his originality) within the Vienna Circle consists in his acquaintance with sciences which do not use a rigorous mathematical equipment, that is with social sciences, and not with natural sciences, to which Neurath – as Jan Sebestik remarks – pays no "attention" ("Vorwort": RPV, 13). And there is no doubt that this background explains very well – as Michael Friedman has recently suggested – the fact that Neurath, unlike the other members of the Vienna Circle, "shows very little interest in either the problem of a priori principles or the problem of elucidating the peculiar position and role of philosophy vis à vis the special sciences"[4].

At any rate, it is quite impossible to understand Neurath's intellectual activity without an accurate account of his extremely broad work as economist, sociologist, social utopian, culture organizer and finally ardent apostle of the *Bildstatistik*. Since recent studies on Neurath are concernend with all these aspects (neglected for too long) it is no wonder that a lot of the contributions we are discussing deal with this fascinating side of his thought and action. Wolfgang Pircher, for example, shows how war economy represents for Neurath (and for other social scientists at the beginning of 20th century too) a crucial phenomenon for the planning of a non-market economy and, more generally, for the building of a "rational world" ("Der Krieg der Vernunft": RPV, 103). In his "Socialism, Ecology and Austrian Economics" John O'Neil proposes a reconstruction of the debate between Neurath and the Austrian school of economics. His paper explores indepth the discussion between Neurath and Ludwig von Mises as regards Neurath's criticism both of the socialist alternatives to the market and of the "algorithmic conception of practical reason", which according to Neurath was only a typical form of "pseudorationalism". "Mises' attack on the possibility of socialism – according to O'Neil – exhibits precisely the kind of pseudorationalism in the domain of practical reason that Neurath had attacked in his earlier writings" (RPV, 131).

In this context, Hegselmann rightly observes that Neurath's studies on war economy constitute "the key" of his late scientific and political conceptions (ONPA, 101). Neurath's "social engineering", on the other hand, meets his "scientific utopia" of a new social order. But Neurath's socialisation plans elaborated during the German social crisis and the brief experience of the Bavarian *Räterepublik* has to be distinguished – as Camilla Nielsen and Thomas E. Uebel suggest in their "Zwei Utopien in einer gescheiterten Revolution. Otto Neurath und Gustav Landauer im Vergleich" (RPV, 62-95) – from the kind of spiritual mission which was preconized by Gustav Landauer. For Neurath, indeed, utopia doesn't mean vision, but "plan": it doesn't signify political system of the councils, but economic system of the councils. Thus utopia and science are always intertwined and Neurath's project of a new republic which has to be interpreted in the framework of the enlightenment, which is at the same time self-critique of reason: that is reason undergoing constant revision and always needing new enlightenment, similarly to the dialectical critique of reason supported by the Frankfurt School (RPV, 89-90). It is also for this reason that many years later, towards the end of his life, Neurath still considered a great philosophical utopia such as Plato's *Republic* as a danger for the present social and political life: it was the danger, as Antonia Soulez points out in her "Comprendre, est-ce tout pardonner? Otto Neurath et la République de Platon en 1944-45" (ON, 147-165), of a "antidemocratic and repressive system", celebrating Greek ethnocentrism against the kind of cosmpolitism professed later by Epicurus. For Neurath, Plato could thus not remain object of detached historical analysis, but became a controversial theme for present debates: the past (that is the Platonic utopia) has not to be understood, since to understand would mean to pardon. And more generally all this was in Neurath's opinion only the last, but illuminating proof that politics can't actually be separated from epistemology (ON, 155). It could still be contended that a similar intrusion of political arguments into the field of epistemology may have, at any rate, beneficial results. One can agree on the matter that this point really represents the core of Neurath's way of thinking; but it is not obvious that we have to accept the link between politics and epistemology as a sound pattern. To examine this aspect more closely will perhaps be useful to a balanced evaluation of Neurath beyond the over-estimation of his legacy, which seems to emerge from the more recent studies on the social-economical thought and the political action of the Vienna Circle's "big locomotive".

III.

Philosophy of science represents certainly the most important item rediscovered in Neurath in the last two decades. By now many aspects of his work stimulating renewed interest have been studied (and shared) within the scientific community. Moreover some of the main theses characterising his historical and holistic approach to scientific theories have contributed to the revision of the traditional image concerning the relationship between logical empiricism and post-empiristic philosophy. Haller has reminded us once more that Neurath seems to be the forerunner of epistemological statements such as those of Feyerabend, Kuhn and Lakatos (cfr. R. Haller, "Neurath et Feyerabend": ONPA, p. 19). On the basis of this judgement, which emerges without substantial variations in other contributions as well (see R. Hegselmann, "Otto Neurath": ONPA, 114 and E. Nemeth, "Les voyageurs égarés de Descartes": ONPA, p. 150), Neurath's sadly forgotten peculiarity within logical empiricism appears now defintively established. This turns out to be clear from Sebestik's comments, particularly when he stresses that more than anything else Neurath's work can bring into question the usual view of "positivism and neo-positivism as bare collection of facts, from which one should automatically derive the corresponding theory" (J. Sebestik, "Vorwort": RPV, 9). Indeed, as everybody knows, if there is any author who seems to be firmly convinced of the "theory-ladenness" so wide-spread within the post-positivistic philosophy (but also in less influential trends of the philosophy of science), it is really Neurath: "this positivist – declares Sebestik – speaks about the relationships between theory and observation like Koyré and Bachelard" ("Raison analytique et pensée globale: Otto Neurath": ON, 44). However, one of the most original and topical themes of Neurath's philosophy of science is his holism and his first steps towards a "naturalization of epistemology", thanks to which he anticipated important aspects of Quine's thought. And no doubts seem to survive on this last point too, as testified by a quite unanimous agreement (see e.g. Th. Mormann, "Neuraths anticartesische Konzeption von Sprache und Wissenschaft": RPV 38, and J. Sebestik, "Raison analytique et pensée globale: Otto Neurath": ON, 67).

While to insist on Neurath's uniqueness within the Vienna Circle has apparently became a commonplace, there is still potential for closer investigations on single aspects or connections to further shed light upon his "naturalized epistemology". Only two issues shall be stressed here. The first concerns Neurath's view of language and his insights into "physicalism". In his "La question du langage universel entre Neurath et Carnap", Mélika Ouelbani gives, for instance, a good survey of the debate between Neurath and Carnap, showing how their conceptions of langauge as well as of protocol-sentences assume a different view as far as the task of philosophy is concerned (for Carnap, unlike Neurath, it is a *logical* one), and a different sight of unified science, too. Moreover, according to Mèlika, Neurath is wrong when he ascribes to protocol sentences as they are conceived by Carnap a definitive character, for such a property should be inconsistent with Carnap's principle of "logical tolerance". All this shows once more that it is important to focus on the specificity of Neurath's conception of language – a point that represents an interesting challenge for the research on Neurath, first of all as regards new insights in the metaphysical background of language, which Neurath discovered very late – according to Sebestik – in spite of his antimetaphysical fury (see "Raison analytique et pensée globale: Otto Neurath": ON, 62). However there is at least another noteworthy question: can it be justified that the leading role attributed to ordinary, everyday language proves Neurath's convergence with Wittgenstein's philosophy of language after the *Tractatus*, even though Neurath was always opposed to "Wittgenstein's metaphysical-mystical tendencies"?[5]

The other issue to be considered here has to do with Neurath's attitude towards American pragmatism. According to Mormann, Neurath shares with Peirce the criticism of Cartesianism, a specific interest in the ways in which belief is testified, and finally a clear refusal of the *tabula rasa* meant as *conditio sine qua non* for analysis of knowledge ("Neuraths anticartesische Konzeption von Sprache und Wissenschaft": RPV, 40-43). Even though Mormann points out that, unlike Peirce, Neurath was concerned with a "finite pragmatism", such an (alleged) affinity raises some doubts. Perhaps it would be better to compare Neurath with William James. First of all because it is plausible that from the very beginning Neurath had a certain acquaintance with James's pragmatism. We should not forget that the German translation of James's famous book on *Pragmatism* was published in 1908 by Wilhelm Jerusalem whose intellectual relationship with Neurath at the beginning of the 20th century deserves to be considered more closely. Secondly, Neurath himself explicitly mentioned James's "pluralistic universe", more specifically within the context of his statements about the famous *Ballungen* or instable concepts of everyday life as well as of his criticism of atomic sentences, which according to Neurath cannot be considered as the definitive basis for science and its language. In Neurath's opinion, indeed, the rationalistic prejudice according to which there is only one "system of the world" must be refused, and this is exactly what James means with his "pluralistic universe", that is, his thesis about a plurality of world views and world interpretations (O. Neurath, "Universal Jargon and Terminology"[6]; see also J. Sebestik, "Raison analytique et pensée globale: Otto Neurath": ON, 66). Finally, it can be claimed that Neurath's conception of truth shows interesting and intrinsic affinities with that one of the great pragmatist James. This point is really a very important one. Insofar as Neurath *doesn't* maintain a coherentistic theory of truth, is it perhaps correct to think, that his own theory seems to converge to a certain degree with a pragmatic theory of truth? If this way of interpreting a central theme of Neurath's thought makes good sense, could James be considered an essential point of reference, once we agree to consider James's theory of truth differently from the banalization of the receveid view?[7]

All these issues constitute in part a new field of research, which has to do with crucial themes (as in the case of the concept of truth) throughout all of Neurath's work. But it seems important to bear in mind that Neurath's epistemological theses are not formulated in a rigorous way, so that it is not easy to draw exhaustive arguments from his often inchoate writings. These limitations are stressed by some essays discussed here. Antonia Soulez, for example, argues that Neurath was the "least argumentative of the Vienna Circle's philosophers" ("Qui ètait Otto Neurath?": ON, 13). Mormann, for his part, remarks that Neurath has never offered a complete elaboration of his ideas regarding local systematization ("Neuraths anticartesische Konzeption von Sprache und Wissenschaft": RPV, 47). And François Schmitz suggests that the image of Neurath as a philosopher of science is more or less "an *a posteriori* reconstruction of Vienna Circle's history" ("La conception neurathienne de la réforme du langage": ONPA, 44). To be sure, these objections are noteworthy. We must always bear them in mind, if we want understand *why* something as the marginalization of Neurath could at all have happened in the past. And this "past" was decisive in influencing the usual view of logical empirism as well as contributing to the paths, followed by the history of analytical philosophy thereafter. It is interesting to note that the most relevant part of Neurath's legacy is not committed to a systematic work of the kind, for instance, of Carnap's *Logical Construction of the World*, but rather to a collective work that he tried to coordinate and direct. This latter role was Neurath's genuine attitude, and the collective work was, obviously, the *Encyclopedia of Unified Science*. But this great project belongs to a new and different stage of Neurath's thought and activity – a stage to be studied more closely, also as regards the emblematic meaning of the encyclopedia: this "boat on the open sea" which need always some repair.

IV.

This is precisely the point to stress. The core of Neurath's conceptions about epistemology and philosophy of science lies in the project and in the partial achievements of his *Encyclopedia of Unified Science*. In this regard Mormann is surely right when he claims that the leitmotiv of Neurath's encyclopedism has to be identified with the possibility to proceed only to *local* "systematization" of the sciences ("Neuraths anticartesische Konzeption von Sprache und Wissenschaft": RPV, 46-48). The very traditional idea of scientific and/or philosophical system must also be contrasted with the revolutionary perspective of the open

encyclopedia. The encyclopedia of unified science is indeed the synthesis of
Neurath's view both about historical development of science and anti-foundationalist patterns in epistemology.

The more specific problem of the ancestors to which Neurath is indebted
seems of minor importance – as far as it may be instructive – compared to the
statement that Neurath conceived for his project (an interesting reference to
Bernard Bolzano's "theory of science" is suggested by Sebestik in his "Raison
analytique et pensée globale": ON, 64). Both of the excellent contributions of
Hans-Joachim Dahms on the relationship between Neurath and the American tradition of pragmatism mainly represented by Charles Morris as well as on the
projected, but never published monographs of the "Encyclopedia" offer a considerable historical-critical overview to all the scholars who are interested in
Neurath's great challenge in the second half of the 1930s (H.-J. Dahms, "Pragmatismus, Enzyklopädieprojekt, Zeichentheorie": N, 25-73, and "Otto Neuraths
'International Encyclopedia of Unified Science' als Torso": RPV, 184-230).
Dahms bases his extensive analyses on Neurath's unpublished correspondence,
items which are extremely valuable reconstructing even the tensions underlying
Neurath's ambition to play the role of a kind of "new D'Alembert". These
tensions led to the exclusion of possible contributors such as Hans Reichenbach
(Neurath – and most of the other logical positivists – didn't agree with Reichenbach's theory of probability) and, above all, to a troubled relationship with
Charles Morris's scientific empiricism. Neurath indeed believed that Morris was
too much indebted, in his *Foundations of the Theory of Signs,* to the "architec-
tonic" spirit of the Kantian philosophy, whereas Morris, for his part, responded
to Neurath stressing the importance not just of the system, but purely of the
"systematization" (N, 58-60). Nevertheless, the fact that, according to Neurath,
Morris's semiotics couldn't aspire to the role of unifying concept of the encyclopedia and that, more generally, within analytic philosophy the "pragmatics" of
Morris has been neglected for a long time or was only marginally considered (as
in the case of Carnap) constitutes the proof – in Dahms's opinion – of the failed
agreement between positivism and pragmatism under the banner of "scientific
empiricism" (N, 63).

But Dahms also deals with another important issue of the "Encyclopedia of
Unified Science". The "enormous" difference (RPV, 185) between Neurath's
original project and its outcome, highly incomplete and belatedly published,
raises the question about the image of logical empiricism and its possible impact
had the original plan of the encyclopedia, enthusiastically sketched, been completed. The reasons why Neurath's project never reached its goal are manifold
and can be connected to the tensions we remembered above between the
"Wiener" and the "Berliner" as well as between European logical empiricists and
American pragmatists. There is yet another prominent aspect: originally the
encyclopedia embraced monographs on the history of sciences and the history of
logic (Enriques and Lukasiewicz were in charge of writing them). Moreover a
monography on the sociology of knowledge had also been planned. In short, the

historical and sociological orientation, usually identified with Thomas Kuhn's book on the structure of scientific revolutions and for a long time conceived as a radical break with the supposedly unhistorical view of science promoted by logical positivism, is in reality the *continuation* of Neurath's unfinished project. So the partial failure of this plan contributed – as Mormann suggests – to establishing the received view of logical empiricism, mainly considered as a rigid set of "dogmas" that has represented the main controversial element of the post-empiristic way of thinking.

From this point of view, Neurath seems to deserve the role of a "philosophical hero" (together with that of a great misunderstood) within the history of logical positivism. Yet this theoretical and historical thesis may be shared only on the condition that, as we as suggested above, the many facets of the story are considered on the basis of a critical, more detached assessment. Philosophical heros have their faults too, and from defeated heros we have nevertheless something to learn. But these stories will be have to wait for another occasion.

NOTES

1. F. Stadler, "Otto Neurath – Encyclopedia and Utopia", in E. Nemeth, F. Stadler (Eds.), *Encyclopedia and Utopia. The Life and Work of Otto Neurath (1882-1945)*, Dordrecht/Boston/London, Kluwer, 1996, p. 3.
2. An English translation appeared in 1989 as *Reflexive Epistemology. The Philosophical Legacy of Otto Neurath* at Kluwer, Dordrecht-Boston.
3. Cfr. R. Haller, H. Rutte, *Preface* to O. Neurath, *Gesammelte philosophische und methodologische Schriften*, Wien, Hölder-Pichler-Tempsky, 1981, vol. I, p. XII.
4. *Reconsidering Logical Positivsm*, Cambridge, Cambridge University Press, 1999, p. 10.
5. See R. Haller, "Otto Neurath – For and Against", in *Encyclopedia and Utopia. The Life and Work of Otto Neurath [1882-1945]*, cit., p. 34.
6. In *Gesammelte philosophische und methodologische Schriften*, cit., vol. II, p. 903.
7. See H. Putnam, "James's theory of truth", in *The Cambridge Companion to William James*, Cambridge, Cambridge University Press, 1997, pp. 166-185.

Università dell'Aquila
Dipartimento di Storia
Via Roma 33
I-67100 L'Aquila
Italy

Home address:
Via Tolmezzo 3
I-20132 Milano

REVIEWS

MICHAEL FRIEDMAN, *Reconsidering Logical Positivism.* Cambridge: Cambridge University Press, 1999.

This volume is a collection of essays Michael Friedman published over a period of fifteen years together with two postscripts in which he explains some important changes in his views in the interim. The title of the book is somewhat misleading, since it does not deal with the whole spectrum of Logical Positivism, not even with the philosophy of the Vienna Circle in its full breadth. Rather, the protagonist of *Reconsidering Logical Positivism* (RLP henceforth) is Carnap, and, to a lesser extent, Schlick and Reichenbach. The radical empiricists's faction of the Vienna Circle, in particular, Neurath, Frank, and Hahn is not treated at all. On the other hand, Friedman takes into account quite a few figures that usually do not appear in the more standard treatises on Logical Positivism, e.g. neo-Kantians such as Rickert, Natorp and Cassirer, and conventionalists such as Poincaré, Weyl and others.

Without further dodging the issue, I'd like to state that RLP is mandatory reading for anybody who is interested in history and interpretation of logical positivism, even if one may not agree with all of Friedman's theses.

According to Friedman the central innovation of Logical Positivism was not a new version of empiricism but rather a new conception of a priori knowledge and its role in empirical knowledge. If this is true, (neo-) Kantianism of some brand becomes a necessary point of reference. Consequently, Friedman is concentrating his interpretative efforts on the German members of the Vienna Circle. Specifically Austrian influences such as Bolzano or Brentano are only mentioned in passing. Hence, exaggerating a little bit, the topic of RLP may be characterized as a sort of "German Neo-Kantian Logical Positivism".

The book consists of three parts: (1) *Geometry, Relativity, and Convention*, (2) *Der Logische Aufbau der Welt*, and (3) *Logico-Mathematical Truth*. In this review, I will follow this order.

A central topic of twentieth century philosophy of science was the invention (or discovery) of non-Euclidean geometries, and, even more importantly, the advent of Einstein's relativity theories. Thus, a good deal of the Logical Empiricists's work was dedicated to the philosophical digestion of Einstein's scientific achievements. Schlick played an important role in this process, already before the Vienna Circle was founded. The first chapter of Part I is a critical commentary of Schlick's *Philosophical Papers*. Friedman intends to give a sort of succinct résumé of Schlick's intellectual development in the first two decades of the century pointing at the various strains and stresses his work exhibits. At the beginning, Schlick may be characterized as a "critical or structuralist realist" understanding knowledge as a system of interconnected judgments whose concepts get their meaning from their mutual relationships within this system. As is well-known, this "idealist" conception may be traced back to Hilbert's account of knowledge of mathematical concepts based on implicit definitions. Since, according to Schlick, direct confrontation with the notorious "given" cannot yield knowledge we are left with the crucial problem of explaining how such a "structural" account of knowledge may lead to knowledge of reality. For this purpose, Schlick relies on his method of "coincidences" inspired by Einstein's application of non-Euclidean geometries to his general theory. Hence, for Schlick, Einstein's general theory of relativity not only destroys the Kantian bridge between thought and reality, namely pure intuition, it also shows us positively, via the method of coincidences, how to restore such a bridge in a radically new form.

The second chapter "Carnap and Weyl on the Foundations of Geometry and Relativity" deals with Carnap's first published work, his dissertation *(Der Raum)* written 1921 in Jena under the direction of the neo-Kantian Bauch. At that time, the philosophy of geometry was in a rather chaotic situation: mathematicians, physicists, and philosophers all intended to contribute to a rather confused debate on the "Raumproblem". As would become typical for him, Carnap attempted to bring some order into the discussion. He proposed to carefully distinguish between three distinct types of space: *formal, intuitive,* and *physical* space. The epistemic enterprises corresponding to these types of space are mathematics, philosophy, and physics, respectively. Remarkably, at that time Carnap considered Husserl's "Wesensschau" as "the" philosophical method of dealing with the genuine philosophical problems of space and time.

The last two chapters of Part I, "Geometry, Convention and the Relativized Apriori: Reichenbach, Schlick, and Carnap" and "Poincaré's Conventionalism and the Logical Positivists", contain a lucid discussion of the entangled topics of holism, conventionalism, and general relativity theory. Friedman comes to the conclusion that Poincaré's conventionalism is incompatible with general relativity theory and that Reichenbach, Schlick, and Carnap in some sense misunderstood it. Poincaré favoured a hierarchical conception of science that pretty well fits the basic scheme of Kantian philosophy: one science presupposes the other and every succeeding science presupposes all preceding sciences. Even though Poincaré is considered as one of the founding fathers of conventionalism, there is not much space for conventional choices in Poincare's science. For him, there are only three possible geometries, to wit, Lobashevsky's, Euclid's, and spherical geometry. In particular, he does not take into account Riemannian geometry with variable curvature, i.e., the geometry of general relativity theory. This means, that the role of Riemannian geometry in the general theory does not fit really into a conventionalism à la Poincaré.

The topic of Part Two is Carnap's first opus magnum *Der Logische Aufbau der Welt.* Its two chapters, "Carnap's Aufbau Reconsidered" and "Epistemology in the Aufbau", have shaped many of the new revisionist interpretations of the *Aufbau* put forward in the last fifteen years. Moreover, this chapter contains a more recent postscript, in which the author indicates important changes in his views concerning the more subtle relations between the different neo-Kantian schools and Carnap's developing philosophical positions.

In the decades since its publication the *Aufbau* has been subjected to many a criticism. Maybe the most incisive ones have been Goodman's and Quine's which pursue quite different lines of attacks. Goodman attacks the *Aufbau* account at its base, so to speak, pointing at some allegedly fatal flaws of the constitutional method of quasi-analysis. Quine's criticism is concerned with the intermediate levels of the constitutional process claiming that Carnap's constitution of the physical world hopelessly founders. Meanwhile some authors have argued that these objections are not as devastating as one used to think.[1] Be this as it may, Friedman launches a new objection against the feasibility of the *Aufbau* program which is quite independent of that of Goodman and Quine. It may be characterized as an argument from the top and is concerned with the problem of the so-called "relation of foundation" (*Fundierungsrelation*). Thus, even if Goodman's and Quine's objections might be overcome somehow and the whole edifice of scientific knowledge had been successfully reconstructed in terms of Carnap's initial language of elementary experiences and the relation of similarity, given the impossibility of characterizing the founding relation Carnap's endeavor is doom to fail, or so Friedman argues.

According to him, the *Aufbau* program aims at a complete formalization of scientific knowledge motivated by a conception of scientific objectivity that seeks to disengage objective meaning from "subjective" ostension. For the sake of the argument we may grant that this has been achieved by building up the constitutional system on the base of a single relation of "recollection of similarity" as a founding relation. Now, as Friedman points out, the question arises as to what is the nature of that "founding relations": do they belong to the class of logical relations, are they empirical relations or yet another kind? Carnap argues, not very convincingly, that the class of founded relations should be considered a primitive notion of logic, and, at the same time, that the founded relations are just the "experienceable, natural" relations. Friedman objects that thereby ostension enters the stage again by the backdoor: "But what can the experienceable, natural relations be except precisely those relations somehow available for ostention?" Against Friedman, one may argue that "experienceable" for Carnap here meant something like "scientifically experiencable". Thereby, the objectivity of the constitutional system and the naturalness of its founding relation depends on its capacity of reconstructing the system of scientific knowledge. In other words, a Carnapian philosopher could subscribe to a sort of Quinean naturalized structuralism according to which the essential point is to "save the structure" and beyond this there is nothing left to do.

Part 3 of RLP concentrates on Carnap's *Logical Syntax of Language*. More precisely, Friedman reads this work as a (transformed and deepened) continuation of the *Aufbau*-program in which Carnap proposes a systematic transformation of the discipline of philosophy as a whole. Instead of being entangled in obscure and fruitless ontological disputes about the "reality" or the "true nature" of some contested class of entities (such as numbers and other mathematical objects) we should engage in precise disputes about the logico-linguistic form in which the language, or maybe better, the languages of science, are to be cast. We should recognize that there is, after all, no genuine ontological import – no implications as to the "objects" and "facts" in the world – in the philosophical questions with which we have hitherto been struggling in vain. For, when we attain Carnapian philosophical self-consciousness, we see that we have actually been concerned with the much more fruitful, albeit purely pragmatic, question of language planning. In this way, Carnap's attempt to transform traditional philosophy into the new enterprise of language planning is intended not only to bring peace and progress to the discipline, but also, as Friedman rightly emphasizes, peace and progress for humankind in general. It is here, where we meet one of the few points where Friedman touches on the broader, not merely philosophical issues of Carnap's thought. As he points out, this program unites the Carnap of the *Aufbau* and the Carnap of *Logical Syntax*.

Although quite sympathetic with it, in "Tolerance and Analyticity in Carnap's Philosophy of Mathematics" Friedman comes to the rather pessimistic assessment that Carnap's proposal has not found a very sympathetic reception among his fellow philosophers. According to him, Gödel, Tarski, and Quine have compellingly criticized essential parts of Carnap's approach: Gödel severely criticized Carnap's philosophy of mathematics since it is based on a strong metalanguage whose existence appeared doubtful to him due to his incompleteness results; Tarski and Quine, for different reasons, came to the conclusion that the analytic/synthetic distinction is untenable. According to Friedman, this led to a form of naturalized pragmatism that abandoned essential distinctions of Carnap's more "Fregean" philosophical enterprise. Maybe Friedman paints a too gloomy picture of the prospects for a Carnapian account here. Recently, it does not seem as obvious as it used to be that an "enlightened" pragmatic empiricism of Quinean provenance is unquestionably superior to Carnap's now classical Logical Empiricism. In some sense, Carnap may even be said to be more pragmatic than Quine. Of course, such a claim needs an elaboration of what is to be understood by "pragmatic" here. This problem is ignored by Friedman completely. Without further explanation, he directly opposes "purely pragmatic" considerations with "contentful" assertions. Thereby, he subscribes to a rather bland version of pragmatism that not necessarily is the best to make sense of the "pragmatic" issues of Carnap's philosophy.

In sum, RLP is a pleasure to read. The author does not hide the sometimes essential changes some of his views have undergone. Hence, RLP exhibits one of the major reinterpretations of the history of analytic philosophy in the making, so to speak. Thus, the reader should make sure not to skip the postscripts that show some of the changes of Friedman's own thinking. In sum, RLP marks a significant step forward in our understanding of Carnap's Logical Positivism.

NOTES

1. Cf. for example J. Proust, 1989, *Questions of Form, Logic and the Analytic Proposition from Kant to Carnap*, Minneapolis, University of Minnesota Press, T. Mormann, 1994, "A Representational Reconstruction of Carnap's Quasianalysis", PSA 1994, vol.1, or A. W. Richardson, 1998, *Carnap's Construction of the World. The Aufbau and the Emergence of Logical Empiricism*, Cambridge, Cambridge University Press.

Thomas Mormann

WESLEY C. SALMON, *Causality and Explanation.* New York - Oxford: Oxford University Press, 1998.

This book is a collection of Salmon's major essays on causality and explanation, spanning over a period of about twenty-five years. As the author says in the Introduction, the collection "is offered in the hope that it will provide new insight into causality and its role in understanding our world" (p. 10). And indeed it does. Its twenty-six papers, seven of which previously unpublished, testify Salmon's relentless efforts to revive the notion of causality and to establish its strong connections with explanation in a probabilistic framework.

The volume is divided into five parts. The first four develop Salmon's perspective on explanation and causality, while the fifth contains six papers, three of which published for the first time, devoted to "Applications to other disciplines: archaeology and anthropology, astrophysics and cosmology, physics". The author has enriched the book with careful editing which includes, in addition to a Preface and a general Introduction, forewords to its five Parts and several additions, like cross-references helping to connect the papers, postscripts and various insertions in square brackets throughout the text, providing comments and references to recent literature.

Though connected for centuries in the philosophical literature, causality and explanation are disjoined in most of the work done by philosophers of science. Hempel's influential account of explanation does not link it to causality, being based on the idea that an event is explained when it can be shown that its occurrence (in given circumstances) was to be expected in the light of a set of scientific laws. Causal explanation becomes one kind of explanation among others, equally acceptable. This attitude is a byproduct of the crisis undergone by the notion of causality after the new physics cast doubt on determinism at the beginning of the 20th century. Against this tendency, Salmon's work is aimed at restoring the "obvious and basic relationship" (Preface, p. ix) between causality and explanation.

The resurgence of the notion of causality from the above-mentioned crisis is linked to its probabilistic interpretation. The origin of probabilistic causality is to be traced back to the work of Salmon's teacher: Hans Reichenbach (*The Direction of Time*, 1956), whose ideas inspired him to a great extent. A probabilistic approach to causality was taken up by I.J. Good ("A Causal Calculus", 1961-62), P. Suppes (*A Probabilistic Theory of Causality*, 1970) and many others, giving rise to an ongoing debate. In the literature on the topic causality and explanation are still disjoined, or at least not strictly connected. The reason for this is lies in a distinction between two types of causality, which is imposed by a probabilistic treatment of the subject. In a probabilistic framework it is necessary to keep separate causal talk referring to population variables (or "property causality") and causality between single events (called "token" or "aleatory" causality). Between these two kinds of causality there is a tension, reflecting that between prediction and explanation in probabilistic contexts. Property causality has predictive power, but predictability can also be obtained from spurious correlations; on the other hand, single events can be unpredictable, but can receive a causal explanation after they occur. The information regarding statistical relationships often does not bear on the explanation of single events. Much of the debate on probabilistic causality revolves around this distinction and the possibility of matching property and token causality. In view of these problems, Good treats the notion of "explicativity" independently from that of causality, and Suppes develops a theory of causality which is not intrinsically linked with that of explanation. An analogous separation can be found in various areas of research, like econometrics, where there is a vast literature on causal models which serve practical purposes of manipulability (directed to political economy) but are not always seen as explanatory.

On the contrary, Salmon wants to work out a unified theory of property and token causality, functional to his main purpose: reviving mechanical explanation in a probabilistic framework. His attempts in this direction date back to his essay "Statistical Explanation" (1970) where he put forward a model of explanation based on the notion of statistical relevance (thereupon called S-R model), meant as an alternative to Hempel's approach. This was followed by a number of articles – many of which are reprinted in *Causality and Explanation* – leading to the volume *Scientific Explanation and the Causal Structure of the World* (1984) containing the most complete formulation of Salmon's idea that explaining means exhibiting the causal mechanisms responsible for the occurrence of phenomena. The debate provoked by this book made Salmon revise some aspects of his notion of causation, as reflected by other papers included in the collection, especially "Causality without Counterfactuals" (1994).

Salmon makes a point of leaving the door open to indeterminism when dealing with explanation and causality. In other words, he holds the view that an adequate theory of explanation should not be committed to either determinism or indeterminism, but be compatible with both. Together with his conviction that genuine explanation is causal, the view in question involves the adoption of a probabilistic notion of causality.

His theory of explanation is articulated on two levels. At the first level we find his S-R model, according to which an event is explained by showing which factors are statistically relevant to its occurrence. The event to be explained is associated with a reference class, which is gradually restricted by "screening off" the irrelevant factors, until homogeneity of the reference class, in the sense of including all and only the relevant factors, is obtained. The event is explained when its place within such a network of statistical relations is specified. Explanation of this kind conveys complete information on the statistical relationships holding among the variables of the population examined.

The second level of explanation is causal in character, and is based on the notion of "causal process", defined as a spatio-temporal continuous entity having "the capacity ... to transmit information, structure and causal influence" (p. 253). Causal processes are responsible for causal propagation, and provide the links between causes and effects. They can be thought of as forming a net, whose knots represent interactions between processes. When processes are modified by such interactions there is causal production, and such modifications are propagated by the process. Causal explanation obtains when phenomena are located at some point within the net of causal processes, thereby identifying the mechanism responsible for their occurrence.

Unlike statistical relevance relationships, which are defined in terms of probability values alone, causal processes and interactions make reference to physical properties. Causal explanation is "ontic" in kind, and informs on how the mechanisms responsible for the occurrence of phenomena work.

Salmon's theory captures our intuitions about causality, explanation and scientific understanding, but it is not free from difficulties. The most problematic feature of the S-R model is connected to the homogeneity of the reference class, a requirement which is much too strong to be satisfied in more than a few cases. Apart from explanations making reference to strong theories like classical and statistical mechanics, that requirement indeed appears far from satisfiable, and even desirable. This has been pointed out among others by J. Woodward, who observed that in fields like the social sciences a reference class is often taken to be explanatory even though finer partitions of it would be possible. Similar considerations apply to causal processes, defined as spatio-temporal continuous entities. For one thing, this notion does not find applicability in vast areas of science (in the first place quantum mechanics, but also the social sciences and psychology); in addition, Salmon's definition of causal processes has raised various objections and is still under debate.

A further issue regards the linkage between the two levels of explanation. Around 1990, when the article "Causal Propensities: Statistical Causality versus Aleatory Causality" (n. 13 in the collection) was published, Salmon thought that only aleatory causality could give "an adequate understanding of causality"

(p. 207). This conviction has been challenged by C. Hitchcock on the claim that, in the absence of information on statistical relationships, Salmon's aleatory causality – namely information on processes and interactions – does not contain any clue as to what properties should be taken as explanatory. On this basis he rejects Salmon's distinction between two levels of explanation and maintains that explanatory relevance should rest on the counterfactual information given by statistical correlations. Hitchcock's remarks have persuaded Salmon to revise his position and to reaffirm a strict connection between the two levels of discourse. Accordingly, causal analysis in terms of processes and interactions is seen as a model that needs to be implemented with information on statistical relevance relations in order to allow recognition of explanatory properties. This is the conclusion attained in "Causality and Explanation: A Reply to Two Critiques" (1997) to which the reader is referred (p. 260 of the collection). The ongoing debate on Salmon's conception of causality testifies to its originality.

For Salmon causal explanation does not exhaust the field of scientific explanation. In addition to the latter, he credits explanation based on the idea of theoretical unification, developed in somewhat different ways by Michael Friedman and Philip Kitcher. Basically, this approach maintains that scientific explanation consists in unifying within the same theoretical framework phenomena which are apparently disparate. It is peculiarly directed towards the explanation of scientific laws by interconnecting them in a wider conceptual structure. These approaches are regarded by Salmon as "mutually compatible and complementary" (p. 10) because they convey a different sort of understanding of phenomena.

As already said, the last part of the collection examines possible applications of Salmon's ideas, including three papers on archaeology, a field largely neglected by philosophers of science. The book closes with the article "Dreams of a Famous Physicist", bearing the subtitle "An Apology for Philosophy of Science", where Salmon argues against Weinberg's attack on philosophy of science, reaffirming at the same time the importance of working out a coherent and powerful conception of scientific explanation and causality.

This is a valuable book, a most welcome addition to the literature on one of the key problems of philosophy of science, written by one of its most outstanding representatives.

REFERENCES

Salmon, W.C. (1970), "Statistical Explanation", in *Nature and Function of Scientific Theories*, ed. by R. Colodny, Pittsburgh: University of Pittsburgh Press, pp. 173-231, also in W.C. Salmon, R.C. Jeffrey and J.G. Greeno, *Statistical Explanation and Statistical Relevance*, Pittsburgh: University of Pittsburgh Press, 1971, pp. 29-87.
Salmon W.C. (1984), *Scientific Explanation and the Causal Structure of the World*, Princeton: Princeton University Press.
Salmon W.C. (1997), "Causality and Explanation: A Reply to Two Critiques", *Philosophy of Science* **64**, pp. 461-477.

Maria Carla Galavotti

KARIN GERNER, *Hans Reichenbach – sein Leben und Wirken.* Osnabrück: Phoebe-Autorenpress 1997.

Hans Reichenbach (1891-1953), the undisputed leader of the Berlin "Gesellschaft für Wissenschaftliche Philosophie" between 1929 and 1933, is widely acknowledged as a pioneer of analytical epistemology and as one of the first philosophers in the German academic context to take the findings of modern physical and mathematical sciences into full account. He can be considered together with Moritz Schlick as one of the main proponents among philosophers of Einstein's Theory of Relativity and perhaps as its most influential populariser. Reichenbach's work on the concept of probability in scientific – especially physical – induction is still read with profit by modern philosophers, his "Elements of Symbolic Logic" (1947) have influenced linguistic work on the philosophy of grammar. In recent years, due to the contributions of Andreas Kamlah in the first place, Reichenbach's work has found attention in many versatile and detailed articles mostly in German, alongside with an annotated edition of his *Collected Works* in nine volumes (*Gesammelte Werke*, Braunschweig/Wiesbaden 1977 ff.). What is still missing is a comprehensive, scholarly biography of Hans Reichenbach that puts his work in the twofold context of time and of the contributions of his contemporaries. Such a project seems particularly welcome and even necessary in the case of those philosophers of science, whose full impact and lasting importance cannot be evaluated other than against the biographical background of their interactions with scientists on the one hand and the public on the other. Writing a fully successful scientific biography of a personality like Reichenbach with his wealth of connections with famous contemporaries such as Einstein, Planck, Bertolt Brecht (along with mathematician Ludwig Bieberbach consistently misspelled by the author), and Max Horkheimer requires, of course, unusual erudition and previous research into the many directions of Reichenbach's work. Thus the following remarks should not merely be read as a critique of Gerner's book but rather as a reflection on the possible value and standing of a scientific biography of a philosopher of science and as a slight admonishment against its underestimation by the community of scholars. The reviewer is convinced that for such a project to become a full success, is required a mature and experienced scholar who is at least in the middle of his career.

The booklet under review, however, was written by Kamlah's student Karin Gerner as a philosophical dissertation, and consequently and quite understandably lacks that degree of maturity. This reviewer has not seen the dissertation itself which may contain further interesting material. In fact, Gerner decided to leave aside from her dissertation a "systematic part on the development of Reichenbach's theory of probability and induction" in order to reach a "broader readership" (p. 3). In view of the remote place of publication and the mixed

popular-scholarly approach of the author, who uses unexplained scientific concepts and repeatedly introduces personalities (such as Heinrich Scholz and Hans
Rademacher) that are not further characterized, it may be questioned whether
that goal of a wide circulation can be attained. Moreover the principal question
arises as to whether a populariser (Reichenbach) can be made popular without
previously recurring in great detail to the contemporary situation of reception of
a science (in this case the theory of relativity) in the 1920s which is far from
being general public knowledge even today. It seems still more difficult to pass
judgment – in a popular manner – on the scientific merits and the epistemic
status of Reichenbach's work, in particular of his "most important book" (Kamlah 1993, p. 251) *Axiomatik der relativistischen Raum-Zeit-Lehre* (1924) and of
his concept of probability, which Reichenbach himself, according to the author
(p. 202), estimated higher than his work in the philosophy of relativity. These
accomplishments need to be analysed both in relation to the work of contemporaries as well as from the standpoint of modern philosophy of science. The
author succeeds in showing – on the basis of unpublished letters – that Reichenbach acted – together with physicists such as Erwin Schrödinger – as a helpful
critic of Einstein's "unified field theory" and was taken seriously by Einstein (p.
79). However, as to the positive scientific (as opposed to the sociological, popu-
larizing) value of Reichenbach's work, Einstein was rather unenthusiastic, as
occasional disputes between the two would show (p. 79ff.). It is obviously both
lack of space and the deliberate attitude to skip a systematic discussion that
prevents the author from going into these points of disaccord more thoroughly
than e.g. on p. 51 where relevant letters of Einstein's to Moritz Schlick are
mentioned but not discussed. As to Reichenbach's concept of probability Gerner
gives an interesting excursus on induction (pp. 165-172), which shows after-
effects of Reichenbach's concept in modern epistemological discussions (Christian Piller). (However, the excursus seems strangely misplaced and should have
been inserted earlier in the booklet after p. 114.) It is exactly Reichenbach's
notion of probability where a more detailed discussion of the opposing stand-
points of mathematicians (Mises, Kolmogorov) and epistemologists (Reichenbach, Popper, Carnap) would have revealed more clearly the aims and functions
of Reichenbach's work. Gerner muses about the early publication date of Reichenbach's article in the *Mathematische Zeitschrift* (1932) as compared to A.
Kolmogorov's path-breaking booklet of 1933 on the axiomatization of probability and maintains an "undeserved neglect" (p. 200) of Reichenbach's work. She
does not mention, however, the technical (mathematical) importance of Kolmogorov's work for the introduction of stochastic processes, which even
surpassed in influence Richard von Mises's work of 1919. The latter would have
been the natural point of reference for a discussion of Reichenbach's concept of
probability, because von Mises had close connections both to the "Berliner
Gesellschaft" and the "Wiener Kreis". Mises shared with Reichenbach several
philosophical convictions, especially a repudiation of an occasional dogmatic
over-zeal, on the part of members of the Wiener Kreis, for the logistic under-

pinning of epistemology. Gerner alludes to Mises's axiomatization of probability (p. 116) but does not clearly explain his notion of "irregularity" of a "collective", nor why Reichenbach's theory could do without it. A very interesting discussion on probability during the Prague meeting of empiricist philosophers in 1929 is mentioned but not the fact that it was in Prague that Mises and Reichenbach presented their different attitudes to the application problem, to mathematical idealization and approximation most clearly. Instead, the author comments mostly on Carnap's position towards Reichenbach's probability concept (p. 99ff.) and comes to the conclusion that Reichenbach's position was "generally accepted" (p. 116, by whom?). This leaves the reader somewhat confused with respect to the later reservations of the Wiener Kreis against Reichenbach's logic of probability which are reported at p. 155ff.

Thus the main purpose and value of the booklet in its present state seems to be to put a hold of parts of the valuable and multifaceted biographical material contained in the Reichenbach Archives in Pittsburgh which has so far not been systematically used in the more specific publications on Reichenbach. Interesting are Gerner's discussions of the reasons for Reichenbach's refusal to accept an appointment in Prague in 1929 (p. 105ff.), the description of the foundation of the journal "Erkenntnis" in 1930 (pp. 57/58, 89ff.) after several failed attempts, and the report on Reichenbach's broadcasting lectures (p. 111ff.), both as source for additional income and for his popularization efforts. Insightful is Gerner's discussion of the reasons for the relative oblivion of the Berliner Gesellschaft as compared to the Wiener Kreis, which she explains against the background of emigration and the relative dominance of one figure, Reichenbach, in Berlin. The discussion of Reichenbach's exile in Turkey (1933-1938), for the most part, follows printed sources (Widmann, Neumark), the ensuing description of the fate of "Erkenntnis" in Nazi-Germany uses interesting letters written by Felix Meiner which show the courage of the publisher of that journal. The section on Reichenbach's years in the U.S., which the author declares as somewhat outside her main focus, sheds light on Reichenbach's relation with the Frankfurter Schule and on the help he extended to them to settle in California. Unfortunately reference to the literature is not always reliable. Dahms' (1994) book on the "Positivismusstreit" (1994) is not quoted, several articles from Danneberg/Kamlah/Schäfer (1994) are only cited with the title, in one case (K. Volkert, p. 76) even leaving out the name of the respective author. Above all it is only rarely clear which of the many letters which are being quoted have already been used in the literature elsewhere.

Several mistakes have to be mentioned mostly in general historical context (confusion of "Akademikerhilfe" with "Notgemeinschaft", p. 46, misrepresentation of "Notverordnungen", p. 108, unawareness of identity of W. Thomson and Kelvin, p. 193), several judgements in this realm show unfamiliarity and sometimes the naiveté with respect to universal history (Reichenbach allegedly "not bearable for the state [sic]" (p. 65) and R. allegedly attacked because he was "against the First World War" (p. 63)).

Summing up, Gerner's short biography of Hans Reichenbach can be considered as a very first step into the direction of a comprehensive scientific biography of Hans Reichenbach which remains a desideratum for historiographical research.

References

Kamlah, A. (1993): "Hans Reichenbach – Leben, Werk und Wirkung"; in: R. Haller and F. Stadler (eds.): *Wien, Berlin, Prag. Der Aufstieg der wissenschaftlichen Philosophie*; Wien: hpt, pp. 238-283.

Dahms, H.-J. (1994): *Positivismusstreit. Die Auseinandersetzung der Frankfurter Schule mit dem logischen Positivismus, dem amerikanischen Pragmatismus und dem kritischen Rationalismus*; Frankfurt am Main: Suhrkamp.

Danneberg, L., A. Kamlah, and L. Schäfer (eds.) (1994): *Hans Reichenbach und die Berliner Gruppe*; Braunschweig/Wiesbaden: Vieweg.

Reinhard Siegmund-Schultze

BARRY SMITH, HERBERT HOCHBERG (Eds.), *Austrian Realism: From Aristotelian Roots to the Vienna Circle* (*Monist,* vol. 83, no. 1, 2000).

The title of the volume (whose general topic originally was supposed to be "The Austrian Tradition: From Bolzano to the Vienna Circle") might be read in at least two ways: (a) Realism *in* Austria; or (b) *Austrian* realism as a particular brand of realism, as opposed to, say, American realism. (Here and in what follows I use "realism" in a very general sense, namely for the view that (i) there is a mind-independent reality and (ii) that we can gain some knowledge about this reality.) The latter reading implies the suggestion that there is such a thing as a genuine *Austrian* realism. This might be interpreted as a variant of the so-called "Neurath-Haller thesis". Roughly, this thesis says "that there exists a separate and internally coherent tradition of *Austrian* philosophy within the German-language philosophy as a whole".[1]

Indeed, realism in the sense explicated above may count, by and large, as a common feature of what is called "Austrian philosophy" (i.e., the tradition of Bolzano, Brentano, Meinong and the Vienna Circle) – although it is rather questionable whether Rudolf Carnap does fit in this picture. (See below, for Johanna Seibt's contribution to the present volume.)

But even leaving aside the case of Carnap, a reader who expects either an introduction into a specific Austrian realism or a survey of contemporary approaches to Austrian philosophy, will be somewhat disappointed. Rather, the volume provides a number of worthwhile but quite disparate essays, some of which are only indirectly or very marginally concerned with the Austrian tradition and some of which are in no way concerned with realism; if at all, they are unified by a sort of Wittgensteinian family resemblance.

Mario Mignucci ("Parts, Quantification and Aristotelian Predication", 3-21) advances an interpretation of Aristotle, according to which predicative sentences are to be understood in terms of the part-whole relation. According to this interpretation, "A is B", e. g., "Socrates is [a] man", is to be understood as "A is part of B" (e. g., "Socrates is part of man"). This becomes comprehensible against the background of the following two claims (both of which are supposed to be held by Aristotle): (a) There holds a particular kind of part-whole relation between a species and its genus (e. g., "Man is part of animal"). (b) The relation between an individual and a species is analogous to the relation between a species and its genus. The latter claim, in its turn, reflects the lack of a fundamental distinction between singular and general terms in Aristotle.

Unfortunately, the author makes no effort to elaborate possible impacts of this alleged Aristotelian interpretation of predications on Austrian philosophy. There might be an interesting connection to Brentano's theory of substances and accidents:[2] According to Brentano, *Socrates* is a substance, but *the wise Socrates* is an accident; and Socrates is part of the wise Socrates, although Socrates's wisdom is not a detachable part of the wise Socrates.

Deborah Brown ("Immanence and Individuation: Brentano and the Scholastics on Knowledge of Singulars", 22-46) sets out "to explain the connection between the theory of immanence in its medieval and Brentanian forms and the problem of individuation." (23) According to Brentano's theory of intentionality, every mental act has an immanent, mental object which serves as medium between the objects we are directed at and the mind. This theory has its roots in Thomas Aquinas. According to Aquinas, the immanent objects are the results of a process of abstraction. What we have in mind when we have knowledge of an object is not the object itself, but the form of the object. But the form is general. This raises the question of how knowledge of individuals is possible.

Jan Berg's article ("From Bolzano's Point of View", 47-67) is a comprehensive description of Bolzano's work in metaphysics, logic, theory of science, theory of probability, and the foundations of mathematics, geometry and physics. It is argued convincingly that Bolzano has anticipated important achievements of later developments, in particular in the field of mathematics and the foundations of geometry and physics.

Peter Simons ("The Four Phases of Philosophy: Brentano's Theory and Austria's History", 68-88) applies Brentano's four-phase model of the development of philosophy to Austrian philosophy. According to Brentano, philosophy repeatedly goes through a four-phase cycle from high-level theoretical philosophy over practical-popular thinking and scepticism to a period of intellectual decay which is characterized by dogmatism and mysticism. Simons divides, following Brentano's model, the Austrian philosophy into four phases connected with Bolzano (phase I), Brentano and his disciples (phase II) and the Vienna Circle (phase III); phase IV is supposed to last from 1938 up to now. Simons argues that to a considerable extent it is Austrian politics that is responsible for the decline of Austrian philosophy from the beginning of the 19th to the end of the 20th century.

As Simons himself notes, Brentano's model is oversimplified and superficial and has hardly any explanatory value. It is at best an overdone wrapping for the prosaic truth that "philosophical quality has its historical ups and downs". (75)

Simons pictures the development of Austrian philosophy from 1945 onwards as follows: Due to the fact that until the 1970s the Education Ministry was in the hands of the conservatives, Austrian philosophy was until then "a purely provincial affair". Only in the seventies, when Hertha Firnberg became Minister of Science, were things getting better: This was the period of "the more outward-looking professors" Röd, Haller, and Weingartner. (81-84) If (as has actually happened after the completion of Simons's paper) Austrian politics turns to the right, "academics might [...] come under new pressure". (85)

It is undoubtedly true (and, at least partly, for obvious reasons) that philosophy in Austria after 1938 was (and is) far from reaching the heights of former epochs. But it is not true, as Simons suggests, that the return to the analytical roots happened only in the seventies. As a matter of fact, Haller was appointed in 1967, Rudolf Freundlich, the oldest of the post-war analytical philosophers in Austria, in 1966, – both of them by conservative ministers. In the fifties, the analytic Arthur Pap taught in Vienna, and Weingartner organized colloquia on analytical themes in the mid-sixties.[3] Simons also strongly overestimates the influence of politics on philosophy in today's Austria. Nowadays, the general financial situation as well as struggles within the universities – surely not peculiarly Austrian problems – play a much bigger role than the political color of the present Minister of Science.

Erwin Tegtmeier ("Meinong's Complexes", 89-100) deals with "the problem of complexity": On the one hand, a complex entity (in brief, a complex) is something that consists of several parts or constituents; on the other hand, a complex is a unity of its own. What brings this unity about? What "bundles together" the constituents such that they form a complex, not just a collection of simple entities? Tegtmeier approaches the problem by a critical consideration of Meinong's theory of complexes. Moreover, he discusses Gustav Bergmann's and Reinhardt Grossmann's views on Meinong's theory. Tegtmeier stresses the role of *facts* for an explication of complexity: Complexity cannot be explained by the assumption of a "connector", because "[i]t is not the connector that creates complexity, but the fact of its holding between the constituents." (100)

Ingvar Johansson ("Determinables as Universals", 101-121) starts with an argument for "immanent realism" in general, i.e., for the view that there are universals *in rebus*, independent of language and mind, but also independent of Platonic entities outside of space and time. Some immanent realists (e. g., D. M. Armstrong) have accepted the existence of *determinate* universals (e. g., a particular shade of red, for instance *scarlet*) but denied the existence of *determinable* universals (e. g., *colored*). Johansson advocates the thesis that there are determinable universals as well as determinate ones.

Among the arguments Johansson offers for this thesis the "gap argument" – an argument from best explanation – is the most convincing: Johannsson observes that there is a fundamental *lack of similarity* (not to be mixed up with *dissimilarity*) between, say, a determinate color and a determinate shape; this "gap" between colors and shapes can be explained best by the assumption "that all color-determinates have something in common, namely the ontological determinable of color. All the shape-determinates have something else in common, namely the ontological determinable of shape; and similarly for volumes. The determinables, in turn, are simply different." (108)

Less convincing, however, is the "argument from patterns" in which Johansson tries to infer "there are color- and shape-determinables" from "we cannot think of a color determinate that is not fused with a shape determinate."

Per Lindström ("Quasi-Realism in Mathematics", 122-149) deals with the question of what the "subject matter" of mathematics is. Lindström rejects the usual answer that mathematics is about "abstract objects, numbers, functions, sets, etc., in other words, eternal and unchangeable objects, existing independently of us" (123), because, he argues, we cannot have any epistemic "contact" to abstract objects. Thus, he suggests considering as the subject matter of first order arithmetic not an abstract structure (i. e., the structure of natural numbers) but an *imagined* structure (of numbers). (Later on, he tries to extend this approach to second and higher order arithmetic and set theory.)

Lindström is aware of the fact that one might ask what the nature and ontological status of "imagined structures" is (he does not consider them as mental objects!), but he simply dismisses these questions as follows: "These are (pseudo?) problems I don't pretend to be able to deal with. But I don't feel that, for my present purposes, it is at all important to deal with them." (126)

As an ontologist, I feel that Lindström has raised an ontological question and stated an ontological hypothesis. It is odd that he considers questions that aim at a clarification of his own thesis as irrelevant for his purpose.

The background of David Armstrong's paper "Difficult Cases in the Theory of Truthmaking" (150-160) is the "view that every truth has a truthmaker. The truthmaker for a particular truth is that object or entity in the world in virtue of which that truth is true." (150) Negative truths, general (universally quantified) truths, and modal truths pose particular problems for a theory of truthmakers. Armstrong's task in this paper is to resolve these problems. Armstrong holds it to be necessary to assume *general facts* as truthmakers for general truths, but he thinks that he can do without negative facts. He believes (contra Wittgenstein) that truthmakers for modal truths are required; but he wants to do without possibility or necessity states of affairs, respectively.

The subject of Johanna Seibt's essay "Constitution Theory and Metaphysical Neutrality: A Lesson for Ontology?" (161-183) is what she calls Carnap's "neutrality thesis", which says, roughly, that it is possible to formulate an (empirical) theory that is metaphysically neutral in the sense (i) that it is compatible with competing ontologies and (ii) that it does not entail any ontological commitments at all. Seibt investigates the possibility of metaphysical neutrality in this sense, in particular Carnap's strategy of "complete structuralization" (i. e., the view that a scientific theory should represent nothing but "pure structures"); and she observes that – since "pure structures are not the kind of stuff of which explanations or justifications are made" (177) a mere description of structures is dissatisfying and insufficient from an ontologist's point of view.*

NOTES

1.   Barry Smith: *Austrian Philosophy. The Legacy of Franz Brentano* (Chicago and La Salle, Illinois: Open Court, 1996), p. 18.
2.   I owe this hint to Prof. Johann C. Marek.
3.   I am indebted to Prof. Heiner Rutte for very detailed historical informations, only a small part of which is conveyed here.
*   I am indebted to the FWF (Projekt F 407) for financial support.

Maria E. Reicher

PAOLO MANCOSU, *From Brouwer To Hilbert: The Debate on the Foundations of Mathematics in 1920s.* New York, Oxford: Oxford University Press, 1998.

JAMES ROBERT BROWN, *Philosophy of Mathematics: An Introduction to the World of Proofs and Pictures.* London, New York: Routledge, 1999.

With an elapse of about one year two books were published in the domain of philosophy of mathematics, which, at first glance, have very little in common but, on closer scrutiny, turn out to be perfectly *complementary*. The first book – compiled by Mancosu – contains a collection of important essays (together with four instructive introductions) which have been directive in the foundation and philosophy of mathematics in the twentieth century. The second book, written by Brown, is an introduction to the philosophy of mathematics of today, which at the same time contemplates the developments since the zenith of the foundational debate during the twentieth. Both books complement each other and together form an excellent basis for a graduate course into philosophy of mathematics. I shall review them in the order of their appearance.

*

Mancosu's book is centred on the foundational debate between Hilbert and Bernays on the one hand, and Brouwer and Weyl on the other, from 1920 to 1930. Although, this seems to be a very limited temporal framework, the selection of the relevant essays is so circumspect and complete, that one gets a profound and detailed impression of this short but unique debate on the proper foundations of mathematics. (Probably a better impression than most of the mathematical, not to speak of the philosophical contemporaries of Hilbert, Bernays, Brouwer and Weyl had.) Furthermore, Mancosu strives to expand the perspective by presenting in the introductory chapters the relevant developments "prior" and "up to" the 1920s – and even sometimes beyond.

Mancosu's book is divided into four parts. The first three parts centre on the mathematical and philosophical positions of the four combatants in the debate about the proper foundations of mathematics during the twentieth: (1) L. E. J. Brouwer; (2) H. Weyl, (3) P. Bernays and D. Hilbert. Part (4) deals with the belated emergence and formalisation of intuitionistic logic by A. Heyting. The papers in the four parts are in each case arranged in chronological order and include not only the most important essays of the four combatants but also some interesting contributions of lesser known figures such as Hölder, Borel, Glivenko and Kolmogorov. The selection of essays is very clever and pleasing from a philosophical perspective, because it reveals that the *dramatis personae* were not just ingenuous, yet narrow-minded mathematicians but true philosophers, if one understands this as referring not so much to the guild of professional philosophers but primarily to persons, who reflect – up to the limits of the conceivable – what the essence of their field of research is. In this context Mancosu deserves special praise for including (besides Brouwer) also Weyl as one of the decisive actors in the foundational drama, because Weyl is, to my mind, much too much neglected, although he was presumably the most contemplative and philosophical head in this group of outstanding mathematicians.

Somewhat less convincing is the selection in part 3 concerning the metamathematical approach of Bernays and Hilbert. Besides three articles by Hilbert himself this part contains four essays by Bernays, which are primarily of an explanatory nature. Bernays sets out to explain Hilbert's achievements regarding the development of a proof theory for a new foundation of mathematics as well as the philosophical underpinnings of this program. Although Bernays' role in the formation of that program can hardly be overemphasised, Mancosu's selection seems to me slightly out of balance. Furthermore there is a certain danger to mingle Hilbert's views with those of Bernays. Obviously, Mancosu values Bernays' contributions to the program, in particular his philosophical reflections to justify the program, more than I do. I agree with Mancosu that Bernays was the better-trained philosopher (in the professional sense of the word) and that Hilbert expressed his views frequently in unclear and sometimes even confused terms. But all this does not mean that Hilbert was not the far more original thinker regarding the epistemological analysis with respect to a firm foundation of mathematics and in this truly analytical sense the more profound philosopher. Finally, the selection of essays for the last section on intuitionistic logic is again very appealing, insofar Mancosu did not only include the fundamental articles of Brouwer and Heyting but also the almost unknown but extremely interesting essays by Borel, Glivenko and Kolmogorov.

A short remark to the four introductions (the first is written by van Stigt, the second and third by Mancosu, and the fourth co-authored by Mancosu and van Stigt). They are all very well written and offer a huge amount of information about the positions of the four main figures and the development of their views before and during the 1920s. The reader gets a lively and detailed impression of the debate and the issues and disagreements at stake. Most readers hardly imagine how much work has been put into these introductions. Although little is new – at least for the expert most points are fairly standard – it's quite a lot of work to bring together all the details and to arrange them in lucid fashion. Here, the authors have done a great job. I can only add two minor points.

In chapter 3.3.2 Mancosu discusses L. Nelson's *Critical Philosophy* in relation to Hilbert's metamathematics and the disputed role of "pure (spatial) intuition" for a proper foundation of mathematics. Mancosu (referring to Peckhaus) correctly points out that Nelson's *Critical Philosophy* became criticised primarily by "two of Hilbert's best collaborators, Bernays and Courant", and not so much by Hilbert himself. But he forgets to mention that the crucial argument against Nelson's claim that the whole of mathematics must be grounded in pure spatial intuition first was formulated by Ackermann in his "Vorbemerkung" to the new edition of Nelson's *Kritische Philosophie und mathematische Axiomatik*. Ackermann correctly points out that Nelson's view cannot be shared by Hilbert in spite of the fact that in Hilbert's account of mathematics *ideal elements* play an essential role. However, the *role* and *meaning* of *ideal elements* cannot be captured by pure intuition. This is the decisive argument against Nelson's claim that the whole of mathematics – including metamathematics – must be grounded in pure intuition. The objections of Bernays to Nelson's view is, by the way, interesting proof that Bernays' and Hilbert's views are not always exactly the same. Bernays boldly denies that "pure intuition" plays any role in mathematics proper (its function is restricted to metamathematics), whereas Hilbert is much more careful in this respect.

In chapter 2.6 Mancosu notes quite correctly that Weyl since 1925 "already attempts to take a middle stand between Hilbert and Brouwer". (p. 80) He then goes on to discuss what that could mean, in particular Weyl's cryptic remark that in theoretical physics "we have before us the great example of a (kind of) knowledge of a completely different character than the common or phenomenological knowledge" and in which Hilbert's axiomatic approach may turn out to prevail over intuitionism. This would mean, as Weyl points out, "a decisive defeat of pure phenomenology, which thus proves to be insufficient ... even in the area of mathematics". Now, what does it mean "to take a middle stand", or more precisely, is such a middle stand at all possible? And, if yes, what does it amount to? Mancosu doesn't tell us, presumably because he did not know that the Mathematical Institute in Göttingen holds a typescript of Weyl from 1930, entitled "Axiomatik", in which Weyl demonstrates what he takes as the reasonable core of Hilbert's axiomatic approach to mathematics as well as to science.

One final remark concerning Mancosu's translation of Hilbert's "Neubegrün-dung der Mathematik" as "The New *Grounding* of Mathematics". Although I am not a native speaker, this translation seems to me even more unsatisfactory than the already inappropriate orthodox translation "New Foundations of Mathematics". If the German expression "Neubegründung" means something more than or beyond "New Foundation" it's more like "New *Justification*" than "New Grounding".

*

Brown's *Philosophy of Mathematics* is – as the subtitle correctly states – an introduction to the world of proofs and pictures. This is meant quite literally, as can be seen already from the titles of the eleven chapters, which contain among others: "Picture-proofs and Platonism" (chapter 3), "Constructive Approaches" (chapter 8), "Proofs, Pictures and Procedures in Wittgenstein" (chapter 9) and "Computation, Proof and Conjecture" (chapter 10).

Brown makes no secret of his sympathy with Platonism, of course, only a mild form of it, because he is aware that original Platonism has to overcome a number of obstacles: First, there is "The problem of Access" to the time- and spaceless numbers. Second, and even more pressing, "The Problem of Certainty" of our knowledge with respect to mathematical statements (descriptions of mathematical facts) and their proofs. These (and many more) topics are discussed in chapter 2, where Brown also scrutinises some of the (more recent) rival views, such as Kitcher's empiricism and Lakatos' fallibilism. The classical modern competitors of Platonism, formalism and constructivism, are discussed later, in chapters 5 and 8, where Brown reveals his secret admiration for Hilbert and his deep-seated distrust for Brouwer and his followers, such as Bishop's Constructivism and Dummett's Anti-realism. (I'll come to that).

The defence of moderate Platonism is, however, not the main goal of the book. Its true intention is the endorsement of the *usefulness* of pictures in two respects: first, in *grasping* mathematical statements in general, and second, in *proofs* of mathematical statements in particular. Because this is a much-debated issue, let me make some short, nonetheless clarifying remarks.

Pasch was the first modern mathematician, who banished all use of pictures whatsoever in *proofs* of mathematical statements. (Unfortunately, Pasch is not mentioned in the book.) But, as every expert of Pasch's *Lectures about Newer Geometry* knows, he did not banish pictures from mathematics altogether, but insisted, on the contrary, on their *indispensability* for a correct and unambiguous understanding of the axioms of geometry. This is, in particular, true for the axioms of incidence in projective geometry. Consequently, Pasch's book entails for each and every incidence axiom a "drawing", which reveals how the axiom and the expressions 'point', 'straight line', 'between' have to be understood.

From this it should be clear that Pasch promotes a *propositional* attitude with respect to *proofs*, but not with respect to our grasping sentences from pre-propositional elements like names of and relations between individuals. Consequently, there is nothing to object to the use of pictures and diagrams in explaining the meaning of first sentences and their grammatical constituents. Only in proofs pictures are not permitted, because proofs are by definition logical deductions of *sentences* from *sentences*. Whether Pasch's stance is 'justified' is another question. Brown contemplates the role of pictures in proofs at length in connection with Wittgenstein's late philosophy (and Kripke's sceptical interpretation of it). Although Brown's intention is, without doubt, to find a legitimate place for pictures in proofs, his efforts in this respect are to my mind somewhat wavering and altogether too anxious. To grant pictures a merely heuristic role in proofs is too little; to grant them the full force of proof is in general too much. But the *crucial* question is, whether some special kind of pictures – let's call them *symbolic* pictures or diagrams – have (almost) the same force of proof as logical deductions of sentences from sentences. I see no reason a priori, why this should be impossible. Brown comes close to this, but he stops just before its affirmation.

Besides presenting many amusing as well as realistic examples of mathematical problems and proofs, the book deserves special mention for its two chapters on 'Applied Mathematics' and 'Definitions' – topics, which in general are too much neglected, although their bearing on questions of philosophy of mathematics can hardly be denied. Brown's approach of comparing again and again the situation in mathematics with what is done in science – in particular in theoretical physics – cannot be praised enough. However, no review without some critical remarks! In this respect, I have to offer two quite different points.

In chapter 5 on "Hilbert and Gödel", Brown deals, inter alia, with Hilbert's program to prove the consistency of arithmetic in an absolute manner, from within so to speak, and Gödel's long-lasting, almost devastating impact on this program. Now, although Brown is very enthusiastic about Hilbert as one of the most ingenious mathematicians of the 20th century, he nonetheless presents Hilbert's program in a way, which does not make clear why this program should have worked at all. What Brown calls an 'absolute' proof of consistency "by exhibiting a concrete (visual) model" (p. 68) is no absolute proof in Hilbert's sense, but just what he calls in his lecture "Foundations of Mathematics" (1921) the method of "Aufweisung" of a 'system of things', which fulfil the axioms of the theory. This method belongs to the old idea of axiomatic. Hilbert's new idea of an absolute consistency proof is, however, completely different. First, we formalise a theory in such a way that all deduction rules are explicitly stated. Next we show that only a finite number of different types of formulae can occur according to the deduction rules. Last but not least we show (type by type) that among all the different types of formulae a certain kind of formula, which represents an inconsistency of the theory, cannot occur. This idea of an 'absolute' consistency proof is similar to Hilbert's famous proof that the set of all possible algebraic invariants has a finite basis. Here, like there, was no reason a priori why it should not work. Gödel's second result was indeed the real surprise. Nobody had imagined – except (perhaps) H. Weyl in his *Consistency in Mathematics* (1929) – that if *T* is consistent it cannot be proved to be consistent.

My second point is quickly stated. I find the presentation of Kripke's interpretation of Wittgenstein's sceptical considerations regarding the very possibility
of "rule following" much too long. Admittedly, this is a question of taste, yet not
only of taste but also of internal coherence. If someone like Brown confesses,
and here I agree with him, that Wittgenstein's scepticism has little relevance for
the working mathematician, why should one spend so much ink. If Wittgenstein's point is, however, that mathematics has an inevitable conventional component, then I can only reply, that this was already recognised and clearly stated
by Hilbert in his public lecture "Natur und mathematisches Erkennen" (1919)
where he explains the *practical* necessity of 'ideal elements' in mathematics as
well as in science.

This remark should, however, not scare away any reader, who is interested in
philosophy of mathematics. The book is written in a very reflected and relaxed
point of view and, what I appreciate most, with a bit of humour. Together with
Mancosu's book it forms an ideal basis for a graduate course in philosophy of
mathematics and its recent history.

Ulrich Majer

JOHN BLACKMORE (Ed.), *Ludwig Boltzmann: Troubled Genius as Philosopher.*
Dordrecht: Kluwer 1999. (*Synthese* 119)

The life and work of Ludwig Boltzmann (1844-1906) is receiving ever more
attention by people interested in philosophy of science and its history. No fewer
than seven talks centrally concerned with Boltzmann are scheduled for this
year's meeting of the History of the Philosophy of Science Working Group
(HOPOS).[1] It is therefore very beneficial that just now, John Blackmore has put
together a set of topical articles and material on Boltzmann's philosophy and
published it as a *Synthese* special issue.

Blackmore himself has not lost his previously confessed enthusiasm about
Boltzmann,[2] which is conspicuously noticeable in a virulent way in his contributions to the subject. One reason for his high appreciation is obviously Boltzmann's courageous campaign for atomism in physics, since Blackmore considers
theories about the reality of atoms and molecules to be "perhaps the most important theories in science." (180) However, the question of whether or to what
extent Boltzmann himself did believe in the reality of atoms is still a subject of
debate, as Blackmore points out in his editorial introduction. (2) That is because
from the 1890s on Boltzmann held a series of varying philosophical positions
concerning the relation between reality and physical theory. There are *Bildtheorie* views not unlike Heinrich Hertz's and pragmatic approaches to theory
evaluation to be found in his writings both published and unpublished. There are

as well philosophy of language sounding proposals to cleanse science from 'meaningless' metaphysical questions. The problems posed by these differing conceptions and their implications for Boltzmann's views on the reality of atoms are competently addressed by several authors in the volume under review, particularly in Blackmore's own contribution on the subject of "Boltzmann and Epistemology."

In Boltzmann's *Bildtheorie*, according to Blackmore's analysis of it, the idea that the scientific picture is a representation of something trans-conscious is abandoned, which in his eyes amounts to Boltzmann's surrender to "extreme epistemology." It is not always clear what 'extreme epistemology', Blackmore's catchword, means, but it is often related to Mach's phenomenalism. In any case, it is analysed as a counter-productive impediment to Boltzmann's defence of atomism, and in its later form, linguistic philosophy, even as a "disaster for philosophy of science." (171) Let us not ponder whether a label like 'extreme epistemology' can do justice to all the diverse thinkers to whom it is applied, like Mach, Ostwald, Duhem and the Neopositivists (who later were mostly physicalists instead of phenomenalists). The aim of Blackmore's story is a different one: He wants to explain how Boltzmann came to flirt with phenomenalism in the first place.

Blackmore answers that the genius surrendered because he was troubled, troubled by the anti-metaphysical attack of Mach, Duhem, Jaumann and others against the (trans-conscious) reality of atoms, which at the same time threatened the core of Boltzmann's physical work and thinking: kinetic theory and physical discontinuity. In order to defend the latter as the most effective and promising way of scientific theorising, he publicly sacrificed its representational character and renounced atomistic realism. Henk Visser, in the same volume, joins in with Blackmore's speculation that Boltzmann's move may have been a mere compromise in order to be in a metaphysically less exposed position to defend atomism. (140)

But could not Boltzmann have adopted 'extreme epistemology' out of the positive conviction that shaking loose of the metaphysical riddles of representational realism is liberating and productive? "Many questions that used to appear unfathomable thus fall away by themselves," as he claimed in an 1899 address.[3] The historical *evidence* that it was all just a desperate act of surrender (cf. esp. 174) is scarce, and sceptics (like the author of this review) are unlikely to be convinced.

However, Blackmore surely does succeed in blocking the way for anyone who would want to tell Boltzmann's epistemological success story as the gradual deliverance from the onerousness of realism. The story is not such a straightforward one. Why, for example, did he display a rather realist stance in his philosophy of nature and a rather anti-realist one in methodology at the same time? Boltzmann's epistemological odyssey is riddled with incoherencies. And in the end, it seems to have brought him (home again?) to realism. Boltzmann's epistemology is sure to attract further attention by scholars. New reference points for this research have been laid out in Blackmore's and other pieces in the *Synthese* issue.

Another point of interest that has already received some attention in the past few years is the relation between Wittgenstein's philosophy and Boltzmann's. Visser inquires further into this problem, concentrating on the picture theory of language in Wittgenstein's early philosophy. I have already mentioned Boltzmann's picture theory, according to which any of a variety of structures can be counted as a picture – a view not unrelated to Wittgenstein's conception of pictures as facts. Visser can point to this and a list of further interesting parallels and aspects of possible influence, especially through Boltzmann's 1905 *Populäre Schriften*. Unfortunately, no progress has yet been made in clarifying how precisely Boltzmann's ideas actually exerted their influence on young Wittgenstein. And since the younger Ludwig nowhere in his work refers to the writings of the older, for the time being, the story remains a speculative one.

But historical detail, it must be emphasised, is not neglected in this volume. Stephen Brush offers a new perspective on the history of gas theory. He portrays the phenomenon that throughout the development of the kinetic theory, its progress was again and again provoked by poignant objections from individual critics. While these "gadflies", as Brush calls them, later came to be forgotten by history, their criticism greatly promoted the research of the "geniuses" who were engaged in the positive development of the theory, above all Boltzmann, "the genius most often stung by gadflies." (37) The story Brush has to tell is important and fascinating to read – obstructed only by the occasional irritating impression that Brush attempts to employ the concept of "genius", not only as an ironic means to give the narration a relaxed and informal character, but as a serious historiographical category.

Brush is not alone with this slightly noticeable undertone of hagiography. When Visser and Blackmore both refer to Galileo's quest for the reality behind the appearances, opposing it to Bellarmine's 'saving the phenomena', and comparing both to Boltzmann's atomism vs. Kirchhoff's mathematical phenomenalism, it becomes a little too clear who the good guys in the tale are. (136, 169, 179 f.)

The persuasiveness of the essays collected in the volume under discussion is most impressive whenever specific questions are addressed. This is the case with Robert Deltete's detailed study of the Lübeck *Naturforscherversammlung* of 1895, where "Ostwald and Boltzmann came to heavy blows" (as Georg Helm reported directly from the scene, 60). As Deltete adroitly reconstructs, Boltzmann's opposition to the energeticism of Helm and Ostwald developed only over time, when the energeticists continued propagating their ability to develop energetics as a simple, accurate and comprehensive theory of nature, but proved unable to stand up to their promises. Boltzmann resolved to not let these inadequacies go unchallenged, which in the end led to his success at Lübeck in delivering a blow to energeticism from which it was not to recover. It is very enjoyable that at least this important part from Deltete's comprehensive dissertation on the history of energeticism is now made more easily available by means of inclusion in this *Synthese* issue.

Michael Stöltzner identifies a historical phenomenon he calls "Vienna Indeterminism." Its evolution is manifest in the work of Mach and Boltzmann, and it culminates in the doctrines of the Viennese physicist Franz Exner. Exner assumed a fundamental indeterminism, on the basis of which he attempted to justify all types of regularities. Stöltzner's discussion of the relation of indeterminism to Mach's and Boltzmann's separation of reality from causality displays a wealth of detail and is not always easy to follow. In any case, he demonstrates very neatly what I consider the central point to his account: Namely, that Mach, Boltzmann and Exner shared an opposition against the Kantian assumption of an absolute causality as a necessary precondition to the understanding of nature. Rather, they redefined causality as functional dependence. The Vienna physicists in question established this tradition long before the concept of causality came to be challenged by the development of quantum physics.

Another opposition of Boltzmann's to the Kantian tradition so dominant in the intellectual life of his time is dealt with in Henk de Regt's study of the views on explanation and understanding in science implied by Boltzmann's picture theory. Like many before and after him, Boltzmann held that explanations based on mechanical pictures are the most intelligible ones. He accounted for this fact by their correspondence to our laws of thought. But these laws, and that is the anti-Kantian part of the idea, are a product of evolution and therefore not necessarily infallible. That is why Boltzmann preferred mechanical explanation but admitted that it might turn out not to suffice for all future scientific purposes. This new aspect of Boltzmann's picture theory, i.e., that it was also meant to account for explanatory purposes of scientific pictures, along with its underlying naturalistic reasoning, should be of interest to many a philosopher of science, given the current attention to 'mental models' and their role in scientific understanding.

Boltzmann's naturalistic laws of thought also give preference to a finitist philosophy of mathematics, which is in turn related to his atomist convictions, as both Setsuko Tanaka and Stöltzner show in some detail.

In addition to the scholarly articles in the volume, the editor has also included some source material in English translation, which is partly published here for the first time. The first portion of this material consists of short informative and entertaining pieces on Boltzmann by three of his assistants, G. Jäger, J. Nabl, and S. Meyer. More importantly, the book contains Boltzmann's notes for the first three of his 1903 lectures on natural philosophy, "expanded" into English. (200, fn1) These notes (for altogether 20 lectures) and an elaborated record of the course by an unknown student have only recently been transcribed and published by Ilse Fasol-Boltzmann.[4] But Boltzmann's notes are sometimes enigmatic and hard to understand, and even harder to translate. For example, in the English 'expansion' of the third lecture, there appears the puzzling image of someone drowning while everything passes before his eyes. (199) But in the German original, *nothing* of this is mentioned but the word *"Revue"*, while the unknown student renders the corresponding passage as "I undertake to let the most simple and clear concepts of our thinking pass before our eyes (*Revue passieren zu lassen*) ..." [5] Very probably, the unknown student's elaboration would lend itself much better to translation than Boltzmann's fragmentary notes.

On the other hand, the lectures on natural philosophy definitely constitute an invaluable source on Boltzmann's philosophy, and the enterprise to make them more easily available to the English-speaking world is surely meritorious. Tanaka has done the interested reader an excellent service by means of her illuminative and detailed precis of lectures 3-18.

That the book is published as a *Synthese* issue is not a disadvantage. It is sure to act both as a stimulus to further research and as a tool to make the philosophy of Ludwig Boltzmann more accessible. As such, it should serve to create a more prevalent and nuanced awareness of Boltzmann's work within the philosophical community.

NOTES

1. I am referring to the contributions by Nadine de Courtenay, David Hyder, Theodor Leiber, Peeter Müürsepp, Matthias Neuber, Michael Stöltzner and myself to the HOPOS 2000 Conference at Vienna.
2. Cf. *Ludwig Boltzmann: His Later Life and Philosophy, 1900–1906*, Book One: A Documentary History, ed. by John Blackmore, Dordrecht: Kluwer 1995, p. vii.
3. "Über die Entwicklung der Methoden der theoretischen Physik in neuerer Zeit", quoted from the translation in Ludwig Boltzmann: *Theoretical Physics and Philosophical Problems: Selected Writings*, ed. by Brian McGuinness, Dordrecht: Reidel 1974, p. 91.
4. Ludwig Boltzmann: *Principien der Naturfilosofi. Lectures on Natural Philosophy 1903–1906*, ed. by Ilse M. Fasol-Boltzmann, Berlin etc.: Springer 1990.
5. *Ibid.*, 83 resp. 159. The irritating image first appears in an English translation of a fragment from the notes in *Ludwig Boltzmann: His Later Life and Philosophy, 1900–1906*, Book One (cf. note 2), p. 134.

Torsten Wilholt

KURT BLAUKOPF, *Unterwegs zur Musiksoziologie. Auf der Suche nach Heimat und Standort,* annotated by REINHARD MÜLLER. Graz-Wien: Nausner&Nausner, 1998. (Bibliothek sozialwissenschaftlicher Emigranten)

It is a special matter of luck that Kurt Blaukopf, the doyen of Austrian sociology of music, had the opportunity shortly before his death to describe his unusual intellectual career. Thus we are in possession not only of a very personal book of one the great figures in Austria's intellectual history, but also of a contemporary document of the lively political and cultural developments between the wars which created the Austrian Republic out of the remnants of the Habsburg Empire. The present volume is appropriately published in a series devoted to social scientists that became emigrants shortly before or during World War II. The editor, Reinhard Müller, has added some useful comments, explanations and historical dates in the footnotes.

The volume opens with an autobiographical text that also figures as an introduction and which contains a number of remarkable facts: Born in 1914 in the Bukowina (today a region in the north of Rumania), Kurt Blaukopf came to Vienna early in his life. He was raised in a well-situated and cultivated family and enjoyed an excellent education which formed an unusually dedicated and self-conscious young man who was eager to learn and to participate both in the political and cultural life of his time. Immediately after finishing high school Blaukopf already devoted himself to the sociology of music, a field which was just coming of age and still far from being one of the subjects studied and taught at universities. He trained himself following his own curriculum and thus made himself familiar with the current intellectual movements. At this time he discovered and became especially fond of Max Weber's essay "Die rationalen und soziologischen Grundlagen der Musik", and of the ideas of the Vienna Circle. He admired Otto Neurath's "Empirische Soziologie", and followed his idea of encyclopedical knowledge in pursuing questions in sociology, law, philosophy, and politics. Blaukopf also took quite seriously some of the methodological requirements put forward in the programmatic essay of the Vienna Circle "Wissenschaftliche Weltauffassung" (1929). He emphasized the logical analysis of the categories used in the cultural studies *(Kulturwissenschaften)*, considered the benefits of the methods and findings of the natural sciences, and tried to apply mathematical and in particular statistical procedures in musicology (p. 20).

Throughout his whole life Blaukopf felt strong emotional ties to his native country Austria. His patriotic attitude, which runs like a thread through his entire life and work, must be understood as a reaction to his time. In contrast to many others he never doubted the cultural autonomy of Austria and its right to exist as a nation of its own. It hurt him deeply when in the admission form to the university his entry for nationality was changed from "Austrian" to "Jewish" with the justification that there is no Austrian nation. He felt expelled, like an inner emigrant in his own country.

Blaukopf became a real emigrant in March 1938 when Austria was occupied by the Nazis. He first escaped to Paris where he arrived with only a few belongings, but with all his "mental goods": the complete typescript of his "Musiksoziologie", excerpts, and commentaries filling two suitcases. In Paris he met a group of Austrian emigrants who tried to combine an intellectual life with political commitment. In this context it seemed a political task to make explicit the peculiar features of the Austrian musical tradition, and Blaukopf was encouraged to publish in the pertinent journals of emigrants like the "Pariser Tageszeitung". Two years later, after a short internment, he obtained a student-visa for Israel in 1940 where he enjoyed a profound musical education at the conservatory. Soon he met the Austrian Society in Palestine, a section of the Free Austrian Movement, which addressed the public with cultural events, leaflets, and a book series called "Österreichische Schriftenreihe".

When he returned to Vienna in 1947 he faced an Austria that showed little interest in coming to grips with its history and showed even less concern for its immediate past. It was hard for Blaukopf to gain a firm foothold. He had to minimize his scientific activities and took up work as a daily writer ("Tagesschriftsteller"). As musical critic and journalist he kept in touch with the current cultural affairs. His interest as a sociologist in the technological progress of music production and reproduction led to his engagement with the journals "phono" and "HiFi Stereophonie". From 1962 on he was again connected with the academic world by giving lectures at the Viennese Conservatory. Blaukopf's personal contributions and his scientific achievements were finally honoured when he first became head of the Institute of Sociology of Music which was founded in 1965, and later was appointed as full professor in 1977. The autobiographical section, which takes up one third of the book, ends with an ungrateful postscript of a grateful person ("Undankbares Nachwort eines Dankbaren") in which he once again emphasizes his very personal concerns in his field of research.

In the second part of this volume the author compiled texts which were originally written in various languages, hardly accessible or still unpublished. They all date from the period of emigration and testify to Blaukopf's search for a reliable portrait of his lost home and its history (p. 9). For the most part these are essayistic contributions to emigrant journals published in Paris, Jerusalem, London, and Vienna. They focus on three subjects: music and society, focusing on Austria, philosophy and aesthetics.

The texts belonging to the field of musicology in a narrower sense document the wide range of Blaukopf's interests. In "Musikwissenschaft und Rassentheorie" he wants to show that racial thinking surfaced not only in the Third Reich, but had its roots already in the former German musicology (p. 25). Blaukopf bluntly lists the names of his now famous colleagues whose racist writings apparently did not hinder their later career. In "Jüdische Musik" we learn about the never realized project of a "Fest der jüdischen Musik" in Jerusalem. Blaukopf traces this project in the unpublished correspondence of prominent composers and music writers, and uncovers the eminent role played by the composer, theoretician, and co-founder of the International Society of Contemporary Music, Rudolf Réti. Several papers document Blaukopf's approach to the musical "heroes" Beethoven and Mozart driven by a broad understanding of the history of ideas. Part of his concern is to defend Mozart as an Austrian composer against attempts to see him as a central figure in German culture. Worthy of note is also the complicated history of the long essay "Musik im Geist der Aufklärung. Mozart und die Eigenheit der österreichischen Kultur". He started this text in 1938/39 in Paris, continued it in the French internment camp, and finished it in Jerusalem in 1940. Two studies dealing with sociological aspects of contemporary music refer to problems of music reception prominent at the time. A further

interest of Blaukopf's musicological work is documented in "Die Enzyklo-
pädisten und die Musik", which was Blaukopf's graduation thesis at the conservatory.

In the section "Mit dem Blick auf Österreich" we find political texts dealing
with Austrian history. An anonymously published leaflet addressed to the Aus-
trians in the British Army reveals Blaukopf's personal political attitude according to which it is mainly the socialist and communist proletariat that fights for a
free and independent Austria. This attitude also informs a number of individual
studies on outstanding figures who were well-disposed toward Austria: Guido
Adler, Michael Hainisch, Friedrich Engels, Ferdinand von Saar, and Josef Fried-
jung. Especially touching is the last text in this section "Ruhe im Hafen", dating
from 1946, in which Blaukopf urges to come to terms with the past. He speaks of
the hearts that need to be cleaned of injustice, from the misery of being mis-
treated, and points out that overcoming the economical misery is not enough for
gaining peace (p. 250).

The volume concludes with five short pieces dealing with various philosophical themes and with a plaidoyer for founding an Austrian Encyclopedia. In this
way the intellectuals, artists, and writers of this country should enable and support the cultural resurrection of Austria. From a theoretical point of view Blaukopf puts forward a materialism and humanism that is rooted in Marxism, and
defends this position against a number of simplifying criticisms. He emphatically
rejects Sartre's existentialism, since it rejects all values and truths (p. 272).

In a letter of Engels Blaukopf finds a valuable hint which expresses a central
motive of his philosophical thinking at that time: The entire history, Engels says,
must be studied from scratch, and the conditions of the existence of various parts
of society must be dealt with one by one, before one can draw any inferences
about the political, legal, aesthetical, philosophical, and religious ideas appropriate to their sociological background (p. 260). In this vein Blaukopf takes a fresh
look at the history of philosophy in Austria. His starting point is the ambiguous
character of Austrian philosophy in the 18th century which he describes as both
aristocratic and bourgeois. Related to that is the intellectual barrier ("geistige
Scheidewand") between Austria and Germany, which results from the
completely different social and political conditions pertaining in these countries.
Whereas in Germany the founders of great philosophical systems gained a
dominating influence, philosophy in Austria developed as a by-product of
reflections on arts, on legal issues, on medicine, on economy, etc. (p. 253ff.). As
the typical characteristic of the Austrian philosophical tradition Blaukopf praises
the materialistic and "naively realistic" attitude, which however could not
prevent that the idealistic doctrines of Ernst Mach became influential also within
the socialist movement.

Certainly Blaukopf did not claim in these texts to put forward a thoroughgoing analysis of philosophical problems or of questions in the history of philosophy. Instead, he wanted to draw attention to the special situation of philosophy
in Austria, which is part of the cultural heritage that needs to be preserved and

taken care of. These suggestive ideas remained unappreciated at the time of their first publication. One hopes all the more that their reprint in this volume helps to keep alive the interest which the Austrian tradition enjoys in the meantime and which bears fruits also thanks to the life work of Kurt Blaukopf.

Andrea Lindmayr-Brandl
and Johannes L. Brandl

CLAUS-DIETER KROHN, PATRIK VON ZUR MÜHLEN, GERHARD PAUL, LUTZ WINCKLER (Eds.), *Handbuch der deutschsprachigen Emigration 1933-1945,* Darmstadt: Wissenschaftliche Buchgesellschaft, 1998.

From the perspective of later centuries the banishment and forced emigration of scholars and scientists from Central Europe during the Nazi period will perhaps be regarded as the single most important event in the history of the 20th century science. Practically all academic disciplines were affected by Nazi purges (and in particular those fields where Germany and Austria played a leading role in the scientific scene). The emigration that followed benefited countries around the world. In light of these dimensions it seems astonishing that research on this phenomenon and its consequences only began in the eighties (preceded by some meritable but isolated private initiatives in earlier decades) in the countries of what once constituted the Großdeutsches Reich. A milestone in that new development was the monumental *Dictionary of Central European Emigrés* (edited by Röder and Strauss) with about 15000 entries on individual émigrés from politics, culture and science. But it was only with the large congress "Vertriebene Vernunft" that was held in Austria in 1987 and the establishment of a special program for research into scientific emigration by the Deutsche Forschungsgemeinschaft in Germany that interest spilled over from earlier studies in political and artistic emigration to scientific emigration.

The monumental handbook to be reviewed here can be seen as taking stock of what has been achieved in the area of exile studies between then and now. And these results are surely impressive. After an introductory part (I) the book studies the following areas:

- single countries of exile (part II),
- exile of politicians (part III),
- scientific emigration (with individual sections for most academic disciplines) (part IV)
- literary and artistic exile (part V).

A survey covering the (later) return of emigrants from exiles to their countries of origin and a description of the history of exile studies as such (in part VI) round off the volume.

To review the entire book in great detail would go beyond the scope of this article. I will restrict my critical remarks to the emigration of philosophers (some of whom like Adorno or Carnap appear again in other sections). The reader immediately notices that philosophy – unlike all the other academic disciplines – is described by three authors: Gunzelin Schmid Noerr presents the "Critical Theory" of the Frankfurt School (805-812) and Friedrich Stadler the Vienna Circle of Logical Empiricists (813-823). These articles are preceded by a tentative summary of emigration in philosophy written by Nikolaus Erichsen (791-804). One can only congratulate the editors that they were able to recruit the best experts in their respective fields: Schmid Noerr, editor and apt commentator of the Collected Works of Max Horkheimer (the founder and leading figure of the Frankfurt School) and Stadler, the founder of the Vienna Circle Institute and, most recently, author of the impressive volume *Studien zum Wiener Kreis*. In my opinion both of their articles offer the best concise accounts on the Frankfurt School and the Vienna Circle to be found in any dictionary or reference book today.

Schmid Noerr begins by distinguishing two senses of "critical theory". In a more general sense, the term (as distinct from "traditional" theory) covers all those philosophical options which conceive their work under the perspective of an improvement of society. But the Frankfurtians also took the term as a brand name for their own philosophical work which – in contrast to more traditional marxist approaches – included theory of culture and theory of subjectivity, in order to obtain a materialist social philosophy and a philosophically oriented social research. In two places (pp. 807 and 811) Schmid Noerr surprises the reader – and in my opinion he is right in doing so – by indicating two positive elements within the generally ambivalent picture of America entertained by the Frankfurt School. One concerns certain tendencies in social research found in the U.S., the other is Horkheimer's stance towards the Vietnam war.

Friedrich Stadler's article covers in condensed form many of the themes treated in his above-mentioned magnum opus. But he also includes new ideas and material, putting the emphasis on the emigration of Logical Empiricists to the U.S. Among other things, the article offers a description of the various institutions and centers from which Viennese philosophy of science entered the American philosophical scene (p. 817f.): The Unity of Science Institute and, subsequently, the Boston Colloquium for the Philosophy of Science, the Minnesota Center for the Philosophy of Science, and the Pittsburgh Center for the Philosophy of Science – all of them with different profiles and directions in their actual philosophical work. In which way the formerly "Viennese" gospel was thereby transformed and merged into analytic philosophy would be an interesting story.

Since I assume that the readers of this Yearbook are more familiar with the themes covered by Schmid Noerr and Stadler, I will devote the rest of this re-

view to Erichsen's presentation. Of course, an attempt at a synthesis of everything known up to now about the exile of philosophers can only be warmly welcomed. All too often philosophers ignore what other schools and options outside
their particular field and affiliation have to offer, and here Erichsen's contribution could help redress this situation. His article also indirectly shows how much
has to be done in order to arrive at a correct summary. To begin with, he presents
lists of names, statistics and some telling diagrams about the emigrants. This, to
be sure, is in itself an achievement. But since some false information has slipped
into the basic data, the statistics built upon them are to a certain extent flawed.
First, the community of philosophers is divided first in four groups: the ordinarii
(tenured professors) (group A), private lecturers (*Privatdozenten*) and extra-
ordinarii (group B), promoted professors (but without habilitation) (group C),
and emigrants who only embarked upon a career in philosophy once they settled
in their country of exile (group D). Group A also includes Philipp Frank, even
though he held a chair in *physics* from 1913 till 1939 as Einstein's successor in
Prague. In group B we find Otto Neurath, even though he actually completed his
habilitation in 1917 in Heidelberg, albeit in *economy* and, what is more important, lost this venia legendi two years later as a consequence of his participation
in the failed Bavarian Soviet Republics. Ernst Bloch and Georg Lukács are also
cited in group B, even though both of them never received a habilitation. As a
consequence the large general statistical diagrams about all emigrants, the date
of their emigration, the countries of exile, their career in these exile countries,
their later return etc. on p. 797 must of course be corrected taking into account
these errors. It also seems some entire lines in that diagram must be changed:
although far more émigré philosophers in group A who went to the USA are
cited, Great Britain is presented as the country which took in 60% of the emigrants in that group. A lot must still be done to arrive at more reliable basic data
and more accurate statistical diagrams.

At least as important as general statistical data is the information on the
different schools, to which the philosophers belonged, the disciplines they represented or their "philosophic milieus" respectively, as Erichsen refers to it (794).
He argues that mainly Jewish philosophers of religion and logical empiricists
were victims of the Nazi purges. His argument is certainly right, but perhaps too
restricted. With respect to Marxism, the reader finds contradicting statements on
the third prominent candidate to be named here. On p. 794 we read that prior to
1933 the Marxists had no chance of attaining an academic position at a German
university (from which it would follow that they could then not be expelled from
there later.) On the following page we learn that "*besides Marxists* and representatives of Jewish philosophy of religion mostly logical empiricists ... were victims
of this sort of prosecution" (emphasis added). But perhaps this contradiction is
only a seeming one and dissolves, once the term "Marxist" is defined more
exactly. Certainly some (politically) more orthodox Marxists like Bloch and
Lukács (to name just two of the most prominent) never got an appointment at a
German-speaking university before 1933 (and therefore could not be expelled

from there after 1933). On the other hand, some more open-minded thinkers like Horkheimer and Adorno got positions in the academic world before 1933 and were then dismissed from Frankfurt University immediately after the Nazis came to power.

At the end of his article Erichsen explores how the philosophical landscape in Germany and Austria changed as a result of the Nazi "purges" and forced emi- gration. He seems to believe that these changes were not so far-reaching. In order to back his claim he draws from the results of two other authors. But Claudia Schorcht's argument that it was largely non-Nazi philosophy that was taught at *Bavarian* universities in the Nazi era does not seem representative for other German universities (on which detailed surveys had been published as in the cases of Frankfurt, Göttingen and Hamburg). And Hans Sluga's argument (which Erichsen also cites to back his own) that "surprisingly, the Nazi revolu- tion was not to make all that much difference in that diversity" (i.e. of philosoph- ical schools and options), applies only to the schools that remained in Germany and Austria after critical theorists, logical empiricists and Jewish philosophers of religion had been forced to emigrate, i.e., the larger part of neo-Kantianism, phe- nomenology and philosophy of life (*Lebensphilosophen*).

In the context of exile research the question of what became of the different emigrant philosophers groups and their options in the countries of their respec- tive exiles is of course more interesting than the question of the fate of philoso- phy in Germany after their emigration. Concerning the USA, to which most of the philosophers sought exile, it is important to see whether and how the immi- grant philosophical schools from Central Europe merged with philosophies al- ready present there, most notably, with American pragmatism. To test the rele- vance of the factor "emigration" for the content of the philosophical develop- ment, Erichsen makes an interesting speculation: what, for instance, would have become of Logical Empiricism, if it had not have been expelled from Central Europe? Would it then have been able to maintain its impetus for social and cul- tural reform or rather would it sooner or later have become "academic" (803), as it actually happened in the USA? The good thing about this thought-experiment is that it need not remain purely speculative, but can be studied relatively well. Logical Empiricists often spoke among themselves about the fate of their philos- ophies in their new environment. But that of course can mainly be unearthed in unpublished letters and archives (especially in the correspondence between Car- nap and Neurath).

All these critical remarks on Erichsen's contribution should of course not de- tract from his achievement: he should be applauded for being the first to venture a synthesis, but at the same time he risks exposing himself more to errors and to criticism.

Hans-Joachim Dahms

ACTIVITIES OF THE INSTITUTE VIENNA CIRCLE

1. ACTIVITIES 2000

Lecture Series

Location: Department of Mathematics, University of Vienna

ELLIOTT SOBER (University of Wisconsin, Madison, USA)
Evolution and the Problem of Other Minds
(March 16, 2000)

MASSIMO FERRARI (Universita d'Aquila, I)
The Young Schlick – A Piece of Intellectual Biography
(March 27, 2000) Together with the Italian Cultural Institute

BARRY LOEWER (Rutgers University, NJ, USA)
From Physics to Physicalism
(April 10, 2000)

GEREON WOLTERS (Universität Konstanz, D)
Carl Gustav Hempel: Pragmatical Empiricist
(June 5, 2000)

International Workshop

Concepts of Knowledge and Economic Thought by and on Otto Neurath and Josef Popper-Lynkeus
(University of Economics and Business Administration, May 3-5, 2000)
Together with the Department of Economics, University of Economics and Business Administration, Department of Philosophy (Working Group for Analytic Philosophy) and Center for Interdisciplinary Research, University of Vienna

Vienna International Summer University

"Scientific World Conception"
2000: Development and Preparation phase
Together with the Center for Interdisciplinary Research at the University of Vienna (CIR)

8th Vienna Circle Lecture and Keynote Lecture of HOPOS 2000

MICHAEL FRIEDMAN (Indiana University, USA)
On the Idea of a Scientific Philosophy?
Bank Austria, Vienna, July 6, 2000

International Conference

HOPOS 2000:
Third International Conference on the History of Philosophy of Science
University Campus Vienna, July 6-9, 2000
Together with the History of Philosophy of Science Working Group (HOPOS) and the University of
Vienna, Center for Interdisciplinary Research (CIR)
Program Committee/Chairs: Michael Heidelberger (Berlin), Friedrich Stadler (Vienna)

Book presentation

*Elemente moderner Wissenschaftstheorie. Zur Interaktion von Philosophie,
Geschichte und Theorie der Wissenschaften.* Ed. by Friedrich Stadler. Vienna-
New York: Springer 2000 (= Publications of the Institute Vienna Circle, vol. 8).
University Campus Vienna, July 8, 2000

2. PREVIEW 2000/2001

International Conference

*Intellectual Migration and Cultural Transformation. The Movement of Ideas
from german-speaking Europe to the Anglo-Saxon World*
University of Sussex/ Brighton, September 25-28, 2000
Together with the Centre for German-Jewish Studies, University of Sussex

European Science Foundation (ESF) – Network

*"Historical and Contemporary Perspectives of Philosophy of Science in
Europe", 2000-2002*
Together with the Universities of Bologna, Constance, Athens, Budapest (Loránd Eötvös University),
Groningen and King's College (London)

Lecture

MARIA-CARLA GALAVOTTI (University of Bologna)
Bruno de Finetti – Radical Probabilist
Department for Mathematics, University of Vienna (December 11, 2000)
Together with the Italian Cultural Institute

Book Presentations

Kunst, Kunsttheorie und Kunstforschung im wissenschaftlichen Diskurs. In memoriam Kurt Blaukopf. Ed. by Martin Seiler and Friedrich Stadler. Vienna: HPT & ÖBV 2000 (= Scientific World Conception and Art, vol. 5).
BM:BWK, Austrian Ministry for Education, Science and Culture, Vienna (t.b.a.)

Gödel: A Life of Logic. By John Casti and Werner DePauli-Schimanovich. Cambridge: Perseus Books 2000.
Department for Mathematics, University of Vienna (November 6, 2000)

Thomas E. Uebel, *Vernunftkritik und Wissenschaft. Otto Neurath und der Erste Wiener Kreis im Diskurs der Moderne – eine verspätete Rückkehr.* Vienna-New York: Springer 1999 (= Publications of the Institute Vienna Circle, Vol. 9).
Department for Mathematics, University Vienna (t.b.a.)

John von Neumann and the Foundations of Quantum Physics. Ed. by Miklós Rédei and Michael Stöltzner. Dordrecht-Boston-London: Kluwer (= Vienna Circle Institute Yearbook 8/2000).
Erwin Schrödinger International Institute for Mathematical Physics, Vienna (November 29, 2000)

Lecture

WALTER THIRRING (University of Vienna)
John von Neumann's Influence in Mathematical Physics

International Anniversary Symposion
(On the occasion of the 10th anniversary of the Institute Vienna Circle)

The Vienna Circle and Logical Empiricism – Re-Evaluation and Future Perspectives of the Research and Historiography
University Campus Vienna, July 12-14, 2001

International Summer University – Scientific World Conceptions

VISU 2001: Unity and Plurality of Science
University Campus Vienna, July 16-28, 2001
Together with the Center for Interdisciplinary Research at the University of Vienna
Main Lecturers: Don Howard (University of Notre Dame, USA)
Elliott Sober (University of Wisconsin, USA)

3. RESEARCH PROJECTS / WORKSHOPS

Research Project

Scientific World Conception and Art:
Art, Theory of Art and Studies in Art in the Scientific Discourse,
4th stage (1999-2001)
Coordination: Martin Seiler and Friedrich Stadler

Research Project

Liberalism and Logical Empiricism
Together with the Center for Interdisciplinary Research at the University of Vienna, and the Institute of Economics, University of Vienna. 1998-2000. Jubilee fonds of the OeNB (Austrian National Bank)
Coordination: Friedrich Stadler, *Project head:* Georg Winckler

4. PUBLICATIONS

Epistemological & Experimental Perspectives on Quantum Physics. Ed. by Daniel Greenberger, Wolfgang L. Reiter and Anton Zeilinger. Dordrecht-Boston-London: Kluwer 1999 (= Vienna Circle Institute Yearbook 7).

Elemente moderner Wissenschaftstheorie. Zur Interaktion von Philosophie, Geschichte und Theorie der Wissenschaften. Ed. by Friedrich Stadler. Vienna-New York: Springer 2000 (= Publications of the Institute Vienna Circle, vol. 8).

Kunst, Kunsttheorie und Kunstforschung im wissenschaftlichen Diskurs. In memoriam Kurt Blaukopf. Ed. by Martin Seiler and Friedrich Stadler. Vienna: HPT & ÖBV 2000 (= Scientific World Conception and Art, vol. 5).

Friedrich Stadler, *The Vienna Circle – Studies in the Origins, Development and Influence of Logical Empiricism.* Vienna-New York: Springer 2000 (= Publications of the Institute Vienna Circle, special volume).

Thomas E. Uebel, *Vernunftkritik und Wissenschaft. Otto Neurath und der Erste Wiener Kreis im Diskurs der Moderne – eine verspätete Rückkehr.* Vienna-New York: Springer 2000 (= Publications of the Institute Vienna Circle, vol. 9).

John von Neumann and the Foundations of Quantum Physics. Ed. by Miklós Rédei and Michael Stöltzner. Dordrecht-Boston-London: Kluwer (= Vienna Circle Institute Yearbook 8/2000).

Logischer Empirismus und Reine Rechtslehre. Beziehungen zwischen dem Wiener Kreis und der Hans Kelsen-Schule. Ed. by Clemens Jabloner and Friedrich Stadler. Vienna-New York: Springer 2001 (= Publications of the Institute Vienna Circle, vol. 10).

Hans-Joachim Dahms, *Neue Sachlichkeit oder sachte Neulichkeit? Moderne Architektur, Kunst und Literatur der 20er und frühen 30er Jahre im Spiegel der zeitgenössischen Philosophie.* Vienna: HPT & ÖBV 2001 (= Scientific World Conception and Art, vol. 5).

Appraising Lakatos - Mathematics, Methodology and the Man. Ed. by Ladislav Kvasz, George Kampis, Michael Stöltzner. Dordrecht-Boston-London: Kluwer 2001 (= Vienna Circle Institute Library 1).

5. LIBRARY AND DOCUMENTATION

–   Expansion of primary sources and secondary literature on the Vienna Circle and its influence.

–   Acquisition of estates and archival materials in Austria and abroad.

–   Archival material (Xerox copies) from the Minnesota Center for Philosophy of Science, the Pittsburgh Center for Philosophy of Science and the Boston Center for Philosophy of Science, Research Library Kurt Blaukopf.

INDEX OF NAMES

Not included are: Figures, Tables, Notes, References

VIENNA CIRCLE INSTITUTE YEARBOOK

The *Vienna Circle Institute* is devoted to the critical advancement of science and philosophy in the broad tradition of the Vienna Circle, as well as to the focussing of cross-disciplinary interest on the history and philosophy of science. The Institute's *Yearbooks* provide a forum for the discussion of exact philosophy, logical and empirical investigations, and analysis of language. Each volume centers around a special topic which is complemented with a permanent section with essays arising from the scientific activities at the Institute and reviews of recent works in the history of philosophy of science or others with a particular relation to the tradition of logical empiricism.

KLUWER ACADEMIC PUBLISHERS – DORDRECHT / BOSTON / LONDON